<< multimedia >>

From Wagner To Virtual Reality

<< multi

From Wagner To

Edited by

Randall Packer
and Ken Jordan

Foreword by

William Gibson

media>>

Virtual Reality

W • W • NORTON & COMPANY
NEW YORK • LONDON

Support for *Multimedia* and the companion Web site at www.artmuseum.net is provided by Intel Corporation.

Printed in the United States of America
First Edition

The text of this book is composed in Bulmer and Bell Gothic
with the display set in Cholia Sans
Composition by Allentown Digital Services Division of R.R. Donnelley & Sons Company
Manufacturing by The Haddon Craftsmen, Inc.
Book design by Rubina Yeh
Production manager: Andrew Marasia

Library of Congress Cataloging-in-Publication Data

Multimedia: from Wagner to virtual reality / edited by Randall Packer and Ken Jordan.
p. cm.
Includes bibliographical references and index.
ISBN 0-393-04979-5
1. Multimedia systems. 2. Multimedia (Art) I. Packer, Randall. II. Jordan, Ken.

QA76.575.M8319 2001
006.7—dc21 00-066994

W.W. Norton & Company, Inc.,
500 Fifth Avenue, New York, N.Y. 10110
www.wwnorton.com

W.W. Norton & Company Ltd.,
Castle House, 75/76 Wells Street, London W1T 3QT

1 2 3 4 5 6 7 8 9 0

"*After more than a century of electric technology, we have extended our central nervous system itself in a global embrace, abolishing both space and time as far as our planet is concerned. Rapidly, we approach the final phase of the extensions of man—the technological simulation of consciousness, when the creative process of knowing will be collectively and corporately extended to the whole of human society, much as we have already extended our senses and our nerves by the various media.*"

—Marshall McLuhan, *Understanding Media*

"*Form is never more than the extension of content.*"

—Robert Creeley

<< contents >>

<< foreword >>

Geeks and Artboys

by William Gibson

I can, these years later, quite clearly see the shape of the chip I had on my shoulder when I wrote and/or assembled the piece that now stands as my contribution to the discourse charted in this collection.

I was an artboy, or had become one in my own eyes, and I felt myself, in that era of unevenly emergent cyberstuff, and quite painfully, to be lumped in with geeks of the first water.

I had not yet been given to recall that I had started out very much a geek myself, or seen that in so many ways I still was one. And I had been out, through much of the late eighties, on what Bruce Sterling dubbed the Virtual Chicken Circuit. Rather than the traditional rubber chicken, the alpha geeks of Virtual Reality and I had been dining on all sorts of delicacies (the bill invariably picked up by some branch of the local bureaucracy charged with trying to figure out the shape of the future). We were invited to dine in Tokyo, Barcelona, Venice, and Linz, because this was in the heyday of the old gloves-and-goggles paradigm of VR, and there was a certain buzz to the effect that this stuff might be "the next television." At these events, I seem to recall, there was relatively little discussion of the Internet, or what its impact might be. The buzz was mainly around the goggles and the gloves, these end-user interface devices. As futurists, I'm afraid, we were scarcely worth the fancy dinners. As futurists so often do, we threw the baby out and brainstormed the marvelous potential of the bath water.

The baby, an evolving and determinedly prehensile multimedia, went right on evolving, and still does, no sign whatever of stopping, in a world made more than a little unrecognizable by the passage of a mere fifteen years.

By the time I could be induced, by whatever sort of editorial suasion, to sit down and arrange words in a row about what I thought at some particular moment about all of this, I had grown desperately tired of being mistaken for someone who knew what he was talking about.

Rather, I felt it imperative that I *not* know what I was talking about. I wanted to be recognized for some subrational, basically shamanistic function that I was convinced I served. I wanted it recognized that I was in fact an artboy. Which meant, at least in my understanding of what it took to qualify, that I did not, must not, know what I was doing. If that which writes novels, as someone once told me Henry James had said, is not that which attends dinner parties, I wanted it recognized that that which envisions "cyberspace" is not that which glazes over at the press conference following a federally funded VR fest.

Consequently I donned the attitudinal equivalent of a black raincoat, took the intertextual machete to a piece of my own fiction (originally commissioned by a Japanese cosmetics firm), and declared my literary descent from William S. Burroughs. I waxed nonlinear, burned brightly at both ends, and hoped to be definitively identified thereby as an artboy.

Had the volume you now hold existed then, I would probably have not been able to do that. Had I read this material, my eyes would have been opened in a very different way. I would have had to come to accept that Ted Nelson and William Burroughs, strange to think, were actually aiming, at various times, in the same direction.

I had had some inkling of this upon first discovering the cut-and-paste universe of word processing: I had seen that one could now do things, at least potentially, that Inspector Lee had scarcely dreamed of; henceforth all text was infinitely plastic, no scissors required. (Only recently have I had the opportunity to examine facsimile pages of Burroughs's cut-up "journals," which are astonishing prefigurations, in paper, of pure hypertext.)

I would have been presented with ample evidence that geeks and artboys have considerable in common, under the skin, and *have* had, with regard to multimedia, from the very start. (The very start being entirely debatable, though I myself have elsewhere described cave paintings as "the first screen.")

I had not read Douglas Engelbart or Vannevar Bush then, although I had read quite a lot of what is offered here on the artboy side. I did not then know, and I was struck by this repeatedly as I recently read or reread the texts collected here, that the rants of artboys tend to date quite early on, whereas the rants of geeks retain a certain timelessness and an enviable intensity. The reason for this, I think, is that artboys of whatever era are required to expend energy in the assumption of attitude, in the donning of that black raincoat, whereas geeks (often quite scarily) are not.

I would have discovered Morton Heilig's Rube Goldberg, peculiarly *Popular Mechanics* visions of VR, neither geek nor artboy but somehow Hollywood Master Techie (and perhaps Hollywood was where the two impulses first fused, cinema having been the brilliant bastard offspring of a union once unthinkable to anyone but a frothing Italian futurist).

I would have discovered, to my great delight, the figure of Billy Klüver, the perfect recombinant sport along the supposed great divide of the Two Cultures: a geek (Bell Laboratories laser systems engineer) whose inner artboy (collaborations with Rauschenberg, Johns, and Warhol) allowed him to unhesitatingly and smilingly describe the great New York power failure of 1966 as "a Happening."

I would have seen a certain impulse toward nonlinearity, which I had been culturally conditioned to think of as the hallmark of the century's avant-garde, manifesting, often crucially, in the theorizing of Ivan Sutherland and Tim Berners-Lee.

I would, I hope, have been rightly humbled.

Who can read Vannevar Bush's "As We May Think" today and not be amazed by the weird acuity of this man's imagination? These visions of hyper-efficient, cigar-smoking technocrats, with walnut-sized "dry" cameras strapped to their foreheads, of glass-topped (and walnut sided?) desks wherein are rear-projected images collected "in the field," are, in their way, so genuinely, so embarrassingly, so rawly, prescient as to make a science fiction writer squirm. (There was a reason, you see, that Roosevelt was advised by Vannevar Bush, and not by Robert Heinlein or Isaac Asimov. Heinlein and Asimov were, relatively speaking, artboys. Bush, who personally put together what we think of today as the military industrial complex, was a *geek*.)

Today, in the opening years of a new century, I suspect that the distinction between the geek and the artboy means less than it ever has. Indeed, I wonder how effectively one can now be either to the exclusion of the other. And, indeed, I sometimes wonder how deep a distinction it ever was. Hippies, someone must somewhere have pointed out, didn't discover LSD. And of all those who partook in that particular version of the lotus, it now looks as though the ones who then shambled into the garage to continue fiddling with what eventually became the personal computer were the ones who most utterly and drastically and truly changed things.

My own early work is haunted by the collage constructions of Joseph Cornell, a prototypical artboy/geek (to the extent, it seems, of a sort of voluntary autism) who fits nowhere at all in the standard art histories, and by certain luxuriantly retrograde impulses in the work of Jorge Luis Borges, a poet who came to regard his later role of librarian as inherently the more romantic of the two callings. Cornell treated the island of Manhattan as an intensely private archive of images, and was said to possess in photographic memory the exact contents of every secondhand shop in New York City, shelf by dusty shelf. Borges regarded the universe itself as a library, infinitely recombinant, infinitely recursive, in which a single text might exist in variorum editions beyond number. I would add both these artists to the company collected here, and many more as well.

Multimedia, in my view, is not an invention but an ongoing discovery of how the mind and the universes it imagines (or vice versa, depending) fit together and interact. Multimedia is where we have always been going. Geeks and artboys, emerging together from the caves of Altamira, have long been about this great work. This book is one start toward a different sort of history, a history cognizant of an impulse that seems to me always to have been with us.

I recommend this book to you with an earnestness that I have seldom felt for any collection of historic texts. This is, in large part, where the bodies (or, rather, the bones of the ancestors) are buried. Assembled this way, in such provocative proximity, these visions give off strange sparks. Think of it, if you like, as a cut-up in Burroughs's best sense, an interleaving of histories intended to open intertextual doors, some of which, given the right reader, have never before been opened. Perhaps you are that reader.

If not, keep it handy: you may be that reader one day, be you geek or artboy (of either gender, please) or (more likely) some evolved hybrid of the two.

Someone, it seems, always has to be.

—Vancouver, August 2000

<< overture >>

by Randall Packer and Ken Jordan

Ivan Sutherland and Sketchpad. Courtesy of Lincoln Laboratory, MIT.

> *"And so the arts are encroaching one upon another, and from a proper use of this encroachment will rise the art that is truly monumental."*
>
> —WASSILY KANDINSKY

Multimedia is emerging as the defining medium of the twenty-first century. The World Wide Web, CD-ROMs, virtual reality arcade games, and interactive installations only hint at the forms of multimedia to come. Yet the concept of integrated, interactive media has its own long history, an evolution that spans more than 150 years. Remarkably, this has been a largely untold story. Discussions of the development of the personal computer and the Internet tend to focus on a few highly successful entrepreneurs, neglecting the less-known work of the engineers and artists who first sought to craft a medium that would appeal to all the senses simultaneously—a medium that would mimic and enhance the creative capacities of the human mind. Here, then, is a "secret history" of multimedia: a narrative that includes the pioneering activities of a diverse group of artists, scientists, poets, musicians, and theorists from Richard Wagner to Ivan Sutherland, from Vannevar Bush to Bill Viola.

Beginning with Wagner, subsequent generations of artists sought, and found, integrated forms and interdisciplinary strategies to express their concern with individual and social consciousness and extreme states of subjective experience. In the years since World War II, scientists have pursued personal computing and human-computer interactivity as vehicles for transforming consciousness, extending memory, increasing knowledge, amplifying the intellect, and enhancing creativity. The idealistic and ideological aspirations of both groups have resulted in a new medium that emphasizes individual choice, free association, and personal expression.

Multimedia: From Wagner to Virtual Reality brings together the major writings by multimedia's pioneers for the first time in order to foster a greater understanding of its precedents, landmarks, aesthetic roots, social impact, and revolutionary potential.

Birth of a New Medium

Douglas Engelbart. Courtesy of Douglas Engelbart, Bootstrap Institute.

New technology has always been used to make media. George Lucas shot his latest *Star Wars* epic with digital cameras, though the audience experience was no different than if it had been shot on celluloid. But while not all computer-based media is multimedia, today's multimedia starts with the computer, and takes the greatest advantage of the computer's capability for personal expression.

Digital computers were initially designed as calculating machines. The first fully electronic computer, the ENIAC, was built by the United States military during World War II to produce ballistics tables for artillery in battle. Computers then were clumsy, hulking devices—the ENIAC had eighteen thousand vacuum tubes, and measured fifty feet by thirty feet—that did calculations for scientific research. Only a handful of scientists considered the possibility of personal computing for creative purposes by nonspecialists.

The first scientist to think seriously of this potential was Vannevar Bush. In his 1945 article "As We May Think," he outlined "a future device for individual use, which is a sort of mechanized private file and library." Before the ENIAC was completed, Bush was already contemplating how information technology could enhance the individual's capability for creative thought. "The human mind . . . operates by association," Bush observed. The device that he proposed, which he named the memex, enabled the associative indexing of information, so that the reader's trail of association would be saved inside the machine, available for reference at a later date. This prefigured the notion of the hyperlink. While Bush never actually built the memex, and while his description of it relied on technology that predated digital information storage, his ideas had a profound influence on the evolution of the personal computer.

In the years immediately after the war, under the shadow of the atomic bomb, the scientific establishment made a concerted effort to apply recent advancements in technology to humanitarian purposes. In this climate, Norbert Wiener completed his groundbreaking theory on cybernetics. While Wiener did not live to see the birth of the personal computer, his book, *The Human Use of Human Beings*, has become *de rigueur* for anyone investigating the psychological and sociocultural implications of human-machine interaction. Wiener understood that the quality of our communication with machines affects the quality of our inner lives. His approach provided the conceptual basis for human-computer interactivity and for our study of the social impact of electronic media.

Bush and Wiener established a foundation on which a number of computer scientists associated with the Advanced Research Projects Agency (ARPA)—a United States government-funded program to support defense-related research in the 1960s—began to build. Leading ARPA's effort to promote the use of computers in defense was the MIT psychologist and computer scientist J.C.R. Licklider, author of the influential article "Man-Computer Symbiosis." Defying the conventional wisdom that computers would eventually rival human intelligence rather than enhance it, Licklider proposed that the computer be developed as a creative collaborator, a tool that could extend human intellectual capability and improve a person's ability to work efficiently.

While at ARPA, Licklider put significant resources toward the pursuit of his vision. Among the scientists he supported was Douglas Engelbart, who since the mid-1950s had been seeking support for the development of a digital information retrieval system inspired by Bush's memex. ARPA funding enabled Engelbart to assemble a team of computer scientists and psychologists at the Stanford Research Institute to create a "tool kit" that would, as he phrased it, "augment human intellect." Dubbed the oNLine System (NLS), its public debut in 1968 at the Fall Joint Computer Conference in San Francisco was a landmark event in the history of computing. Engelbart unveiled the NLS before a room of three thousand computer scientists, who sat in rapt attention for nearly two hours while he demonstrated some of his major innovations, including the mouse, windows for text editing, and electronic mail. Engelbart was making it possible, for the first time, to reach virtually through a computer's interface to manipulate information. Each of his innovations was a key step toward an interface that allowed for intuitive interactivity by a nonspecialist. At the end of his presentation, he received a standing ovation.

However, the contributions of the NLS went beyond innovation regarding the computer interface. Engelbart and his colleagues also proposed that creativity could be enhanced by the sharing of ideas and information through computers used as communications devices. The oNLine System had its computers wired into a local network, which enabled them to be used for meaningful collaboration between co-workers. Engelbart understood that the personal computer would not only augment intelligence but augment communication as well. In 1969 his research in on-line networking came to fruition with the creation of ARPANET, the forerunner of the Internet. The first message on the ARPANET, in fact, was sent to Engelbart's lab at SRI from UCLA.

Engelbart's NLS pioneered some of the essential components necessary for the personal computer, but it would be up to a new generation of engineers to advance computing so it could embrace multimedia. As a graduate student in the late 1960s, Alan Kay wrote a highly influential Ph.D. thesis proposing a personal information management device that, in many ways, prefigured the laptop. In 1970, as research in information science was shifting from East Coast universities and military institutions to private digital companies in Silicon Valley, Kay was invited to join the new Xerox PARC (Palo Alto Research Center) in California. PARC's mandate was no less than to create "the architecture of information for the future."

At PARC, Alan Kay pursued his idea for the Dynabook—a notebook-sized computer that enabled hyperlinking, was fully interactive, and integrated all media. With the Dynabook, digital multimedia came into being. Echoing Licklider, Engelbart, and colleagues at PARC, Kay declared the personal computer a medium in its own right. It was a "meta-medium," as he described it in his 1977 essay "Personal Dynamic Media," capable of being "all other media." While the Dynabook remained a prototype that was never built, the work that came from its development, including the invention of the Graphical User Interface (GUI) and subsequent breakthroughs in dynamic computing, was incorporated into the first true multimedia computer, the Xerox Alto.

The Integration of the Arts

Pepsi Pavilion. *Photo by Harry Shunk. Courtesy of Billy Klüver, E.A.T.*

By proposing that the Dynabook be a "meta-medium" that unifies all media within a single interactive interface, Alan Kay had glimpsed into the future. But he may not have realized that his proposal had roots in the theories of the nineteenth-century German opera composer, Richard Wagner.

In 1849, Wagner introduced the concept of the *Gesamtkunstwerk,* or Total Artwork, in an essay called "The Artwork of the Future." It would be difficult to overstate the power of this idea, or its influence. Wagner's description of the *Gesamtkunstwerk* is one of the first attempts in modern art to establish a practical, theoretical system for the comprehensive integration of the arts. Wagner sought the idealized union of *all* the arts through the "totalizing," or synthesizing, effect of music drama—the unification of music, song, dance, poetry, visual arts, and stagecraft. His drive to embrace the full range of human experience, and to reflect it in his operas, led him to give equal attention to every aspect of the final production. He was convinced that only through this integration could he attain the expressive powers he desired to transform music drama into a vehicle capable of affecting German culture.

Twentieth-century artists have continued the effort to heighten the viewer's experience of art by integrating traditionally separate disciplines into single works. Modern experience, many of these artists believed, could be evoked only through an art that contained within itself the complete range of perception. "Old-fashioned" forms limited to words on a page, paint on a canvas, or music from an instrument were considered inadequate for capturing the speed, energy, and contradictions of contemporary life.

In their 1916 manifesto "The Futurist Cinema," F. T. Marinetti and his revolutionary cohorts declared film to be the supreme art because it embraced all other art forms through the use of (then) new media technology. Only cinema, they claimed, had a "totalizing" effect on human consciousness. Less than a decade later, in his 1924 essay describing the theater of the Bauhaus, "Theater, Circus, Variety," László Moholy-Nagy called for a theater of abstraction that shifted the emphasis away from the actor and the written text, and brought to the fore every other aspect of the theatrical experience. Moholy-Nagy declared that only the synthesis of the theater's essential formal components—space, composition, motion, sound, movement, and light—into an organic whole could give expression to the full range of human experience.

The performance work of John Cage was a significant catalyst in the continuing breakdown of traditional boundaries between artistic disciplines after World War II. In the late 1940s, during a residency at Black Mountain College in North Carolina, Cage organized a series of events that combined his interest in collaborative performance with his use of indeterminacy and chance operations in musical composition. Together with choreographer Merce Cunningham and artists Robert Rauschenberg and Jasper Johns, Cage devised theatrical experiments that furthered the dissolution of borders between the arts. He was particularly attracted to aesthetic methods that opened the door to greater participation of the audience, especially if these methods encouraged a heightened awareness of subjective experience. Cage's use of indeterminacy and chance-related technique shifted responsibility for the outcome of the work away from the artist, and weakened yet another traditional boundary, the divide between artwork and audience.

Cage's work proved to be extremely influential on the generation of artists that came of age in the late 1950s. Allan Kaprow, Richard Higgins, and Nam June Paik were among the most prominent of the artists who, inspired by Cage, developed nontraditional performance techniques that challenged accepted notions of form, categorization, and composition, leading to the emergence of genres such as the Happening, electronic theater, performance art, and interactive installations.

Kaprow, who coined the term "Happening," was particularly interested in blurring the distinction between artwork and audience. The ultimate integrated art, he reasoned, would be without an audience, because every participant would be an integral part of the work. As he wrote in his 1966 primer, "Untitled Guidelines for Happenings," "The line between art and life should be kept as fluid, and perhaps indistinct, as possible." This approach led to a performance style that pioneered deliberate, aesthetically conceived group interactivity in a composed environment. Happenings artists devised formal elements that allowed participants the freedom to make personal choices and collective decisions that would affect the performance.

In this climate, artists became increasingly interested in integrating technology into their work. While technology clearly played a significant role in twentieth-century arts (such as photography, film, and video, as well as various fine arts genres), it was not until Bell Labs scientist Billy Klüver placed the potential of advanced engineering into the hands of artists in New York that integrated works of art and technology began to flourish. Klüver conceived the notion of equal collaboration between artist and engineer. He pioneered forms of art and technology that would have been unimaginable to the artist without the engineer's cooperation and creative involvement. With Robert Rauschenberg, Klüver created several of the earliest artworks to integrate electronic media and to encourage a participatory role for the audience, including *Oracle* (1963–1965) and *Soundings* (1968).

In 1966 Klüver and Rauschenberg, along with Robert Whitman and Fred

Waldhauer, cofounded E.A.T. (Experiments in Art and Technology) to bring artists and engineers together to create new works. E.A.T.'s most ambitious production was the Pepsi-Pavilion, designed for the Osaka Expo '70 in Japan—a tremendously ambitious collaborative, multimedia project that involved more than seventy-five artists and engineers. As Klüver explained, audience participation was at the heart of their interests: "The initial concern of the artists who designed the Pavilion was that the quality of the experience of the visitor should involve choice, responsibility, freedom, and participation. The Pavilion would not tell a story or guide the visitor through a didactic, authoritarian experience. The visitor would be encouraged as an individual to explore the environment and compose his own experience."

During this period, the British artist and theorist Roy Ascott began to explore the use of computers in artistic expression. One of the first theoretical attempts to integrate the emerging fields of human-computer interactivity and cybernetics with artistic practice was Ascott's article "Behavioral Art and the Cybernetic Vision," from 1966–1967. Ascott noted that the computer was "the supreme *tool* that . . . technology has produced. Used in conjunction with synthetic materials it can be expected to open up paths of radical change in art." Ascott saw that human-computer interaction would profoundly affect aesthetics, leading artists to embrace collaborative and interactive modes of experience.

When Alan Kay arrived at Xerox PARC in 1970, the foundation was in place for a multimedia that synthesized all the existing art forms, and presented them in an environment that allowed for meaningful interactivity. With the Dynabook, the interactive *Gesamkunstwerk* was brought into the digital realm, and put online.

Through the Looking Glass

The fantasy of being transported into another world, to be taken wholly into an imaginary realm, is a primal desire. With computer-based multimedia, encounters with immersive, virtual worlds will soon become commonplace. Virtual reality, after all, is a logical extension of the integration of the arts. It is also an ideal environment for applying our knowledge of human-computer interactivity.

There is an evocative echo in virtual environments of the earliest known form of human expression—the prehistoric cave paintings found at such sites as the caves of Lascaux in the south of France. These immersive environments, dating from 15,000 B.C., are thought by scholars to have been theaters for the performance of rituals that integrated all forms of media and engaged all the senses.

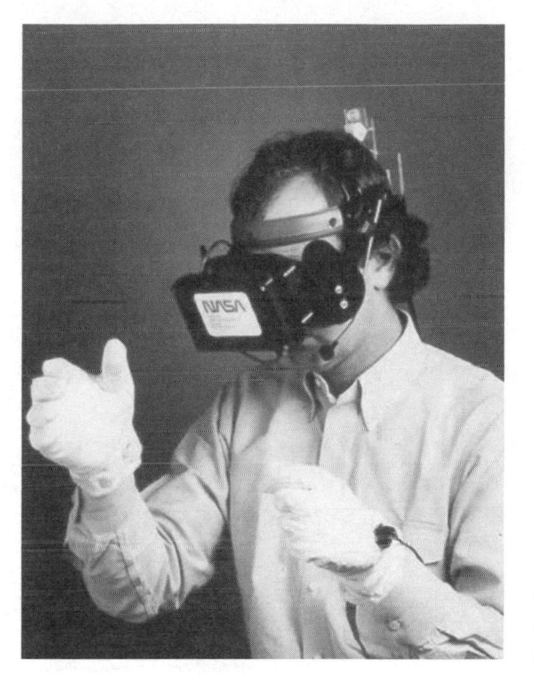

Scott Fisher. VIEW Head-Mounted Display. Courtesy of Scott Fisher, NASA–Ames Research Center.

On the walls were painted animal images and shamanist scrawls, but the environment in which the paintings appeared was surely as significant as the paintings themselves. As Joseph Campbell described it, "these magical spots occur far from the natural entrances of the grottos, deep within the dark, wandering chill corridors and vast chambers, so that before reaching them one has to experience the full force of the mystery of the cave itself." To encounter the cave paintings was to immerse the self in an otherworldly domain, which would heighten consciousness and trigger altered states of perception.

There are many examples of immersion in the history of art. The Dyonisian rituals of Greek theater and the great cathedrals of Europe are two obvious examples. Richard Wagner's *Gesamtkunstwerk* was driven by a vision of theater in which the audience loses itself in the veracity of the drama, creating an immersive experience. As he wrote in *The Artwork of the Future,* "the spectator transplants himself upon the stage, by means of all his visual and aural faculties." To facilitate his vision, Wagner reinvented the conventions of the opera house, and in 1876 opened the Festpielhaus Theater in Bayreuth, Germany, with the first complete production of *The Ring* cycle. The Festpielhaus, with its employment of Greek amphitheatrical seating, surround-sound acoustics, the darkening of the house, and the placement of musicians in an orchestra pit, focused the audience's undivided attention on the dramatic action. His intent was to maximize the suspension of disbelief, to draw the viewer into an illusionary world staged within the proscenium arch.

In the 1950s, a similar vision inspired the American cinematographer Morton Heilig to propose a "cinema of the future" that would surround the audience with facsimiles of life so convincing they would believe themselves to be transported to another domain. Such a cinema, he wrote, "would faithfully reproduce man's outer world as perceived in his consciousness, it will eventually learn to create totally new sense materials for each of the senses . . . [that] they have never known before, and to arrange them into forms of consciousness never before experienced by man in his contact with the outer world."

While Heilig devised a theoretical framework that applied the technologies of his day toward the achievement of virtual experience, it was only as a consequence of advances in computer science that the immersion his work suggested became possible.

By 1965, Ivan Sutherland had already achieved legendary status among computer scientists as the inventor of Sketchpad, the first interactive graphics software. In a short paper published that year, Sutherland mused over the options available to the engineer to display computer data, to create a "a looking glass" into what he described as a "mathematical wonderland." It seemed reasonable for him to suggest that "The Ultimate Display" (as the paper was titled) would represent this data in three-dimensional form, allowing the construction of entirely believable three-dimensional, computer-controlled, virtual worlds. However, like Heilig before him, Sutherland took this suggestion one step further. "The ultimate display," he wrote, "would . . . be a room within which the computer can control the existence of matter." He suggested that such a display could present realities heretofore only imagined, as if seen through Alice's looking glass. Sutherland's proposal was startling, but it launched an entire field of scientific inquiry.

It also fueled the imagination of a generation of artists. One of the first to consider the possibilities of digitally constructed virtual experiences was Myron Krueger. In the early 1970s, Krueger created the pioneering works *Metaplay* and *Videoplace* to explore the potential of computer-mediated interactivity. These works were interactive artistic environments, influenced by Happenings, designed to give participants freedom of choice and opportunities for personal expression. *Videoplace* also connected participants in different locations through networked technologies, creating the illusion of shared space. As Krueger later wrote about the piece, "our teleconference created a place that consisted of the information we both shared . . . a world in which full physical participation would be possible. This world is an 'artificial reality.' "

During the 1970s and 1980s, several engineering projects pursued virtual environment display systems that could represent such an artificial reality. Perhaps the most significant of these in the mid-1980s was led by Scott Fisher at the NASA-Ames Research Center. Fisher's intent was to engage the entire nervous system in a multisensory presentation of virtual space—extending multimedia beyond the screen. The Ames VIEW system (an acronym for Virtual Interface Environmental Workstation) included a headset with two small liquid crystal display screens, a microphone for speech recognition, earphones for surround-sound effects, a head-tracking device, and a dataglove to recognize the user's gestures and place them within the virtual environment. The direction this work pointed in was clear. As Fisher wrote in his 1989 article "Virtual Interface Environments," "with full body tracking capability, it would also be possible for users to be represented in this [virtual] space by life-size virtual representations of themselves in whatever form they choose." Immersive environments, Fisher observed, could give birth to a new form of participatory, interactive electronic theater.

The possibility of such a theater had already taken hold of the public's imagination. In *Neuromancer,* his widely read novel from 1984, William Gibson described in palpable detail a future in which virtual reality was a fact of life. Echoing Myron Krueger's notion that teleconferencing created a "place" that consisted of shared information, Gibson's characters inhabited a virtual environment made possible by the networking of computers, which he named "cyberspace." Gibson's cyberspace provided the first literary definition for the computers, hubs, servers, and databases that make up the matrices of the network. His discussion of cyberspace was so tangible—and seductive, with its suggestion that any computer hacker could "jack-in to the matrix" for an encounter with a sexy avatar—it became a touchstone for every engineer, artist, and theorist working in the field.

Marcus Novak took Gibson's description of virtual environments as the starting point for his own theoretical and artistic explorations. In his essay from 1991, "Liquid Architecture in Cyberspace," he follows the pioneering work of Sutherland, Fisher, and Gibson, et al., to its logical conclusion, and notes its profound implications for architecture, our notions of space, and our attitudes toward the organization of information. He notes that in cyberspace, since all structure is programmable, all environments can be fluid. The artist who designs these immersive digital habitats will be able to transcend the laws of the physical world. As a consequence, architectural forms built in cyberspace can respond to the viewer, encouraging provocative and illuminating interactions. In cyberspace, architecture becomes a form of poetry.

While most research in virtual reality aims to project the viewer into a digital environment by means of a head-mounted display—as in the work of such media artists as Jason Lanier, Char Davies, and Jenny Holzer—some engineers have taken an alternative approach. In the early 1990s, Daniel Sandin and Thomas DeFanti conceived of a virtual reality system that places the human body directly inside a computer-generated environment. They describe their system, called the CAVE (an acronym for Cave Automatic Virtual Environment) in their article "Room with a View": "Unlike users of the video-arcade type of virtual reality system, CAVE 'dwellers' do not need to wear helmets, which would limit their view of and mobility in the real world . . . to experience virtual reality." Instead, participants in the CAVE are surrounded by an immersive, digital "cave painting"—which brings the evolution of immersion full circle, back to the prehistoric caves of Lascaux, and humankind's earliest efforts at personal expression.

The Integrated Datawork

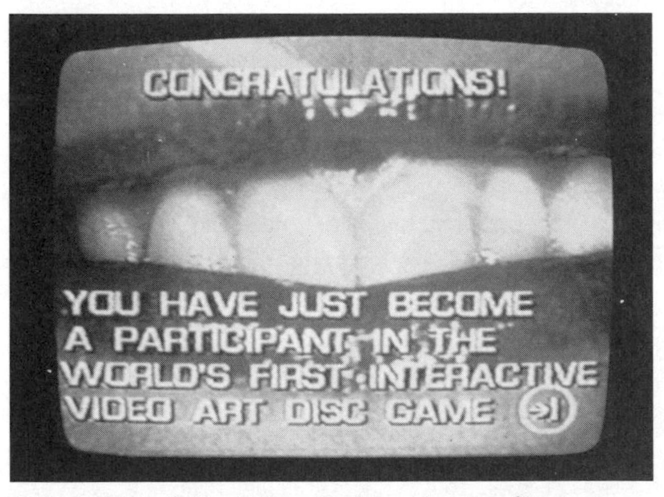

Lynn Hershman, Lorna. *© Lynn Hershman. Courtesy of Lynn Hershman.*

It was Vannevar Bush who, in 1945, determined the chief narrative characteristic of multimedia by proposing a mechanical device that operated literally "as we may think." The challenge, as he saw it, was to create a machine that supported the mind's process of free association. Bush noted how ideas tend to evolve in a nonlinear, idiosyncratic fashion. His memex would be a tool that could supplement this aspect of human creativity by organizing its media elements to reflect the dynamics of the mind at play.

Douglas Engelbart expanded on Bush's premise. His quest to "augment human intellect," as he aptly phrased it, was based on the insight that the open flow of ideas and information between collaborators was as important to creativity as private free association was. The personal computer, as he envisioned it, would not only allow for the arrangement of data in idiosyncratic, nonlinear formats, but by connecting workstations to a data-sharing network and turning them into communications devices, Engelbart's oNLine System allowed for a qualitative leap in the collaboration between individuals—almost as if colleagues could peer into one another's minds as part of the creative process. In the early 1960s, experiments with networked personal computing promised the nonlinear organization of information on a grand scale.

While few recognized this possibility at the time, it inspired a series of influential theoretical writings by the rogue philosopher Ted Nelson. Working outside the academic and commercial establishments, following his own strongly held convictions, Nelson devised an elaborate system for the sharing of information across computer networks. Called Xanadu, this system would maximize a computer's creative potential. Central to Nelson's approach was the "hyperlink," a term he coined in 1963, inspired by Bush's notion of the memex's associative trails. Hyperlinks, he proposed, could connect discrete texts in nonlinear sequences. Using hyperlinks, Nelson realized, writers could create "hypertexts," which he described as "nonsequential writing" that let the reader make decisions about how the text could be read in other than linear fashion. As he observed in his landmark book from 1974, *Computer Lib/Dream Machines,* "the structures of ideas are not sequential." With hypertext, and its multimedia counterpart, "hypermedia," writers and artists could create works that encouraged the user to leap from one idea to the next in a series of provocative juxtapositions that presented alternatives to conventional hierarchies.

Nelson's insights were paralleled by experiments in the literary avant-garde that challenged traditional notions of linear narrative. In his book, he refers to Vladimir Nabokov's *Pale Fire* and Julio Cortázar's *Hopscotch* as two novels that use unconventional branching structures to encourage the reader's active collaboration in the construction of the story. As we have already seen, experimental performances inspired by John Cage—including Happenings, interactive installations, and performance art—also gave rise to a variety of nonlinear narrative strategies. But perhaps the most prescient explorer of this terrain was the novelist William S. Burroughs.

Like Ted Nelson, Burroughs was deeply suspicious of established hierarchies. He was especially interested in writing techniques that suggest the spontaneous, moment-by-moment movement of the mind, and how nonlinear writing might expand the reader's perception of reality. Through his use of the cut-up and fold-in techniques, which he described in his 1964 essay "The Future of the Novel," Burroughs treated the reading experience as one of entering into a multidirectional web of different voices, ideas, perceptions, and periods of time. He saw the cut-up as a tool that let the writer discover previously undetected connections between things, with potentially enlightening and subversive results. With the cut-up, Burroughs prefigured the essential narrative strategy of hypertext and its ability to allow readers to leap across boundaries in time and space.

Since the invention of the electric telegraph by Samuel Morse in the 1830s, commentators have been noting the transformation of our concepts of space and time by wired technology. From the telegraph to the telephone to television to satellite communications, modern telecommunications has eradicated geographic borders, and made speed a central factor in modern life. This effect was commonly acknowledged as long ago as 1868, when, at a banquet held in honor of Morse's life achievement, he was toasted for having "annihilated both space and time in the transmission of intelligence. The breadth of the Atlantic, with all its waves, is as nothing."

Artists have grappled with the implications of this technology since its inception; the narrative experiments of literary authors reflect this current in modern art. Ted Nelson's concept of hypertext represented a profound effort to put this technology toward the service of personal, idiosyncratic expression. Nelson became an evangelist for hypertext, publishing articles, speaking at conferences, spreading the gospel wherever he could. One of those places was Brown University, which during the 1980s became a hotbed of literary explorations of the form. At Brown, the literary critic George Landow and his colleagues developed hypertext tools, such as Intermedia, which allowed authors with little experience in programming to invent new genres of creative writing. In his own work, Landow applied a trained critical eye to the formal aspects of hypertext, making connections to the poststructural textual analysis of critics like Roland Barthes and Jacques Derrida. Just as academic theoretical discourse was questioning the centrality of the author in the production of texts, hypermedia suggested that, in a future of networked digital media, responsibility would shift from author to reader, actively encouraging the reader's collaboration in shaping a narrative.

During this period, media artists whose roots lay in performance and video also began investigating hypermedia as a means of exploring new forms for telling stories. Artists such as Lynn Hershman and Bill Viola were drawn to the computer's ability to break down linear narrative structures. Viola approached the medium as a repository for evocative images that could be projected on screens in installations, "with the viewer wandering through some three-dimensional, possibly life-sized field of prerecorded or simulated scenes evolving in time," as he described it. Hershman was among the first to create digital artworks using interactive media, in such pieces as *Deep Contact,* from 1989. She introduced interactivity into her work to combat the loss of intimacy and control brought about by the dominance of media such as radio and television. Her use of hypermedia allowed the viewer to choose directions inside the artwork's complex branching structure, and shape a personal experience from it.

By the late 1980s, multimedia, which had been at the fringe of the arts and sciences, reached critical mass and went mainstream. Marc Canter, who developed the first commercial multimedia authoring system, was a chief catalyst. Canter pioneered software tools that artists and designers used to create multimedia on their personal computers. His authoring systems synthesized text, images, animation, video, and sound into a single integrated work, using hyperlinks and other hypermedia techniques to connect its various elements.

In 1974, Ted Nelson had declared that "The real dream is for 'everything' to be in the hypertext." It was a proposal that echoed Marshall McLuhan's influential observation that "after more than a century of electric technology, we have extended our central nervous system itself in a global embrace, abolishing both space and time as far as our planet is concerned." But despite Nelson's best efforts, and many advances in the fields of hypermedia and computer-based telecommunications, the global hypermedia library he envisioned remained more dream than reality. Most innovations in hypermedia focused on closed systems, such as the CD-ROM and interactive installations, rather than on open systems using a computer network.

In 1989 Tim Berners-Lee, a young British engineer working at CERN, the particle physics laboratory in Geneva, Switzerland, circulated a proposal for an in-house on-line document sharing system, which he described modestly as "a 'web' of notes with links." After getting a grudging go-ahead from his superiors, Berners-Lee dubbed this system the World Wide Web. The Web, as he designed it, combined the communications language of the Internet with Nelson's hypertext and hypermedia, enabling links between files to extend across a global network. It became possible to link every document, sound file, or graphic on the Web in an infinite variety of nonlinear paths through the network. And instead of being created by a single author, links could be written by anyone participating in the system. Not only did the open nature of the Web lend itself to a wide array of interactive, multimedia experiences, but by hewing to a nonhierarchical structure and open protocols, Berners-Lee's invention became enormously popular, and led to an explosion in the creation of multimedia. By 1993 the Web had truly become an international phenomenon.

The success of the Web seemed to confirm the intuition of artists engaging in digital media that in the future, a global media database would inspire new forms of expression. Roy Ascott, for example, had already been exploring the creative possibilities of networking since the 1980s. He was interested in the notion of "dataspace," a territory of information in which all data exists in a continual present outside the traditional definitions of time and space available for use in endless juxtapositions. Ascott considers dataspace a new type of *Gesamtkunstwerk,* or a *Gesamtdatenwerk,* as he calls it, in which networked information is integrated into the artwork. In such an environment, Ascott writes, "meaning is not something created by the artist, distributed through the network, and received by the observer. Meaning is the product of interaction between the observer and the system, the content of which is in a state of flux, of endless change and transformation."

This notion of the artwork as a territory for interaction, as a locus of communications for a community, echoes the Happenings of a previous generation. On-line role-playing games have become laboratories for exploring this form of interactivity. As the social theorist Sherry Turkle has pointed out, on-line communities, such as Multi-User Dungeons (MUD), "are a new genre of collaborative writing, with things in common with performance art, street theater, improvisation theater, commedia dell'arte, and script writing." Pavel Curtis created one of the earliest MUDs, LambdaMOO, in 1990 at Xerox PARC. Though it consisted only of text, its interactive quality, made possible through intricate storytelling devices via the Internet, gave participants the illusion of immersion in a virtual environment. Interaction in the on-line environment, Curtis claimed, creates a kind of social behavior which "in some ways . . . is a direct mirror of behavior in real life."

Throughout history, art has often been referred to as a mirror of life. But by building upon the twin notions of association and collaboration, computer-based multimedia may well become more than a mirror of life. Already we have seen how multimedia blurs the boundaries between life and art, the personal and the mediated, the real and the virtual. The implications of these tendencies we are only now beginning to grasp.

The Future Is Under Construction

Marcos Novak, Voice 3 + 4 Maze Blue. *© Marcos Novak. Courtesy of Marcos Novak.*

The breadth and potential of multimedia lends itself to utopian proposals. The French media theorist Pierre Lévy describes multimedia as belonging to a trajectory of planetary evolution that runs from DNA to cyberspace—an arc that follows pure information as it reaches toward its most evolved form of expression. He proposes that today's global networks will usher in an era of "collective intelligence," and suggests that "cyberspace constitutes a vast, unlimited field ... designed to interconnect and provide interface for the various methods of creation, recording, communication and simulation." His enthusiastic perspective evokes a world of intriguing possibilities.

At the same time, we are all aware of the dystopian qualities of the 24/7 infotainment juggernaut that is being delivered across the globe through an ever more sophisticated telecommunications network. We read daily about the new media's encroachment on privacy, its opportunity for abuse, and the specter of centralized control that it might make possible. These dangers are real. There is a tension between opposing positions at the heart of the Internet—between those who prize its potential for an open, freewheeling exchange of art and ideas and those who see its pervasiveness as an opportunity to expand upon the marketing-driven broadcast model of twentieth-century media—and it is not at all clear whether utopian or dystopian visions will ultimately prevail.

The articles in this book serve as a poignant reminder of the intentions of multimedia's pioneers. Their words, typically written during the heat of invention, convey a passionate involvement with higher ideals. To a remarkable degree, these scientists, artists, and theorists share a commitment to forms of media and communications that are nonhierarchical, open, collaborative, and reflective of the free movement of the mind at play. It is, in sum, an extraordinary vision. But whether we will achieve it is an unresolved question.

Toward a Definition of Multimedia

Many critics of today's multimedia shy away from attempts to identify a dominant theme behind the emergence of this new medium. They say that the subject is too various, that it resists a neat historical frame. In fact, there is a tendency among critics to celebrate the elusive nature of the subject. Multimedia, by its very nature, is open, democratic, nonhierarchical, fluid, varied, inclusive—a slippery domain that evades the critic's grasp just on the verge of definition. But these qualities did not evolve by happenstance. They were the product of deliberate intent on the part of multimedia's pioneers, who were aiming for quite coherent goals.

Just as there are many possible paths through a network, there are many potential readings of multimedia's history. In ours, the key characteristics intrinsic to computer-based multimedia are defined as integration, interactivity, immersion, hypermedia, and narrativity. These five characteristics determine the scope of multimedia's capabilities for expression; they establish its full potential. The purpose of this book is to show how these characteristics evolved more or less simultaneously, each following its own tradition and trajectory and yet inextricably interwoven with the others in a web of mutual influence.

For our purposes, the following definitions apply:

Integration: the combining of artistic forms and technology into a hybrid form of expression.

Interactivity: the ability of the user to manipulate and affect her experience of media directly, and to communicate with others through media.

Hypermedia: the linking of separate media elements to one another to create a trail of personal association.

Immersion: the experience of entering into the simulation or suggestion of a three-dimensional environment.

Narrativity: aesthetic and formal strategies that derive from the above concepts, which result in nonlinear story forms and media presentation.

With the Dynabook in the early 1970s, Alan Kay invented a machine that incorporated all five of these characteristics for the first time, giving birth to digital multimedia. However, much computer-based media today does not reflect these characteristics. *A Tale of Two Cities* screened on a personal digital assistant remains a nineteenth-century novel; not everything digital is multimedia. Yet with the addition of animated illustrations, hyperlinked text, an on-line discussion group, or, more ambitiously, a MUD-like role-playing game that takes place in a virtual Paris at the time of the Revolution, Dickens's text might—for good or for ill—become the basis for a multimedia work. Just as movies have been based on novels since the silent era, we are seeing multimedia adaptations of classic works, some of which may one day become classics themselves. Given the wide array of forms that multimedia currently takes—Internet art, virtual reality installations, graphical on-line chat spaces, and real-time networked performance, and many more to come—it might be fair to say that the medium's only defining element is its mutability, that we can't possibly predict the variety of its various manifestations. Perhaps multimedia's most consistent quality will be its relentlessly changing nature.

In this book, one further attempt at definition has guided all the others: Multimedia is the form that makes the most complete use of the computer's potential for personal expression.

<< part I >>

Integration

Richard Wagner

<< 1 >>

"Outlines of the Artwork of the Future," *The Artwork of the Future* (1849)

Richard Wagner.

"Whereas the public, that representation of daily life, forgets the confines of the auditorium, and lives and breathes now only in the artwork which seems to it as Life itself, and on the stage which seems the wide expanse of the whole World."

<< For Richard Wagner, the "Artwork of the Future" represented a rejection of lyric opera, which the German composer considered hopelessly superficial, a tired showcase for pompous divas. Yet the implications of this landmark publication go well beyond the transformation of opera. Wagner believed that the future of music, music theater, and all the arts lay in an embrace of the "collective art-work," a fusion of the arts that had not been attempted on this scale since the classic Greeks. In this essay, we find the first comprehensive treatise arguing for the synthesis of the arts, or, as Wagner defines it, the *Gesamtkunstwerk,* the total artwork.

Wagner was convinced that the separate branches of art—music, architecture, painting, poetry, and dance—would attain new poetic heights when put to the service of the drama, which he viewed as the ideal medium for achieving his vision. His totalizing approach to music theater also foreshadowed the experience of virtual reality. Scenic painting, lighting effects, and acoustical design were intended to render an entirely believable "virtual" world, in which the proscenium arch serves as the interface to the stage environment. In 1876, twenty-seven years after *The Artwork of the Future,* Wagner took this approach even further when he opened the famous Festpielhaus Theater in Bayreuth, Germany, where his theatrical innovations included darkening the house, surround-sound reverberance, the orchestra pit, and the revitalization of the Greek amphitheatrical seating to focus audience attention onto the stage.

While Wagner's hyperromantic rhetoric may seem archaic today, there is acumen and foresight in his approach to the *Gesamtkunstwerk* and the synthesis of the arts, which illuminates contemporary notions of multimedia. >>

. . . Artistic Man can only fully content himself by uniting every branch of Art into the *common* Artwork: in every *segregation* of his artistic faculties he is *unfree,* not fully that which he has power to be; whereas in the *common* Artwork he is *free,* and fully that which he has power to be.

The *true* endeavour of Art is therefore all-embracing: each unit who is inspired with a true *art-instinct* develops to the highest his own particular faculties, not for the glory of these special faculties, but for the glory of *general Manhood in Art.*

The highest conjoint work of art is the *Drama:* it can only be at hand in all its *possible* fulness, when in it each *separate branch of art* is at hand in *its own utmost fulness.*

The true Drama is only conceivable as proceeding from a *common urgence of every art* towards the most direct appeal to a *common public.* In this Drama, each separate art can only bare its utmost secret to their common public through a mutual parleying with the other arts; for the purpose of each separate branch of art can only be fully attained by the reciprocal agreement and co-operation of all the branches in their common message.

Architecture can set before herself no higher task than to frame for a fellowship of artists, who in their own persons portray the life of Man, the special surroundings necessary for the display of the Human Artwork. Only that edifice is built according to Necessity, which answers most befittingly an aim of man: the highest aim of man is the artistic aim; the highest artistic aim—the Drama. In buildings reared for daily use, the builder has only to answer to the lowest aim of men: beauty is therein a luxury. In buildings reared for luxury, he has to satisfy an unnecessary and unnatural need: his fashioning therefore is capricious, unproductive, and unlovely. On the other hand, in the construction of that edifice whose every part shall answer to a common and artistic aim alone,—thus in the building of the *Theatre,* the master-builder needs only to comport himself as *artist,* to keep a single eye upon the *art-work.* In a perfect theatrical edifice, Art's need alone gives law and measure, down even to the smallest detail. This need is twofold, that of *giving* and that of *receiving,* which reciprocally pervade and condition one another. The *Scene* has firstly to comply with all the conditions of "space" imposed by the joint *(gemeinsam)* dramatic action to be displayed thereon: but secondly, it has to fulfil those conditions in the sense of bringing this dramatic action to the eye and ear of the spectator in intelligible fashion. In the arrangement of the *space for the spectators,* the need for optic and acoustic understanding of the artwork will give the necessary law, which can only be observed by a union of beauty and fitness in the proportions; for the demand of the collective *(gemeinsam)* audience is the demand for the *artwork,* to whose comprehension it must be distinctly led by everything that meets the eye.[1] Thus the spectator transplants himself upon the stage, by means of all his visual and aural faculties; while the performer becomes an artist only by complete absorption into the public. Everything, that breathes and moves upon the stage, thus breathes and moves alone from eloquent desire to impart, to be seen and heard within those walls which, however circumscribed their space, seem to the actor from his scenic standpoint to embrace the whole of humankind; whereas the public, that representative of daily life, forgets the confines of the auditorium, and lives and breathes now only in the artwork which seems to it as Life itself, and on the stage which seems the wide expanse of the whole World. . . .

Here *Landscape-painting* enters, summoned by a common need which she alone can satisfy. What the painter's expert eye has seen in Nature, what he now, as artist, would fain display for the artistic pleasure of the full community, he dovetails into the united work of all the arts, as his own abundant share. Through him the scene takes on complete artistic truth: his drawing, his colour, his glowing breadths of light, compel Dame Nature to serve the highest claims of Art. That which the landscape-painter, in his struggle to impart what he had seen and fathomed, had erstwhile forced into the narrow frames of panel-pictures,—what he had hung up on the egoist's secluded chamber-walls, or had made away to the inconsequent, distracting medley of a picture-barn,—*therewith* will he henceforth fill the ample framework of the Tragic stage, calling the whole expanse of scene as witness to his power of re-creating Nature. The illusion which his brush and finest blend of colours could only hint at, could only distantly approach, he will here bring to its consummation by artistic practice of every known device of optics, by use of all the art of "lighting." The apparent roughness of his tools, the seeming grotesqueness of the method of so-called "scene-painting," will not offend him; for he will reflect that even the finest camel's-hair brush is but a humiliating instrument, when compared with the perfect Artwork; and the artist has no right to *pride* until he is *free, i.e.,* until his artwork is completed and alive, and *he,* with all his helping tools, has been absorbed into it. But the finished artwork that greets him from the *stage* will, set within this frame and held before the common gaze of full publicity, immeasurably more content him than did his earlier work, accomplished with more delicate tools. He will not, forsooth, repent the right to use this scenic space to the benefit of such an artwork, for sake of his earlier disposition of a flat-laid scrap of canvas! For as, at the very worst, his work remains the same no matter what the frame from which it looks, provided only it bring its subject to intelligible show: so will his artwork, in *this* framing, at any rate effect a livelier impression, a greater and more universal understanding, than the whilom landscape picture.

The organ for all understanding of Nature, is Man: the landscape-painter had not only to impart to men this understanding, but to make it for the first time plain to them by depicting Man in the midst of Nature. Now by setting his artwork in the frame of the Tragic stage, he will expand the individual man, to whom he would address himself, to the associate manhood of full publicity, and reap the satisfaction of having spread his understanding out to that, and made it partner in his joy. But he cannot fully bring about this public understanding until he allies his work to a joint and all-intelligible aim of loftiest Art; while this aim itself will be disclosed to the common understanding, past all mistaking, by the actual bodily man with all his warmth of life. Of all artistic things, the most directly understandable is the Dramatic-Action *(Handlung)*, for reason that its art is not complete until every helping artifice be cast behind it, as it were, and genuine life attain the faithfullest and most intelligible show. And thus each branch of art can only address itself to the *understanding* in proportion as its core—whose relation to Man, or derivation from him, alone can animate and justify the artwork—is ripening toward the *Drama*. In proportion as it passes over into Drama, as it pulses with the Drama's light, will each domain of Art grow all-intelligible, completely understood and justified.[2]

On to the stage, prepared by architect and painter, now steps *Artistic Man,* as Natural Man steps on the stage of Nature. What the statuary and the historical painter endeavoured to limn on *stone* or *canvas,* they now limn upon *themselves,* their form, their body's limbs, the features of their visage, and raise it to the consciousness of full artistic life. The same sense that led the sculptor in his grasp and rendering of the human figure, now leads the *Mime* in the handling and demeanour of his actual body. The same eye which taught the historical painter, in drawing and in colour, in arrangement of his drapery and composition of his groups, to find the beautiful, the graceful and the characteristic, now orders the whole breadth of *actual human show.* Sculptor and painter once freed the Greek Tragedian from his cothurnus and his mask, upon and under which the real man could only move according to a certain religious convention. With justice, did this pair of plastic artists annihilate the last disfigurement of pure artistic man, and thus prefigure in their stone and canvas the tragic Actor of the Future. As they once descried him in his undistorted truth, they now shall let him pass into reality and bring his form, in a measure sketched by them, to bodily portrayal with all its wealth of movement.

Thus the illusion of plastic art will turn to truth in Drama: the plastic artist will reach out hands to the *dancer,* to the *mime,* will lose himself in them, and thus become himself both mime and dancer.—So far as lies within his power, he will have to impart the inner man his feeling and his will-ing, to the eye. The breadth and depth of scenic space belong to him for the plastic message of his stature and his motion, as a single unit or in union with his fellows. But where his power ends, where the fulness of his will and feeling impels him to the *utter*ing of the inner man by means of *Speech,* there will the Word proclaim his plain and conscious purpose: he becomes a *Poet* and, to be poet, a *tone-artist (Tonkünstler).*

But as dancer, tone-artist, and poet, he still is one and the same thing: nothing other than *executant, artistic Man, who, in the fullest measure of his faculties, imparts himself to the highest expression of receptive power.*

It is in him, the immediate executant, that the three sister-arts unite their forces in one collective operation, in which the highest faculty of each comes to its highest unfolding. By working in common, each one of them attains the power to be and do the very thing which, of her own and inmost essence, she longs to do and be. Hereby: that each, where her own power ends, can be absorbed within the other, whose power commences where her's ends,—she maintains her own purity and freedom, her independence as *that* which she is. The *mimetic dancer* is stripped of his impotence, so soon as he can sing and speak; the creations of *Tone* win all-explaining meaning through the mime, as well as through the poet's word, and that exactly in degree as Tone itself is able to transcend into the motion of the mime and the word of the poet; while the *Poet* first becomes a Man through his translation to the flesh and blood of the *Performer:* for though he metes to each artistic factor the guiding purpose which binds them all into a common whole, yet this purpose is first changed from "will" to "can" *by the poet's Will descending to the actor's Can.*

Not one rich faculty of the separate arts will remain unused in the United Artwork of the Future; in *it* will each attain its first complete appraisement. Thus, especially, will the manifold developments of Tone, so peculiar to our instrumental music, unfold their utmost wealth within this Artwork; nay, Tone will incite the mimetic art of Dance to entirely new discoveries, and no less swell the breath of Poetry to unimagined fill. For Music, in her solitude, has fashioned for herself an organ which is capable of the highest reaches of expression. This organ is the *Orchestra.* The tone-speech of Beethoven, introduced into Drama by the orchestra, marks an entirely fresh departure for the dramatic artwork. While Architecture and, more especially, scenic Landscape-painting have power to set the executant dramatic Artist in the surroundings of physical Nature, and to dower him from the exhaustless stores of natural phenomena with an ample and significant background—so in the Orchestra, that pulsing body of many-coloured harmony, the personating individual Man is given, for his support, a stanchless elemental spring, at once artistic, natural, and human.

The Orchestra is, so to speak, the loam of endless, universal Feeling, from which the individual feeling of the separate actor draws power to shoot aloft to fullest height of growth: it, in a sense, dissolves[3] the hard immobile ground of the actual scene into a fluent, elastic, impressionable æther, whose unmeasured bottom is the great sea of Feeling itself. Thus the Orchestra is like the *Earth* from which Antæus, so soon as ever his foot had grazed it, drew new immortal life-force. By its essence diametrically opposed to the scenic landscape which surrounds the actor, and therefore, as to locality, most rightly placed in the deepened foreground outside the scenic frame, it at like time forms the perfect complement of these surroundings; inasmuch as it broadens out the exhaustless *physical* element of Nature to the equally exhaustless *emotional* element of artistic Man. These elements, thus knit together, enclose the performer as with an atmospheric ring of Art and Nature, in which, like to the heavenly bodies, he moves secure in fullest orbit, and whence, withal, he is free to radiate on every side his feelings and his views of life—broadened to infinity, and showered, as it were, on distances as measureless as those on which the stars of heaven cast their rays of light. . . .

The place in which this wondrous process comes to pass, is the *Theatric stage;* the collective art-work which it brings to light of day, the *Drama.* But to force his own specific nature to the highest blossoming of its contents in this *one* and highest art-work, the separate artist, like each several art, must quell each selfish, arbitrary bent toward untimely bushing into outgrowths unfurthersome to the whole; the better then to put forth all his strength for reaching of the highest common purpose, which cannot indeed be realised without the unit, nor, on the other hand, without the unit's recurrent limitation.

This purpose of the Drama, is withal the only true artistic purpose that ever can be fully *realised;* whatsoever lies aloof from that, must necessarily lose itself in the sea of things indefinite, obscure, unfree. This purpose, however, the separate art-branch will never reach *alone,*[4] but only *all together;* and therefore the most *universal* is at like time the only real, free, the only universally *intelligible* Art-work.

—Translated by William Ashton Ellis

F. T. Marinetti, Bruno Corra,
Emilio Settimelli, Arnaldo Ginna, Giacomo Balla,
Remo Chiti

<< 2 >> "The Futurist Cinema" (1916)

F. T. Marinetti. *Photo by Coletti.*

"We shall set in motion the words-in-freedom that smash the boundaries of literature as they march towards painting, music, noise-art, and throw a marvelous bridge between the word and the real object."

<< The Italian Futurists set out to attack the heart of high culture. Denouncing archaic artistic and social structures was key to the Futurist sensibility. The movement was born in Paris with the publication of the first Futurist manifesto in *Le Figaro,* a popular daily newspaper, in 1909. Its author, the Italian poet Filippo Tommaso Marinetti, chose the Parisian public as the target for his manifesto of "incendiary violence," calling for an end to all art that refused to embrace the social transformation brought by technology in the new century.

It was in cinema that Marinetti and his colleagues saw the potential for a form of expression that reflected the speed and energy of the times. Cinema may still have been considered a novelty entertainment, but the Futurists treated it as a legitimate art form. In this essay, they contrast the flexibility of film against the linearity, rigidity, and canonical aura of the book. Cinema, they declared, could be the most dynamic of human expressions because of its ability to synthesize all of the traditional arts, unleashing a form that was totally new. The Futurist cinema would free words from the fixed pages of the book and "smash the boundaries of literature," while it would enable painting to "break out of the limits of the frame." Such an enterprise called for the integration of technology into the arts: the mechanical tools of filmmaking, the Futurists claimed, would produce the "simultaneity and interpenetration of different times and places," foreshadowing later developments in nonlinear narrative found in interactive media. >>

The book, a wholly passéist means of preserving and communicating thought, has for a long time been fated to disappear like cathedrals, towers, crenellated walls, museums, and the pacifist ideal. The book, static companion of the sedentary, the nostalgic, the neutralist, cannot entertain or exalt the new Futurist generations intoxicated with revolutionary and bellicose dynamism.

The conflagration is steadily enlivening the European sensibility. Our great hygienic war, which should satisfy *all* our national aspirations, centuples the renewing power of the Italian race. The Futurist cinema, which we are preparing, a joyful deformation of the universe, an alogical, fleeting synthesis of life in the world, will become the best school for boys: a school of joy, of speed, of force, of courage, and heroism. The Futurist cinema will sharpen, develop the sensibility, will quicken the creative imagination, will give the intelligence a prodigious sense of simultaneity and omnipresence. The Futurist cinema will thus cooperate in the general renewal, taking the place of the literary review (always pedantic) and the drama (always predictable), and killing the book (always tedious and oppressive). The necessities of propaganda will force us to publish a book once in a while. But we prefer to express ourselves through the cinema, through great tables of words-in-freedom and mobile illuminated signs.

With our manifesto "The Futurist Synthetic Theatre," with the victorious tours of the theatre companies of Gualtiero Tumiati, Ettore Berti, Annibale Ninchi, Luigi Zoncada, with the two volumes of *Futurist Synthetic Theatre* containing eighty theatrical syntheses, we have begun the revolution in the Italian prose theatre. An earlier Futurist manifesto had rehabilitated, glorified, and perfected the Variety Theatre. It is logical therefore for us to carry our vivifying energies into a new theatrical zone: the *cinema.*

At first look the cinema, born only a few years ago, may seem to be Futurist already, lacking a past and free from traditions. Actually, by appearing in the guise of *theatre without words,* it has inherited all the most traditional sweepings of the literary theatre. Consequently, everything we have said and done about the stage applies to the cinema. Our action is legitimate and necessary in so far as the cinema up to now *has been and tends to remain profoundly passéist,* whereas we see in it the possibility of an eminently Futurist art and *the expressive medium most adapted to the complex sensibility of a Futurist artist.*

Except for interesting films of travel, hunting, wars, and so on, the film-makers have done no more than inflict on us the most backward-looking dramas, great and small. The same scenario whose brevity and variety may make it seem advanced is, in most cases, nothing but the most trite and pious *analysis.* Therefore all the immense *artistic* possibilities of the cinema still rest entirely in the future.

The cinema is an autonomous art. The cinema must therefore never copy the stage. The cinema, being essentially visual, must above all fulfill the evolution of painting, detach itself from reality, from photography, from the graceful and solemn. It must become antigraceful, deforming, impressionistic, synthetic, dynamic, free-wording.

ONE MUST FREE THE CINEMA AS AN EXPRESSIVE MEDIUM in order to make it the ideal instrument *of a new art,* immensely vaster and lighter than all the existing arts. We are convinced that only in this way can one reach that *polyexpressiveness* towards which all the most modern artistic researches are moving. Today the *Futurist cinema* creates precisely the POLYEXPRESSIVE SYMPHONY that just a year ago we announced in our manifesto "Weights, Measures, and Prices of Artistic Genius." The most varied elements will enter into the Futurist film as expressive means: from the slice of life to the streak of colour, from the conventional line to words-in-freedom, from chromatic and plastic music to the music of objects. In other words it will be painting, architecture, sculpture, words-in-freedom, music of colours, lines, and forms, a jumble of objects and reality thrown together at random. We shall offer new inspirations for the researches of painters, which will tend to break out of the limits of the frame. We shall set in motion the words-in-freedom that smash the boundaries of literature as they march towards painting, music, noise-art, and throw a marvellous bridge between the word and the real object.

Our films will be:

1. **Cinematic analogies** that use reality directly as one of the two elements of the analogy. Example: If we should want to express the anguished state of one of our protagonists, instead of describing it in its various phases of suffering, we would give an equivalent impression with the sight of a jagged and cavernous mountain.

   The mountains, seas, woods, cities, crowds, armies, squadrons, aeroplanes, will often be our formidable expressive words: THE UNIVERSE WILL BE OUR VOCABULARY. Example: We want to give a sensation of strange cheerfulness: we show a chair cover flying comically around an enormous coat stand until they decide to join. We want to give the sensation of anger: we fracture the angry man into a whirlwind of little yellow balls. We want to give the anguish of a hero who has lost his faith and lapsed into a dead neutral scepticism: we show the hero in the act of making an inspired speech to a great crowd; suddenly we bring on Giovanni Giolitti who treasonably stuffs a thick forkful of macaroni into the hero's mouth, drowning his winged words in tomato sauce.

   We shall add colour to the dialogue by swiftly, simultaneously showing every image that passes through the actors' brains. Example: representing a man who will say to his woman: "You're as lovely as a gazelle," we shall show the gazelle. Example: if a character says, "I contemplate your fresh and luminous smile as a traveller after a long rough trip contemplates the sea from high on a mountain," we shall show traveller, sea, mountain.

   This is how we shall make our characters as understandable *as if they talked.*
2. **Cinematic poems, speeches, and poetry.** We shall make all of their component images pass across the screen.

   Example: "Canto dell'amore" [Song of Love] by Giosuè Carducci:

> In their German strongholds perched
> Like falcons meditating the hunt

We shall show the strongholds, the falcons in ambush.

> From the churches that raise long marble
> arms to heaven, in prayer to God
> From the convents between villages and towns
> crouching darkly to the sound of bells
> like cuckoos among far-spaced trees
> singing boredoms and unexpected joys . . .

We shall show churches that little by little are changed into imploring women, God beaming down from on high, the convents, the cuckoos, and so on.

Example: "Sogno d'Estate" [Summer's Dream] by Giosuè Carducci:

> Among your ever-sounding strains of battle, Homer, I am conquered by the warm hour: I bow my head in sleep on Scamander's bank, but my heart flees to the Tyrrhenian Sea.

We shall show Carducci wandering amid the tumult of the Achaians, deftly avoiding the galloping horses, paying his respects to Homer, going for a drink with Ajax to the inn, The Red Scamander, and at the third glass of wine his heart, whose palpitations we ought to see, pops out of his jacket like a huge red balloon and flies over the Gulf of Rapallo. This is how we make films out of the most secret movements of genius.

Thus we shall ridicule the works of the passéist poets, transforming to the great benefit of the public the most nostalgically monotonous weepy poetry into violent, exciting, and highly exhilarating spectacles.

3. **Cinematic simultaneity and interpenetration** of different times and places. We shall project two or three different visual episodes at the same time, one next to the other.
4. **Cinematic musical researches** (dissonances, harmonies, symphonies of gestures, events, colours, lines, etc.).
5. **Dramatized states of mind on film.**
6. **Daily exercises in freeing ourselves from mere photographed logic.**
7. **Filmed dramas of objects.** (Objects animated, humanized, baffled, dressed up, impassioned, civilized, dancing—objects removed from their normal surroundings and put into an abnormal state that, by contrast, throws into relief their amazing construction and nonhuman life.)

8. **Show windows of filmed ideas, events, types, objects, etc.**
9. **Congresses, flirts, fights and marriages of funny faces, mimicry, etc.** Example: a big nose that silences a thousand congressional fingers by ringing an ear, while two policemen's moustaches arrest a tooth.
10. **Filmed unreal reconstructions of the human body.**
11. **Filmed dramas of disproportion** (a thirsty man who pulls out a tiny drinking straw that lengthens umbilically as far as a lake and dries it up *instantly*).
12. **Potential dramas and strategic plans of filmed feelings.**
13. **Linear, plastic, chromatic equivalences, etc.**, of men, women, events, thoughts, music, feelings, weights, smells, noises (with white lines on black we shall show the inner, physical rhythm of a husband who discovers his wife in adultery and chases the lover—rhythm of soul and rhythm of legs).
14. **Filmed words-in-freedom in movement** (synoptic tables of lyric values—dramas of humanized or animated letters—orthographic dramas—typographical dramas—geometric dramas—numeric sensibility, etc.).

    Painting + sculpture + plastic dynamism + words-in-freedom + composed noises [*intonarumori*] + architecture + synthetic theatre = Futurist cinema.

THIS IS HOW WE DECOMPOSE AND RECOMPOSE THE UNIVERSE ACCORDING TO OUR MARVELLOUS WHIMS, to centuple the powers of the Italian creative genius and its absolute pre-eminence in the world.

11 SEPTEMBER 1916

—TRANSLATED BY R. W. FLINT

László Moholy-Nagy

<< 3 >> "Theater, Circus, Variety," *Theater of the Bauhaus* (1924)

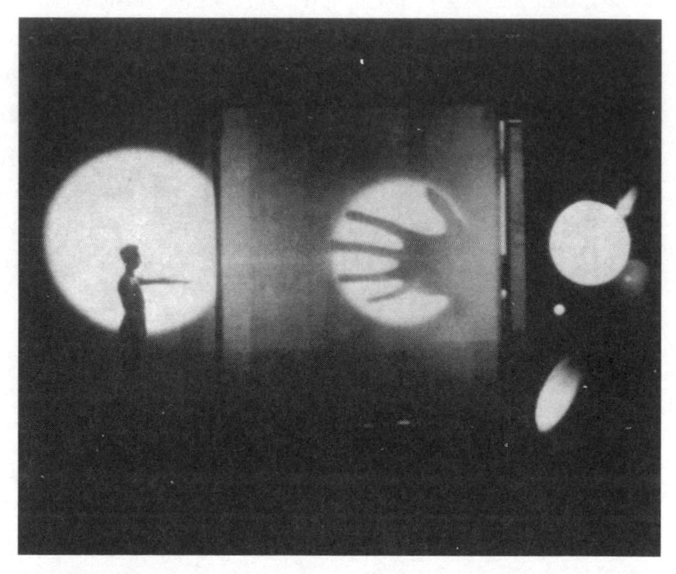

Light Play from Bauhaus Theater. *Photo by Lux Feininger. Courtesy of Prakapas Gallery.*

"The Theater of Totality with its multifarious complexities of light, space, plane, form, motion, sound, man—and with all the possibilities for varying and combining these elements—must be an ORGANISM."

<< Founded in Weimar, Germany, in 1919, the Bauhaus School is best known for its contribution to industrial design. But artists such as László Moholy-Nagy, Oskar Schlemmer, Paul Klee, and Wassily Kandinsky undertook a wide range of aesthetic investigations, using the school as a laboratory to examine the formal principles of abstraction in painting, photography, and sculpture. They also explored the influence of technology, which had a profound impact on their work and ideas. These experiments led Moholy-Nagy and Schlemmer to develop a new kind of theater based on these principles. Their work was an attempt to synthesize the theater's essential components—space, composition, motion, sound, movement, and light—into a fully integrated, abstract form of artistic expression.

Moholy-Nagy referred to this idea as the Theater of Totality, a reinterpretation of Wagner's concept of "total theater." Moholy-Nagy's approach to the synthesis of the arts reduced the importance of the written word and the presence of the actor, placing them on an equal plateau with stage design, lighting, music, and visual composition. His concept of the "Mechanized Eccentric" injected the qualities of machinery into every aspect of the stage performance, including the costume design, props, and movement of the actors, resulting in a theater that emphasized the physical rather than the literary, and that reflected the speed, dynamism, and precision of state-of-the-art technology. It is notable that Moholy-Nagy integrated mechanical motifs in all his work—not only theater but also painting, photography, film, and sculpture.

In addition, Moholy-Nagy's theater challenged the relationship between the spectator and the performance. He called for the use of techniques that would alter the theatrical space, removing the traditional "fourth wall" that separated the audience from the stage. He envisioned mechanical devices that would move across a multiplaned stage, a reorganization of the space that would literally immerse spectators in the action. As Moholy-Nagy put it, the Bauhaus Theater would challenge the passivity of the audience, and "actually allow them to fuse with the action on the stage at the peak of cathartic ecstasy." >>

1. The Historical Theater

The historical theater was essentially a disseminator of information or propaganda, or it was an articulated concentration of action *(Aktionskonzentration)* derived from events and doctrines in their broadest meaning—that is to say, as "dramatized" legend, as religious (cultist) or political (proselytizing) propaganda, or as compressed action with a more or less transparent purpose behind it.

The theater differed from the eyewitness report, simple storytelling, didactic moralizing, or advertising copy through its own particular synthesis of the elements of presentation: SOUND, COLOR (LIGHT), MOTION, SPACE, FORM (OBJECTS AND PERSONS).

With these elements, in their accentuated but often uncontrolled interrelationships, the theater attempted to transmit an articulated experience.

In early epic drama *(Erzählungsdrama)* these elements were generally employed as illustration, subordinated to narration or propaganda. The next step in this evolution led to the drama of action *(Aktionsdrama),* where the elements of dynamic-dramatic movement began to crystallize: the theater of improvisation, the *commedia dell'arte.* These dramatic forms were progressively liberated from a central theme of logical, intellectual-emotional action which was no longer dominant. Gradually their moralizing and their tendentiousness disappeared in favor of an unhampered concentration on action: Shakespeare, the opera.

With August Stramm, drama developed away from verbal context, from propaganda, and from character delineation, and toward explosive activism. Creative experiments with MOTION AND SOUND (speech) were made, based on the impetus of human sources of energy, that is, the "passions." Stramm's theater did not offer narrative material, but action and tempo, which, unpremeditated, sprang almost AUTOMATICALLY and in headlong succession from the human impulse for motion. But even in Stramm's case action was not altogether free from literary encumbrance.

"Literary encumbrance" is the result of the unjustifiable transfer of intellectualized material from the proper realm of literary effectiveness (novel, short story, etc.) to the stage, where it incorrectly remains a dramatic end in itself. The result is nothing more than literature if a reality or a potential reality, no matter how imaginative, is formulated or visually expressed without the creative forms peculiar only to the stage. It is not until the tensions concealed in the utmost economy of means are brought into universal and dynamic interaction that we have creative stagecraft *(Bühnengestaltung).* Even in recent times we have been deluded about the true value of creative stagecraft when revolutionary, social, ethical, or similar problems were unrolled with a great display of literary pomp and paraphernalia.

2. Attempts at a Theater Form for Today

a) *Theater of Surprises: Futurists, Dadaists, Merz*

In the investigation of any morphology, we proceed today from the all-inclusive functionalism of goal, purpose, and materials.

From this premise the FUTURISTS, EXPRESSIONISTS, and DADAISTS (MERZ) came to the conclusion that phonetic word relationships were more significant than other creative literary means, and that the logical-intellectual content *(das Logisch-Gedankliche)* of a work of literature was far from its primary aim. It was maintained that, just as in representational painting it was not the content as such, not the objects represented which were essential, but the interaction of colors, so in literature it was not the logical-intellectual content which belonged in the foreground, but the effects which arose from the word-sound relationships. In the case of some writers this idea has been extended (or possibly contracted) to the point where word relationships are transformed into exclusively phonetic sound relationships, thereby totally fragmenting the word into conceptually disjointed vowels and consonants.

This was the origin of the Dadaist and Futurist "Theater of Surprises," a theater which aimed at the elimination of logical-intellectual (literary) aspects. Yet in spite of this, man, who until then had been the sole representative of logical, causal action and of vital mental activities, still dominated.

b) *The Mechanized Eccentric* (Die mechanische Exzentrik)

As a logical consequence of this there arose the need for a MECHANIZED ECCENTRIC, a concentration of stage action in its purest form *(eine Aktionskonzentration der Bühne in Reinkultur)*. Man, who no longer should be permitted to represent himself as a phenomenon of spirit and mind through his intellectual and spiritual capacities, no longer has any place in this concentration of action. For, no matter how cultured he may be, his organism permits him at best only a certain range of action, dependent entirely on his natural body mechanism.

The effect of this body mechanism *(Körpermechanik)* (in circus performance and athletic events, for example) arises essentially from the spectator's astonishment or shock at the potentialities of his *own* organism as demonstrated to him by others. This is a subjective effect. Here the human body is the sole medium of configuration *(Gestaltung)*. For the purposes of an objective *Gestaltung* of movement this medium is limited, the more so since it has constant reference to sensible and perceptive (i.e., again literary) elements. The inadequacy of "human" *Exzentrik* led to the demand for a precise and fully controlled organization of form and motion, intended to be a synthesis of dynamically contrasting phenomena (space, form, motion, sound, and light). This is the Mechanized Eccentric.

3. The Coming Theater: Theater of Totality

Every form process or *Gestaltung* has its general as well as its particular premises, from which it must proceed in making use of its specific media. We might, therefore, clarify theater production *(Theatergestaltung)* if we investigated the nature of its highly controversial media: the human *word* and the human action, and, at the same time, considered the endless possibilities open to their creator—man.

The origins of MUSIC as conscious composition can be traced back to the melodic recitations of the heroic saga. When music was systematized, permitting only the use of HARMONIES (KLÄNGE) and excluding so-called SOUNDS (GERÄUSCHE), the only place left for a special sound form *(Geräuschgestaltung)* was in literature, particularly in poetry. This was the underlying idea from which the Expressionists, Futurists, and Dadaists proceeded in composing their sound-poems *(Lautgedichte)*. But today, when music has been broadened to admit sounds of all kinds, the sensory-mechanistic effect of sound interrelationships is no longer a monopoly of poetry. It belongs, as much as do harmonies *(Töne),* to the realm of music, much in the same way that the task of painting, seen as color creation, is to organize clearly primary (apperceptive) color effect. Thus the error of the Futurists, the Expressionists, the Dadaists, and all those who built on such foundations becomes clear. As an example: the idea of an *Exzentrik* which is ONLY mechanical.

It must be said, however, that those ideas, in contradistinction to a literary-illustrative viewpoint, have unquestionably advanced creative theater precisely because they were diametrically opposed. They canceled out the predominance of the exclusively logical-intellectual values. But once the predominance has been broken, the associative processes and the language of man, and consequently man himself in his totality as a formative medium for the stage, may not be barred from it. To be sure, he is no longer to be pivotal—as he is in traditional theater—but is to be employed ON AN EQUAL FOOTING WITH THE OTHER FORMATIVE MEDIA.

Man as the most active phenomenon of life is indisputably one of the most effective elements of a dynamic stage production *(Bühnengestaltung)*, and therefore he justifies on functional grounds the utilization of his totality of action, speech, and thought. With his intellect, his dialectic, his adaptability to any situation by virtue of his control over his physical and mental powers, he is—when used in any concentration of action *(Aktionskonzentration)*—destined to be primarily a configuration of these powers.

And if the stage didn't provide him full play for these potentialities, it would be imperative to create an adequate vehicle.

But this utilization of man must be clearly differentiated from his appearance heretofore in traditional theater. While there he was only the interpreter of a literarily conceived individual or type, in the new THEATER OF TOTALITY he will use the spiritual and physical means at his disposal PRODUCTIVELY and from his own INITIATIVE submit to the over-all action process.

While during the Middle Ages (and even today) the center of gravity in theater production lay in the representation of the various *types* (hero, harlequin, peasant, etc.), it is the task of the FUTURE ACTOR to discover and activate that which is COMMON to all men.

In the plan of such a theater the traditionally "meaningful" and causal interconnections can NOT play the major role. In the consideration of stage setting as an *art form,* we must learn from the creative artist that, just as it is impossible to ask what a man (as organism) is or stands for, it is inadmissible to ask the same question of a contemporary nonobjective picture which likewise is a *Gestaltung,* that is, an organism.

The contemporary painting exhibits a multiplicity of color and surface interrelationships, which gain their effect, on the one hand, from their conscious and logical statement of problems, and on the other, from the unanalyzable intangibles of creative intuition.

In the same way, the Theater of Totality with its multifarious complexities of light, space, plane, form, motion, sound, man—and with all the possibilities for varying and combining these elements—must be an ORGANISM.

Thus the process of integrating man into creative stage production must be unhampered by moralistic tendentiousness or by problems of science or the INDIVIDUAL. Man may be active only as the bearer of those functional elements which are organically in accordance with his specific nature.

It is self-evident, however, that all *other* means of stage production must be given positions of effectiveness equal to man's, who as a living psychophysical organism, as the producer of incomparable climaxes and infinite variations, demands of the conformative factors a high standard of quality.

4. How Shall the Theater of Totality Be Realized?

One of two points of view still important today holds that theater is the concentrated activation *(Aktionskonzentration)* of sound, light (color), space, form, and motion. Here man as coactor is not necessary, since in our day equipment can be constructed which is far more capable of executing the *purely mechanical* role of man than man himself.

The other, more popular view will not relinquish the magnificent instrument which is man, even though no one has yet solved the problem of how to employ him as a creative medium on the stage.

Is it possible to include his human, logical functions in a present-day concentration of action on the stage, without running the risk of producing a copy from nature and without falling prey to Dadaist or Merz characterization, composed of an eclectic patchwork whose seeming order is purely arbitrary?

The creative arts have discovered pure media for their constructions: the primary relationships of color, mass, material, etc. But how can we integrate a sequence of human movements and thoughts on an equal footing with the controlled, "absolute" elements of sound, light (color), form, and motion? In this regard only summary suggestions can be made to the creator of the new theater *(Theatergestalter)*. For example, the REPETITION of a thought by many actors, with identical words and with identical or varying intonation and cadence, could be employed as a means of creating synthetic (i.e., unifying) creative theater.

(This would be the CHORUS—but not the attendant and passive chorus of antiquity!) Or mirrors and optical equipment could be used to project the gigantically enlarged faces and gestures of the actors, while their voices could be amplified to correspond with the visual MAGNIFICATION. Similar effects can be obtained from the SIMULTANEOUS, SYNOPTICAL, and SYNACOUSTICAL reproduction of thought (with motion pictures, phonographs, loud-speakers), or from the reproduction of thoughts suggested by a construction of variously MESHING GEARS *(eine* ZAHNRADARTIG INEINANDERGREIFENDE *Gedankengestaltung).*

Independent of work in music and acoustics, the literature of the future will create its own "harmonies," at first primarily adapted to its own media, but with far-reaching implications for others. These will surely exercise an influence on the word and thought constructions of the stage.

This means, among other things, that the phenomena of the subconscious and dreams of fantasy and reality, which up to now were central to the so called "INTIMATE ART THEATER" ("KAMMERSPIELE"), may no longer be predominant. And even if the conflicts arising from today's complicated social patterns, from the world-wide organization of technology, from pacifist-utopian and other kinds of revolutionary movements, can have a place in the art of the stage, they will be significant only in a transitional period, since their treatment belongs properly to the realms of literature, politics, and philosophy.

We envision TOTAL STAGE ACTION (GESAMTBÜHNENAKTION) as a great dynamic-rhythmic process, which can compress the greatest clashing masses or accumulations of media—as qualitative and quantitative tensions—into elemental form. Part of this would be the use of simultaneously interpenetrating sets of contrasting relationships, which are of minor importance in themselves, such as: the tragicomic, the grotesque-serious, the trivial-monumental; hydraulic spectacles; acoustical and other "pranks"; and so on. Today's CIRCUS, OPERETTA, VAUDEVILLE, the CLOWNS in America and elsewhere (Chaplin, Fratellini) have accomplished great things, both in this respect and in eliminating the subjective—even if the process has been naïve and often more superficial than incisive. Yet it would be just as superficial if we were to dismiss great performances and "shows" in this genre with the word *Kitsch.* It is high time to state once and for all that the much disdained masses, despite their "academic backwardness," often exhibit the soundest instincts and preferences. Our task will always remain the creative understanding of the true, and not the imagined, needs.

5. The Means

Every *Gestaltung* or creative work should be an unexpected and new organism, and it is natural and incumbent on us to draw the material for surprise effects from our daily living. Nothing is more effective than the exciting new possibilities offered by the familiar and yet not properly evaluated elements of modern life—that is, its idiosyncracies: individuation, classification, mechanization. With this in mind, it is possible to arrive at a proper understanding of stagecraft through an investigation of creative media other than man-as-actor himself.

In the future, SOUND EFFECTS will make use of various acoustical equipment driven electrically or by some other mechanical means. Sound waves issuing from unexpected sources—for example, a speaking or singing arc lamp, loudspeakers under the seats or beneath the floor of the auditorium, the use of new amplifying systems—will raise the audience's acoustic surprise-threshold so much that unequal effects in other areas will be disappointing.

COLOR (LIGHT) must undergo even greater transformation in this respect than sound.

Developments in painting during the past decades have created the organization of absolute color values and, as a consequence, the supremacy of pure and luminous chromatic tones. Naturally the monumentality and the lucid balance of their harmonies will not tolerate the actor with indistinct or splotchy make-up and tattered costuming, a product of misunderstood Cubism, Futurism, etc. The use of precision-made metallic masks and costumes and those of various other composition materials will thus become a matter of course. The pallid face, the subjectivity of expression, and the gestures of the actor in a colored stage environment are therefore eliminated without impairing the effective contrast between the human body and any mechanical construction. Films can also be projected onto various surfaces and further experiments in space illumination will be devised. This will constitute the new ACTION OF LIGHT, which by means of modern technology will use the most intensified contrasts to guarantee itself a position of importance equal to that of all other theater media. We have not yet begun to realize the potential of light for sudden or blinding illumination, for flare effects, for phosphorescent effects, for bathing the auditorium in light synchronized with climaxes or with the total extinguishing of lights on the stage. All this, of course, is thought of in a sense totally different from anything in current traditional theater.

From the time that stage objects became mechanically movable, the generally traditional, horizontally structured organization of movement in space has been enriched by the possibility of vertical motion. Nothing stands in the way of making use of complex APPARATUS such as film, automobile, elevator, airplane, and other machinery, as well as optical instruments, reflecting equipment, and so on. The current demand for dynamic construction will be satisfied in this way, even though it is still only in its first stages.

There would be a further enrichment if the present isolation of the stage could be eliminated. In today's theater, STAGE AND SPECTATOR are too much separated, too obviously divided into active and passive, to be able to produce creative relationships and reciprocal tensions.

It is time to produce a kind of stage activity which will no longer permit the masses to be silent spectators, which will not only excite them inwardly but will let them *take hold and participate*—actually allow them to fuse with the action on the stage at the peak of cathartic ecstasy.

To see that such a process is not chaotic, but that it develops with control and organization, will be one of the tasks of the thousand-eyed NEW DIRECTOR, equipped with all the modern means of understanding and communication.

It is clear that the present peep-show stage is not suitable for such organized motion.

The next form of the advancing theater—in cooperation with future authors—will probably answer the above demands with SUSPENDED BRIDGES AND DRAWBRIDGES running horizontally, diagonally, and vertically within the space of the theater; with platform stages built far into the auditorium; and so on. Apart from rotating sections, the stage will have movable space constructions and DISLIKE AREAS, in order to bring certain action moments on the stage into prominence, as in film "close-ups." In place of today's periphery of orchestra loges, a runway joined to the stage could be built to establish—by means of a more or less caliper-like embrace—a closer connection with the audience.

The possibilities for a VARIATION OF LEVELS OF MOVABLE PLANES on the stage of the future would contribute to a genuine organization of space. Space will then no longer consist of the interconnections of planes in the old meaning, which was able to conceive of architectonic delineation of space only as an enclosure formed by opaque surfaces. The new space originates from free-standing surfaces or from linear definition of planes (WIRE FRAMES, ANTENNAS), so that the surfaces stand at times in a very free relationship to one another, without the need of any direct contact.

As soon as an intense and penetrating concentration of action can be functionally realized, there will develop simultaneously the corresponding auditorium ARCHITECTURE. There will also appear COSTUMES designed to emphasize function and costumes which are conceived only for single moments of action and capable of sudden transformations.

There will arise an enhanced *control* over all formative media, unified in a harmonious effect and built into an organism of perfect equilibrium.

—TRANSLATED BY ARTHUR S. WENSINGER

Richard Higgins

"Intermedia" (1966)

<< 4 >>

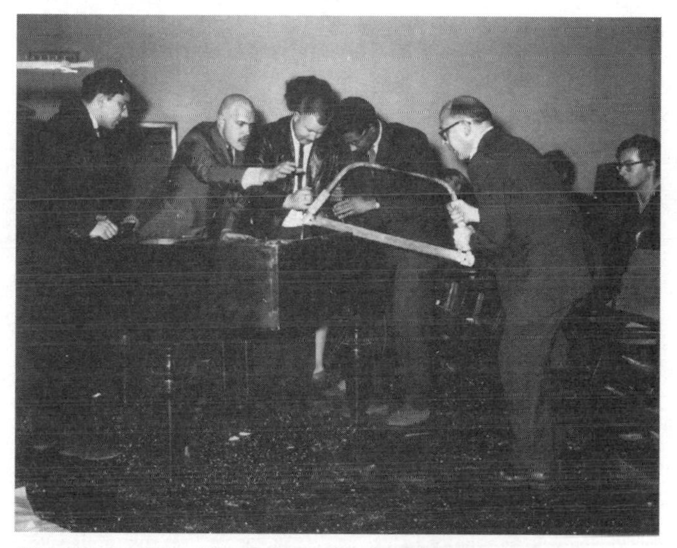

Dick Higgins (second from left) performing Philip Corner's "Piano Activities." Courtesy of Archiv Sohm, Staatsgalerie Stuttgart.

"The Happening developed as intermedium, an uncharted land that lies between collage, music, and the theater. It is not governed by rules; each work determines its own medium and form according to its needs."

<< Fluxus artist Richard Higgins describes a branch of American postmodernism from the 1960s that reflects the tumultuous social atmosphere of the era and its impact on the arts. Higgins's attention is focused on intermedia, a myriad of emerging genres that spilled across the boundaries of traditional media. In the interstices between the arts, mixed-media forms coalesced: Happenings, performance art, kinetic sculpture, electronic theater, as well as a variety of deliberately uncategorizable works.

Higgins cites Marcel Duchamp's readymades, including the infamous urinal *Fountain,* as the first challenge to the European tradition of the sacred aura of the artwork. Following the example of Duchamp, Higgins suggests that artists explore the territory that lies between "the general area of art media and those of life media." He calls for unusual combinations of art, including, for instance, the mixing of painting and shoes (as in the art of Claes Oldenberg). With intermedia, any available object or experience can be incorporated into the artwork. This was the notion at the core of the Happening, a form of performance pioneered by Oldenberg, Allan Kaprow, Jim Dine, and Robert Whitman that was the pinnacle of intermedia art. Their works took place in staged environments, presented as collages of action, music, and found objects that the audience encountered in an order typically without linear development, and without the through line of traditional narrative.

The art of "mixed-means" was defined by its tendency to establish an open system without rules or boundaries. This open-endedness allowed Higgins and his colleagues to extend the Happening to incorporate the audience as an additional, composed element of the work; as he put it: "live people as part of the collage." This shift in the role of the spectator had profound implications for the evolution of interactive media art, and its influence is felt in the changing role of the viewer in human-computer interaction. >>

Much of the best work being produced today seems to fall between media. This is no accident. The concept of the separation between media arose in the Renaissance. The idea that a painting is made of paint on canvas or that a sculpture should not be painted seems characteristic of the kind of social thought—categorizing and dividing society into nobility with its various subdivisions, untitled gentry, artisans, serfs and landless workers—which we call the feudal conception of the Great Chain of Being. This essentially mechanistic approach continued to be relevant throughout the first two industrial revolutions, just concluded, and into the present era of automation, which constitutes, in fact, a third industrial revolution.

However, the social problems that characterize our time, as opposed to the political ones, no longer allow a compartmentalized approach. We are approaching the dawn of a classless society, to which separation into rigid categories is absolutely irrelevant. This shift does not relate more to East than West or vice versa. Castro works in the cane fields. New York's Mayor Lindsay walks to work during the subway strike. The millionaires eat their lunches at Horn and Hardart's. This sort of populism is a growing tendency rather than a shrinking one.

We sense this in viewing art which seems to belong unnecessarily rigidly to one or another form. We view paintings. What are they, after all? Expensive, handmade objects, intended to ornament the walls of the rich or, through their (or their government's) munificence, to be shared with the large numbers of people and give them a sense of grandeur. But they do not allow any sense of dialogue.

Pop art? How could it play a part in the art of the future? It is bland. It is pure. It uses elements of common life without comment, and so, by accepting the misery of this life and its aridity so mutely, it condones them. Pop and Op are both dead, however, because they confine themselves, through the media which they employ, to the older functions of art, of decorating and suggesting grandeur, whatever their detailed content or their artists' intentions. None of the ingenious theories of the Mr. Ivan Geldoway combine can prevent them from being colossally boring and irrelevant. Milord runs his Mad Avenue gallery, in which he displays his pretty wares. He is protected by a handful of rude footmen who seem to feel that this is the way Life will always be. At his beck and call is Sir Fretful Callous, a moderately well-informed high priest, who apparently despises the Flame he is supposed to tend and therefore prefers anything that titillates him. However, Milord needs his services, since he, poor thing, hasn't the time or the energy to contribute more than his name and perhaps his dollars; getting information and finding out what's going on are simply toooooo exhausting. So, well protected and advised, he goes blissfully through the streets in proper Louis XIV style.

This scene is just not characteristic of the painting world as an institution, however. It is absolutely natural to (and inevitable in) the concept of the pure medium, the painting or precious object of any kind. That is the way such objects are marked since that is the world to which they belong and to which they relate.

The sense of "I am the state," however, will shortly be replaced by "After me the deluge," and, in fact, if the High Art world were better informed, it would realize that the deluge has already begun.

Who knows when it began? There is no reason for us to go into history in any detail. Part of the reason that Duchamp's objects are fascinating while Picasso's voice is fading is that the Duchamp pieces are truly between media, between sculpture and something else, while a Picasso is readily classifiable as a painted ornament. Similarly, by invading the land between collage and photography, the German John Heartfield produced what are probably the greatest graphics of our century, surely the most powerful political art that has been done to date.

The ready-made or found object, in a sense an intermedium since it was not intended to conform to the pure media, usually suggests this, and therefore suggests a location in the field between the general area of art media and those of life media. However, at this time, the locations of this sort are relatively unexplored, as compared with media between the arts. I cannot, for example, name work that has consciously been placed in the intermedium between painting and shoes. The closest thing would seem to be the sculpture of Claes Oldenburg, which falls between sculpture and hamburgers or Eskimo Pies, yet it is not the sources of these images themselves. An Oldenburg Eskimo Pie may look something like an Eskimo Pie, yet it is neither edible or cold. There is still a great deal to be done in this direction in the way of opening up aesthetically rewarding possibilities.

In the middle 1950s many painters began to realize the fundamental irrelevance of Abstract Expressionism, which was the dominant mode at the time. Such painters as Allan Kaprow and Robert Rauschenberg in the United States and Wolf Vostell in Germany turned to collage or, in the latter's case, dé-collage in the sense of making work by adding or removing, replacing and substituting or altering components of a visual work. They began to include increasingly incongruous objects in their work. Rauschenberg called his constructions *combines* and went so far as to place a stuffed goat—spattered with paint and a rubber tire around its neck—onto one. Kaprow, more philosophical and restless, meditated on the relationship of the spectator and the work. He put mirrors into his things so the spectator could feel included in them. That wasn't physical enough, so he made enveloping collages which surrounded the spectator. These he called *environments*. Finally, in the spring of 1958, he began to include live people as part of the collage, and this he called a *happening*.

The proscenium theater is the outgrowth of seventeenth-century ideals of social order. Yet there is remarkably little structural difference between the dramas of D'Avenant and those of Edward Albee, certainly nothing comparable to the difference in pump construction or means of mass transportation. It would seem that the technological and social implications of the first two industrial revolutions have been evaded completely. The drama is still mechanistically divided: there are performers, production people, a separate audience and an explicit script. Once started, like Frankenstein's monster, the course of affairs is unalterable, perhaps damned by its inability to reflect its surroundings. With our populistic mentality today, it is difficult to attach importance—other than what we have been taught to attach—to this traditional theater. Nor do minor innovations do more than provide dinner conversation: this theater is round instead of square, in that one the stage revolves, here the play is relatively senseless and whimsical (Pinter is, after all, our modern J. M. Barrie—unless the honor belongs more properly to Beckett). Every year fewer attend the professional Broadway theaters. The shows get sillier and sillier, showing the producers' estimate of our mentality (or is it their own that is revealed?). Even the best of the traditional theater is no longer found on Broadway but at the Judson Memorial Church, some miles away. Yet our theater schools grind out thousands on thousands of performing and production personnel, for whom jobs will simply not exist in twenty years. Can we blame the unions? Or rents and real estate taxes? Of course not. The subsidized productions, sponsored at such museums as New York's Lincoln Center, are not building up a new audience so much as recultivating an old one, since the medium of such drama seems weird and artificial in our new social milieu. We need more portability and flexibility, and this the traditional theater cannot provide. It was made for Versailles and for the sedentary Milords, not for motorized life-demons who travel six hundred miles a week. Versailles no longer speaks very loudly to us, since we think at eighty-five miles an hour.

In the other direction, starting from the idea of theater itself, others such as myself declared war on the script as a set of sequential events. Improvisation was no help; performers merely acted in imitation of a script. So I began to work as if time and sequence could be utterly suspended, not by ignoring them (which would simply be illogical) but by systematically replacing them as structural elements with change. Lack of change would cause my pieces to stop. In 1958 I wrote a piece, *Stacked Deck*, in which any event can take place at any time, as long as its cue appears. The cues are produced by colored lights. Since the colored lights could be used wherever they were put, and audience reactions were also cuing situations, the performance-audience separation was removed and a happening situation was established, though less visually oriented in its use of its environment and imagery. At the same time, Al Hansen moved into the area from graphic notation experiments, and Nam June Paik and Benjamin Patterson (both in Germany at the time) moved in from varieties of music in which specifically musical events were frequently replaced by nonmusical actions.

Thus the Happening developed as an intermedium, an uncharted land that lies between collage, music, and the theater. It is not governed by rules; each work determines its own medium and form according to its needs. The concept itself is better understood by what it is not, rather than what it is. Approaching it, we are pioneers again, and shall continue to be so as long as there's plenty of elbow room and no neighbors around for a few miles. Of course, a concept like this is very disturbing to those whose mentality is compartmentalized. *Time, Life* and the High Priests have been announcing the death of Happenings, regularly since the movement gained momentum in the late fifties, but this says more about the accuracy of their information than about the liveliness of the movement.

We have noted the intermedia in the theater and in the visual arts, the Happening, and certain varieties of physical constructions. For reasons of space we cannot take up here the intermedia between other areas. However, I would like to suggest that the use of intermedia is more or less universal throughout the fine arts, since continuity rather than categorization is the hallmark of our new mentality. There are parallels to the Happening in music, for example, in the work of such composers as Philip Corner and John Cage, who explore the intermedia between music and philosophy, or Joe Jones, whose self-playing musical instruments fall into the intermedium between music and sculpture. The constructed music of Emmet Williams and Robert Filliou certainly constitute an intermedium between poetry and sculpture. Is it possible to speak of the use of intermedia as a huge and inclusive movement of which dada, futurism, and surrealism are early phases preceding the huge ground-swell that is taking place now? Or is it more reasonable to regard the use of intermedia an irreversible historical innovation, more comparable to the development of instrumental music than, for example, to the development of romanticism?

Billy Klüver

"The Great Northeastern Power Failure" (1966)

<< 5 >>

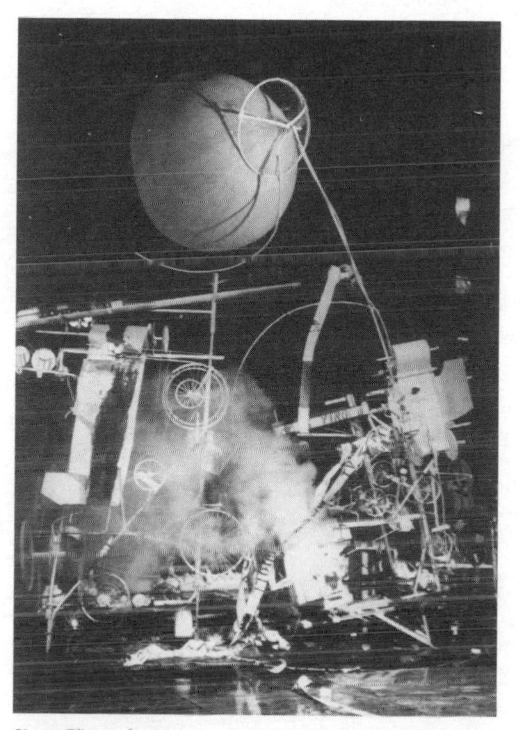

Jean Tinguely, Homage to New York. *Photo by David Gahr. Courtesy of Billy Klüver.*

"The new interface I will define is one in which the artist makes active use of the inventiveness and skills of an engineer to achieve his purpose. The artist could not complete his intentions without the help of an engineer. The artist incorporates the work of the engineer in the painting or the sculpture or the performance."

<< In the late 1950s, the Swedish-born engineer Billy Klüver worked on laser systems for Bell Laboratories in Murray Hill, New Jersey. Klüver, however, was not satisfied by purely scientific pursuits, and his interests led him to explore the artistic milieu that gave rise to pop art, minimalism, and Happenings a short drive away in New York City. Klüver befriended the leading exponents of the new arts, which led to collaborations with such key figures as Robert Rauschenberg, Jasper Johns, Andy Warhol, and John Cage. He became the chief catalyst for the art and technology movement that was launched dramatically in the spring of 1960, at the Museum of Modern Art, with Jean Tinguely's infamous self-destructing kinetic sculpture, *Homage to New York.* Klüver's participation in this work, with its paint bombs, chemical stinks, noisemakers, and fragments of scrap metal, inspired a generation of artists to imagine the possibilities of technology, as the machine destroyed itself, in Klüver's words, "in one glorious act of mechanical suicide."

In the 1960s, Klüver, more than anyone, saw the potential for the integration of art and technology. Inspired by Aristotle's notion of Techne—in which there was no differentiation between the practice of art and science—Klüver proposed the active and *equal* participation of the artist and engineer in the creation of the artwork. In this collaboration, he believed that the engineer required the participation of the artist, who, as a "visionary about life" and an active agent of social change, involved the engineer in meaningful cultural dialogue. At the same time, he felt that the artist, in the spirit of Rauschenberg's famous credo "to close the gap between art and life," had an obligation to incorporate technology as an element in the artwork, since technology had become inseparable from our lives.

Shortly after writing this article, Klüver cofounded the now legendary Experiments in Art and Technology (E.A.T.) with Rauschenberg. E.A.T. encouraged the collaboration of artists and engineers across the country in interdisciplinary technology-based art projects. Klüver's work takes on increasing significance as the notion of the synthesis of art and technology has assimilated into the contemporary arts. >>

Well, to begin with, the title of my talk is not going to be entirely unrelated to what I am going to say. What I will discuss is a new mode of interaction between science and technology on the one hand and art and life on the other. To use a scientific jargon that is currently in, I will try to define a new interface between these two areas.

Technology has always been closely tied in to the development of art. For Aristotle, Techne means both art and technology. As they became different subjects they still fed on each other. New technological discoveries were taken up and used by artists and you are all familiar with the contributions of artists to technology. The contemporary artist reads with ease the technical trade magazines. The new chemical material is hardly developed before it gets used by an artist. Today the artist tends to adopt the new material or the new industrial process as his insignia. We talk about artists in terms that he works in such and such a way or that he uses such and such materials. We hear about artists being poisoned and hurt in their work. In this century, artists have also embraced technology as subject matter: the enthusiasm of the Futurists, the experiments of Dada, the optimism of the Bauhaus movement and the Constructivists, all have looked at technology and science and found material for the artists. But for all this interest, art remains a passive viewer of technology. Art has only been interested in the fallout, so to speak, of science and technology. The effect of technology on art can apparently be even a negative one: the invention of the camera helped kill off representational painting, and we are now witnessing how the computer is about to take care of music and non-representational painting.

The new interface I will define is one in which the artist makes active use of the inventiveness and skills of an engineer to achieve his purpose. The artist could not complete his intentions without the help of an engineer. The artist incorporates the work of the engineer in the painting or the sculpture or the performance. A characteristic of this kind of interaction is that generally only one work of art results. In other words, the engineer is not just inventing a new and special process for the use of the artist. He does not just teach the artist a new skill which the artist can use to extract new aesthetic variations. Technology is well aware of its own beauty and does not need the artist to elaborate on this. I will argue that the use of the engineer by the artist is not only unavoidable but necessary.

Before I try to justify why I believe that this interface exists and why the interaction between artists and engineers will become stronger, let me give you a few simple examples of what I mean in terms of works that already exist. I shall be modest and limit myself to use examples from my own experience. But there exist several others.

You probably have heard about Jean Tinguely's self-destroying machine, *Hommage to New York,* which more or less destroyed itself on March 17, 1960, in the sculpture garden of the Museum of Modern Art in New York. In retrospect I think my modest contribution to the machine was to visit garbage dumps in

New Jersey to pick up bicycle wheels and to truck them to 53rd Street. However, there were a few technical ideas hidden in Tinguely's machine which incidentally were mainly the contributions of my technical assistant at the time, Harold Hodges. There were about eight electrical circuits in the machine which closed successively as the machine progressed toward its ultimate fate. Motors would start, smoke would come out, smaller machines would leave the big one to escape. In order to make the main structure collapse, Harold had devised a scheme using supporting sections of Wood's metal which would melt from the heat of overheated resistors. At another point this method was used to light a candle. Contrary to what I hear frequently said, Tinguely's machine did not contain many of these technical links. It was mostly Tinguely's motors that did it.

A better example is two neon light power supplies that we made for two paintings by Jasper Johns. In one case, the light was the letter A, in the other the letter R. What was new was that Johns wanted no cords to the painting. To stack up batteries to 1200 volts would have been messy, dangerous and impractical. So wc started out with 12 volts of rechargeable batteries and devised a multivibrator circuit which, together with a transformer, would give us 1200 volts. The technical equipment, all 400 dollars' worth of it, was mounted behind John's painting.

My final example is Rauschenberg's large sculpture *Oracle,* which was shown in New York last year. It was the result of work carried out over three years during which time two complete technical systems were finished and junked. The final system enables the sound from five AM radios to be heard from each of the five sculptures in the group, but with each radio being controlled from a central control unit, in one of the sculptures. There are no connecting wires between the sculptures and they are all freely movable, on wheels.

All these examples have one thing in common: they are ridiculous from an engineer's point of view. Why would anyone want to spend 9000 dollars to be able to control five AM radios simultaneously, in one room? I want to emphasize that the examples contain very simple engineering and should not be taken as very original. But each of the projects required an engineer or a technically skilled person to achieve what the artist wanted. And an important point is that the artist could not be quite sure about the outcome.

We have been taught by Robert Rauschenberg that the painting is an object among other objects, subjected to the same psychological and physical influences as other objects. During a musical piece by John Cage, we are forced to accept the equality of all the sounds we hear as part of the composition. In his happenings, Claes Oldenburg lets the actors play themselves although in most instances the actors are unaware of this. He writes his Happenings with a particular person in mind, allowing the specific shyness, nervousness, sensuality of the person to become part of the Happening. The tradition in art can, therefore, not tell us anything else but that the technical elements involved in the works I have described are just as much a part of the work of art as the paint in the painting. It is impossible to treat the sound as part of *Oracle* and not the radios. Jasper Johns has already shown us the backside of the canvas and I am afraid he will have to accept the not-so-elegant backsides of *Field Painting* and *Zone* as well. But if the radios and the amplifiers are part of the work—what about the engineer who designs them? In the same way as Oldenburg works with the peculiarities of people in his Happenings, the artist has to work with the peculiarities and the foreign mode of operation of the engineer. On the basis of this observation, I hereby declare myself to be a work of art—or rather an integral part of the works of art I have just described. I am definitely not a violin player who interprets and feels for the work of his master. I know nothing about art or the artists involved. I am an engineer and as such, only raw material for the artist.

But how can I claim that this new interface between art and technology does in fact exist? Maybe I wanted to become a work of art and devised this ingenious scheme for my own ends? Well, I think that we don't have to look too far. We all know how technology has become part of our lives. And now we can see absolutely no reason why it should not become more so. No sound has been heard from another culture to oppose Western technology. The faster the underdeveloped countries can have it, the faster they want it. On the other end of the spectrum, we now have systems where we don't know quite where the machine ends and the human being begins. I am thinking of the space program which has introduced the new and maybe inhuman objective: the system has to work, no failures are allowed, no personal emotions or mistake may interfere with the success of the project. The space program is developing a new managerial type which is totally responsible. I read recently that President Johnson has let the contract to solve the Appalachian problem go to the electronics industry. We are now getting the fallout from Cape Kennedy and can expect more.

The great initiator of all this technological soul-searching is the computer. Laboriously we are translating every aspect of human activity into computer language. In fact, I believe the computer will turn out to be the greatest psychoanalyst of all times. Now where does all this leave us? The engineers may be psychoanalysts but they are not visionaries. John Cage has recently written a wonderful article called "How to Improve the World." As a blind engineer, one of his observations gave me a real jolt. Cage points out that there exist systems of interaction between human beings which work without any police or power structure whatsoever. In fact, there are hundreds of agreements between the countries of the world that work perfectly well. In particular, technological questions are dealt with without any complications. It seems that technology breeds agreement. This is such a simple observation that it frightens you that you did not think of it. I believe that Cage's discovery fully justifies the statement that technology will force the solution of such problems as food distribution and housing. There is no other stable optimum but to give people food and housing. The Dadaists' suggestion of free food and Buckminster Fuller's suggestion of free housing for the people of the world will happen. But the alternatives that the engineer can imagine for the full use of the fantastic capacity of technology are even so few and limited. He is, as I said, no visionary about life. But the artist is a visionary about life. Only he can create disorder and still get away with it. Only he can use technology to its fullest capacity. John Cage has suggested: Let the engineer take care of order and art (in the traditional sense) and let the artists take care of disorder and life. And I am adding technology. This to sum up: First the artists have to create with technology because technology is becoming inseparable from our lives. "Technology is the extension of our nervous system," as McLuhan says. Second, the artists should use technology because technology needs the artists. Technology needs to be revealed and looked at—much like we undress a woman.

The artist's work is like that of a scientist. It is an investigation which may or may not yield meaningful results; in many cases we only know many years later. What I am suggesting is that the use of the engineer by the artist will stimulate new ways of looking at technology and dealing with life in the future.

What about power failure? I wish we knew more about what happened. We heard a lot about how people became friendly and helped each other out. The whole thing could have been an artist's idea—to make us aware of something. In the future there will exist technological systems as complicated and as large as the Northeastern power grid whose sole purpose will be to intensify our lives through increased awareness.

Nam June Paik

"Cybernated Art" (1966)
"Art and Satellite" (1984)

<< 6 >>

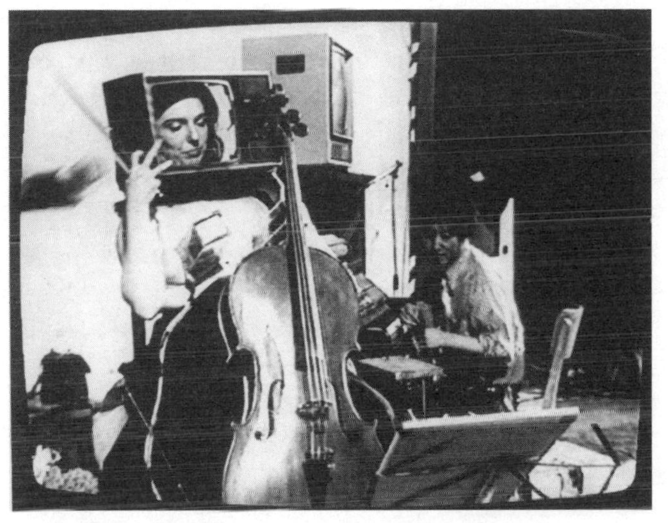

Nam June Paik, Opera Sextronique. *Photo by Ludwig Winterhalter. Courtesy of Nam June Paik.*

"There is no rewind button on the Betamax of life."

<< The Korean born Nam June Paik's formative years were grounded in the study of rigorous post-Webern, twelve-tone music composition. After meeting the iconoclastic American composer John Cage at the Darmstadt Summer School of Music in 1958, Paik turned to performance art and mixed-media works that challenged his classical roots. Most important, Paik embraced the medium of television, and became the founding father of video art. His long and prolific relationship with electronic media began notably with the cellist Charlotte Moorman, in controversial performance works such as *Opera Sextronique* (1967). Paik's oeuvre later included television sculpture, satellite art, robotic devices, and giant video walls with synthesized imagery pulsating from stacks of cathode-ray tubes.

In the brief, mischievously elfin manifesto "Cybernated Art," Paik suggests that art should embrace the technologies of an information society. Paik presents himself as artist-shaman, synthesizing art and technology in an effort to exorcise the demons of a mass-consumer, technology-obsessed society. Paik uses rejected media artifacts in his work, such as vintage television sets; his video works, with their liberal doses of "cybernated shock and catharsis," are poignantly cynical pieces that comment on an American techno-culture dominated by starry-eyed optimists.

In the latter essay, "Art and Satellite," he explains that the material of the information age—the raw data, the continuous flow of content—is becoming an intrinsic element of his artwork. He sees in the satellite, the most advanced communications technology of its day, an opportunity to connect minds and encourage new ways of thinking. Through the connectivity of a "cybernated society," a work of art can become dynamic, always changing, as data flows through the wired network and across our screens. As the artwork incorporates the unending flow and restless nature of information itself, transcending geographical boundaries, it brings about a "synthesis" of all cultures across borders. >>

Cybernated Art (1966)

℞ Cybernated art is very important, but art for cybernated life is more important, and the latter need not be cybernated.
(Maybe George Brecht's simplissimo is the most adequate.)

☠ But if Pasteur and Robespierre are right that we can resist poison only through certain built-in poison, then some specific frustrations, caused by cybernated life, require accordingly cybernated shock and catharsis. My everyday work with videotape and the cathode-ray tube convinces me of this.

Cybernetics, the science of pure relations, or relationship itself, has its origin in karma. Marshall McLuhan's famous phrase "Media is message" was formulated by Norbert Wiener in 1948 as "The signal, where the message is sent, plays equally important role as the signal, where message is not sent."

As the Happening is the fusion of various arts, so cybernetics is the exploitation of boundary regions between and across various existing sciences.

Newton's physics is the mechanics of power and the unconciliatory two-party system, in which the strong win over the weak. But in the 1920s a German genius put a tiny third-party (grid) between these two mighty poles (cathode and anode) in a vacuum tube, thus enabling the weak to win over the strong for the first time in human history. It might be a Buddhistic "third way," but anyway this German invention led to cybernetics, which came to the world in the last war to shoot down German planes from the English sky.

The Buddhists also say

Karma is samsara

Relationship is metempsychosis

We are in open circuits

Art and Satellite (1984)

At the turn of our century, the French mathematician Henri Poincaré said the following thing. . . . (Yes, it was in the midst of so-called material progress and the discovery of new Things. . . .) Poincaré pointed out that what was being discovered was not new THINGS but merely the new RELATIONSHIPS between things already existing.

We are again in the fin de siècle . . . this time we are discovering much new software . . . which are not new things but new thinks . . . and again we are discovering and even weaving new relationships between many thinks and minds . . . we are already knee-deep in the post industrial age. The satellite, especially the live two-way satellite, is a very powerful tool for this human Videosphere. . . .

It is said that all the sciences can trace their roots to Aristotle: but the science of cosmic aesthetics started with Sarutobi Sasuke, a famous *ninja* (a samurai who mastered many fantastic arts, including that of making himself invisible, chiefly to spy upon an enemy). The first step for a *ninja* is learning how to shorten distances by shrinking the earth, that is, how to transcend the law of gravity. For the satellite, this is a piece of cake. So, just as Mozart mastered the newly invented clarinet, the satellite artist must compose his art from the beginning suitable to physical conditions and grammar. Satellite art in the superior sense does not merely transmit existing symphonies and operas to other lands. It must consider how to achieve a two-way connection between opposite sides of the earth; how to give a conversational structure to the art; how to master differences in time; how to play with improvisation, in-determinism, echos, feedbacks, and empty spaces in the Cagean sense; and how to instantaneously manage the differences in culture, preconceptions, and common sense that exist between various nations. Satellite art must make the most of these elements (for they can become strengths or weaknesses), creating a multitemporal, multispatial symphony. . . .

There is no rewind button on the BETAMAX of life. An important event takes place only once. The free deaths (of Socrates, Christ, Bo Yi and Shu Qi) that became the foundations for the morality of three civilizations occurred only once. The meetings of person and person, of person and specific era are often said to take place "one meeting-one life," but the *bundle* of *segments* of this existence (if *segments* can come in *bundles*) has grown much thicker because of the satellite. The thinking process is the jumping of electrical sparks across the synapses between brain cells arranged in multilayered matrices. Inspiration is a spark shooting off in an unexpected direction and landing on a point in some corner of the matrix. The satellite will accidentally and inevitably produce unexpected meetings of person and person and will enrich the synapses between the brain cells of mankind. Thoreau, the author of *Walden, Life in the Woods,* and a nineteenth-century forerunner of the hippies, wrote, "The telephone company is trying to connect Maine and Tennessee by telephone. Even if it were to succeed, though, what would the people say to each other? What could they possibly find to talk about?" Of course, history eventually answered Thoreau's questions (silly ones, at that). There developed a feedback (or, to use an older term, dialectic) of new contacts breeding new contents and new contents breeding new contacts. . . .

Thanks to the satellite, the mysteries of encounters with others (chance meetings) will accumulate in geometric progression and should become the main nonmaterial product of post-industrial society. God created love to propagate the human race, but, unawares, man began to love simply to love. By the same logic, although man talks to accomplish something, unawares, he soon begins to talk simply to talk. . . .

<< part II >>

Interactivity

Norbert Wiener

"Cybernetics in History," *The Human Use of Human Beings* (1954)

<< 7 >>

Norbert Wiener at MIT. Courtesy of MIT Computer Museum.

"Society can only be understood through a study of the messages and the communication facilities which belong to it; and that in the future development of these messages and communication facilities, messages between man and machines, between machines and man, and between machine and machine, are destined to play an ever increasing part."

<< Norbert Wiener, a Hungarian-born mathematician, electrical engineer, and communications specialist, wrote about the nature of human-machine communication in his landmark treatise of 1948, *Cybernetics.* In an attempt to make his theories on information science and machine control accessible to a broader public, two years later he published *The Human Use of Human Beings,* drawing from sociology to theorize about how people might interact and coexist with computers, then revised the book for a new edition in 1954.

Wiener claimed that the quality of man-machine communication influences man's inner well-being. His theory of cybernetics was meant to improve the quality of life in a technological society, where people are increasingly reliant on machines, and where interactions with machines are the norm.

Wiener defined cybernetics as the science of transmitting messages between man and machine, or from machine to machine. The term *cybernetics* has its roots in the Greek word for "steersman" or "governor," and Wiener's use of it suggests how people interact with machines through a controlling device, such as a steering mechanism. Driving a car, passing through an automatic door, or clicking a mouse are all cybernetic actions. Wiener's remarkable insight, which is the premise behind all human-computer interactivity and interface design, is that human communication should be a model for human-machine and machine-to-machine interactions. Wiener points out that the quality of data transmission between man and machine is affected by such factors as feedback, noise, and entropy. As he succinctly puts it, "the fact that the signal in its intermediate stages has gone through a machine rather than through a person is irrelevant." >>

Since the end of World War II, I have been working on the many ramifications of the theory of messages. Besides the electrical engineering theory of the transmission of messages, there is a larger field which includes not only the study of language but the study of messages as a means of controlling machinery and society, the development of computing machines and other such automata, certain reflections upon psychology and the nervous system, and a tentative new theory of scientific method. This larger theory of messages is a probabilistic theory, an intrinsic part of the movement that owes its origin to Willard Gibbs.

Until recently, there was no existing word for this complex of ideas, and in order to embrace the whole field by a single term, I felt constrained to invent one. Hence "Cybernetics," which I derived from the Greek word *kubernētēs,* or "steersman," the same Greek word from which we eventually derive our word "governor." Incidentally, I found later that the word had already been used by Ampère with reference to political science, and had been introduced in another context by a Polish scientist, both uses dating from the earlier part of the nineteenth century.

I wrote a more or less technical book entitled *Cybernetics* which was published in 1948. In response to a certain demand for me to make its ideas acceptable to the lay public, I published the first edition of *The Human Use of Human Beings* in 1950. Since then the subject has grown from a few ideas shared by Drs. Claude Shannon, Warren Weaver, and myself into an established region of research. Therefore, I take this opportunity occasioned by the reprinting of my book to bring it up to date, and to remove certain defects and inconsequentialities in its original structure.

In giving the definition of Cybernetics in the original book, I classed communication and control together. Why did I do this? When I communicate with another person, I impart a message to him, and when he communicates back with me he returns a related message which contains information primarily accessible to him and not to me. When I control the actions of another person, I communicate a message to him, and although this message is in the imperative mood, the technique of communication does not differ from that of a message of fact. Furthermore, if my control is to be effective I must take cognizance of any messages from him which may indicate that the order is understood and has been obeyed.

It is the thesis of this book that society can only be understood through a study of the messages and the communication facilities which belong to it; and that in the future development of these messages and communication facilities, messages between man and machines, between machines and man, and between machine and machine are destined to play an ever-increasing part.

When I give an order to a machine, the situation is not essentially different from that which arises when I give an order to a person. In other words, as far as my consciousness goes I am aware of the order that has gone out and of the signal of compliance that has come back. To me, personally, the fact that the signal in its intermediate stages has gone through a machine rather than through a person is irrelevant and does not in any case greatly change my relation to the signal. Thus the theory of control in engineering, whether human or animal or mechanical, is a chapter in the theory of messages.

Naturally there are detailed differences in messages and in problems of control, not only between a living organism and a machine, but within each narrower class of beings. It is the purpose of Cybernetics to develop a language and techniques that will enable us indeed to attack the problem of control and communication in general, but also to find the proper repertory of ideas and techniques to classify their particular manifestations under certain concepts.

The commands through which we exercise our control over our environment are a kind of information which we impart to it. Like any form of information, these commands are subject to disorganization in transit. They generally come through in less coherent fashion and certainly not more coherently than they were sent. In control and communication we are always fighting nature's tendency to degrade the organized and to destroy the meaningful; the tendency, as Gibbs has shown us, for entropy to increase.

Much of this book concerns the limits of communication within and among individuals. Man is immersed in a world which he perceives through his sense organs. Information that he receives is coordinated through his brain and nervous system until, after the proper process of storage, collation, and selection, it emerges through effector organs, generally his muscles. These in turn act on the external world, and also react on the central nervous system through receptor organs such as the end organs of kinaesthesia; and the information received by the kinaesthetic organs is combined with his already accumulated store of information to influence future action.

Information is a name for the content of what is exchanged with the outer world as we adjust to it, and make our adjustment felt upon it. The process of receiving and of using information is the process of our adjusting to the contingencies of the outer environment, and of our living effectively within that environment. The needs and the complexity of modern life make greater demands on this process of information than ever before, and our press, our museums, our scientific laboratories, our universities, our libraries and textbooks, are obliged to meet the needs of this process or fail in their purpose. To live effectively is to live with adequate information. Thus, communication and control belong to the essence of man's inner life, even as they belong to his life in society. . . .

Messages are themselves a form of pattern and organization. Indeed, it is possible to treat sets of messages as having an entropy like sets of states of the external world. Just as entropy is a measure of disorganization, the information carried by a set of messages is a measure of organization. In fact, it is possible to interpret the information carried by a message as essentially the negative of its entropy, and the negative logarithm of its probability. That is, the more probable the message, the less information it gives. Clichés, for example, are less illuminating than great poems. . . .

Let us consider the activity of the little figures which dance on the top of a music box. They move in accordance with a pattern, but it is a pattern which is set in advance, and in which the past activity of the figures has practically nothing to do with the pattern of their future activity. The probability that they will diverge from this pattern is nil. There is a message, indeed; but it goes from the machinery of the music box to the figures, and stops there. The figures themselves have no trace of communication with the outer world, except this one-way stage of communication with the preestablished mechanism of the music box. They are blind, deaf, and dumb, and cannot vary their activity in the least from the conventionalized pattern.

Contrast with them the behavior of man, or indeed of any moderately intelligent animal such as a kitten. I call to the kitten and it looks up. I have sent it a message which it has received by its sensory organs, and which it registers in action. The kitten is hungry and lets out a pitiful wail. This time it is the sender of a message. The kitten bats at a swinging spool. The spool swings to its left, and the kitten catches it with its left paw. This time messages of a very complicated nature are both sent and received within the kitten's own nervous system through certain nerve end-bodies in its joints, muscles, and tendons; and by means of nervous messages sent by these organs, the animal is aware of the actual position and tensions of its tissues. It is only through these organs that anything like a manual skill is possible.

I have contrasted the prearranged behavior of the little figures on the music box on the one hand, and the contingent behavior of human beings and animals on the other. But we must not suppose that the music box is typical of all machine behavior.

The older machines, and in particular the older attempts to produce automata, did in fact function on a closed clockwork basis. But modern automatic machines such as the controlled missile, the proximity fuse, the automatic door opener, the control apparatus for a chemical factory, and the rest of the modern armory of automatic machines which perform military or industrial functions possess sense organs; that is, receptors for messages coming from the outside. These may be as simple as photoelectric cells which change electrically when a light falls on them, and which can tell light from dark, or as complicated as a television set. They may measure a tension by the change it produces in the conductivity of a wire exposed to it, or they may measure temperature by means of a thermocouple, which is an instrument consisting of two distinct metals in contact with one another through which a current flows when one of the points of contact is heated. Every instrument in the repertory of the scientific-instrument maker is a possible sense organ, and may be made to record its reading remotely through the intervention of appropriate electrical apparatus. Thus the machine which is conditioned by its relation to the external world, and by the things happening in the external world, is with us and has been with us for some time.

The machine which acts on the external world by means of messages is also familiar. The automatic photoelectric door opener is known to every person who has passed through the Pennsylvania Station in New York, and is used in many other buildings as well. When a message consisting of the interception of a beam of light is sent to the apparatus, this message actuates the door, and opens it so that the passenger may go through.

The steps between the actuation of a machine of this type by sense organs and its performance of a task may be as simple as in the case of the electric door; or it may be in fact of any desired degree of complexity within the limits of our engineering techniques. A complex action is one in which the data introduced, which we call the *input,* to obtain an effect on the outer world, which we call the *output,* may involve a large number of combinations. These are combinations, both of the data put in at the moment and of the records taken from the past stored data which we call the *memory.* These are recorded in the machine. The most complicated machines yet made which transform input data into output data are the high-speed electrical computing machines, of which I shall speak later in more detail. The determination of the mode of conduct of these machines is given through a special sort of input, which frequently consists of punched cards or tapes or of magnetized wires, and which determines the way in which the machine is going to act in one operation, as distinct from the way in which it might have acted in another. Because of the frequent use of punched or magnetic tape in the control, the data which are fed in, and which indicate the mode of operation of one of these machines for combining information, are called the *taping.*

I have said that man and the animal have a kinaesthetic sense, by which they keep a record of the position and tensions of their muscles. For any machine subject to a varied external environment to act effectively it is necessary that information concerning the results of its own action be furnished to it as part of the information on which it must continue to act. For example, if we are running an elevator, it is not enough to open the outside door because the orders we have given should make the elevator be at that door at the time we open it. It is important that the release for opening the door be dependent on the fact that the elevator is actually at the door; otherwise something might have detained it, and the passenger might step into the empty shaft. This control of a machine on the basis of its *actual* performance rather than its *expected* performance is known as *feedback,* and involves sensory members which are actuated by motor members and perform the function of *tell-tales* or *monitors*—that is, of elements which indicate a performance. It is the function of these mechanisms to control the mechanical tendency toward disorganization; in other words, to produce a temporary and local reversal of the normal direction of entropy.

I have just mentioned the elevator as an example of feedback. There are other cases where the importance of feedback is even more apparent. For example, a gun-pointer takes information from his instruments of observation, and conveys it to the gun, so that the latter will point in such a direction that the missile will pass through the moving target at a certain time. Now, the gun itself must be used under all conditions of weather. In some of these the grease is warm, and the gun swings easily and rapidly. Under other conditions the grease is frozen or mixed with sand, and the gun is slow to answer the orders given to it. If these orders are reinforced by an extra push given when the gun fails to respond easily to the orders and lags behind them, then the error of the gun-pointer will be decreased. To obtain a performance as uniform as possible, it is customary to put into the gun a control feedback element which reads the lag of the gun behind the position it should have according to the orders given it, and which uses this difference to give the gun an extra push.

It is true that precautions must be taken so that the push is not too hard, for if it is, the gun will swing past its proper position, and will have to be pulled back in a series of oscillations, which may well become wider and wider, and lead to a disastrous instability. If the feedback system is itself controlled—if, in other words, its own entropic tendencies are checked by still other controlling mechanisms—and kept within limits sufficiently stringent, this will not occur, and the existence of the feedback will increase the stability of performance of the gun. In other words, the performance will become less dependent on the frictional load; or what is the same thing, on the drag created by the stiffness of the grease.

Something very similar to this occurs in human action. If I pick up my cigar, I do not will to move any specific muscles. Indeed in many cases, I do not know what those muscles are. What I do is to turn into action a certain feedback mechanism; namely, a reflex in which the amount by which I have yet failed to pick up the cigar is turned into a new and increased order to the lagging muscles, whichever they may be. In this way, a fairly uniform voluntary command will enable the same task to be performed from widely varying initial positions, and irrespective of the decrease of contraction due to fatigue of the muscles. Similarly, when I drive a car, I do not follow out a series of commands dependent simply on a mental image of the road and the task I am doing. If I find the car swerving too much to the right, that causes me to pull it to the left. This depends on the actual performance of the car, and not simply on the road; and it allows me to drive with nearly equal efficiency a light Austin or a heavy truck, without having formed separate habits for the driving of the two. . . .

It is my thesis that the physical functioning of the living individual and the operation of some of the newer communication machines are precisely parallel in their analogous attempts to control entropy through feedback. Both of them have sensory receptors as one stage in their cycle of operation: that is, in both of them there exists a special apparatus for collecting information from the outer world at low energy levels, and for making it available in the operation of the individual or of the machine. In both cases these external messages are not taken *neat,* but through the internal transforming powers of the apparatus, whether it be alive or dead. The information is then turned into a new form available for the further stages of performance. In both the animal and the machine this performance is made to be effective on the outer world. In both of them, their *performed* action on the outer world, and not merely their *intended* action, is reported back to the central regulatory apparatus. This complex of behavior is ignored by the average man, and in particular does not play the role that it should in our habitual analysis of society; for just as individual physical responses may be seen from this point of view, so may the organic responses of society itself. I do not mean that the sociologist is unaware of the existence and complex nature of communications in society, but until recently he has tended to overlook the extent to which they are the cement which binds its fabric together.

We have seen in this chapter the fundamental unity of a complex of ideas which until recently had not been sufficiently associated with one another, namely, the contingent view of physics that Gibbs introduced as a modification of the traditional, Newtonian conventions, the Augustinian attitude toward order and conduct which is demanded by this view, and the theory of the message among men, machines, and in society as a sequence of events in time which, though it itself has a certain contingency, strives to hold back nature's tendency toward disorder by adjusting its parts to various purposive ends.

J.C.R. Licklider

"Man-Computer Symbiosis" (1960)

<< 8 >>

J.C.R. Licklider. Courtesy of MIT Computer Museum.

"The hope is that, in not too many years, human brains and computing machines will be coupled together very tightly, and that the resulting partnership will think as no human brain has ever thought and process data in a way not approached by the information-handling machines we know today."

<< In the late 1950s, J.C.R. Licklider was one of the few scientists who saw the computer's potential as a collaborative partner in the creative process. In his book *Libraries of the Future* he proposed how computers networked in a system could lead to a new kind of "thinking center." During his tenure as director of the United States government's Advanced Research Projects Agency (ARPA) in the 1960s, Licklider had the vision to support controversial but critical research that led to the rise of human-computer interactivity and the personal computer. Licklider's own research on computer networking was a catalyst for the creation of the Internet (originally known as "ARPANET") in 1969.

With this article, Licklider fundamentally changed how we interact with computers by proposing the novel idea of a symbiotic relationship between man and machine. The computer's then typical role was as a subservient device that performed data operations and mechanical calculations. Licklider suggested that a computer could be more effective as a collaborator, and that this interaction would yield results well beyond what people could achieve on their own. He saw the potential for a dialogue between man and machine, a symbiotic partnership that would unleash tremendous creative potential, made possible by the ease, immediacy, and flexibility of a keyboard and real-time graphics display. This foresight was extraordinary considering that "Man-Computer Symbiosis" was written when computers were excruciatingly slow and clumsy, with mainframe systems using punch card input and teletype output the norm. >>

Summary

Man-computer symbiosis is an expected development in cooperative interaction between men and electronic computers. It will involve very close coupling between the human and the electronic members of the partnership. The main aims are (1) to let computers facilitate formulative thinking as they now facilitate the solution of formulated problems, and (2) to enable men and computers to cooperate in making decisions and controlling complex situations without inflexible dependence on predetermined programs. In the anticipated symbiotic partnership, men will set the goals, formulate the hypotheses, determine the criteria, and perform the evaluations. Computing machines will do the routinizable work that must be done to prepare the way for insights and decisions in technical and scientific thinking. Preliminary analyses indicate that the symbiotic partnership will perform intellectual operations much more effectively than man alone can perform them. Prerequisites for the achievement of the effective, cooperative association include developments in computer time sharing, in memory components, in memory organization, in programming languages, and in input and output equipment.

1 Introduction

1.1 Symbiosis

The fig tree is pollinated only by the insect *Blastophaga grossorun.* The larva of the insect lives in the ovary of the fig tree, and there it gets its food. The tree and the insect are thus heavily interdependent: the tree cannot reproduce without the insect; the insect cannot eat without the tree; together, they constitute not only a viable but a productive and thriving partnership. This cooperative "living together in intimate association, or even close union, of two dissimilar organisms" is called symbiosis.[1]

"Man-computer symbiosis" is a subclass of man-machine systems. There are many man-machine systems. At present, however, there are no man-computer symbioses. The purposes of this paper are to present the concept and, hopefully, to foster the development of man-computer symbiosis by analyzing some problems of interaction between men and computing machines, calling attention to applicable principles of man-machine engineering, and pointing out a few questions to which research answers are needed. The hope is that, in not too many years, human brains and computing machines will be coupled together very tightly, and that the resulting partnership will think as no human brain has ever thought and process data in a way not approached by the information-handling machines we know today.

1.2 Between "Mechanically Extended Man" and "Artificial Intelligence"

As a concept, man-computer symbiosis is different in an important way from what North[2] has called "mechanically extended man." In the man-machine systems of the past, the human operator supplied the initiative, the direction, the integration, and the criterion. The mechanical parts of the systems were mere extensions, first of the human arm, then of the human eye. These systems certainly did not consist of "dissimilar organisms living together . . ." There was only one kind of organism—man—and the rest was there only to help him.

In one sense of course, any man-made system is intended to help man, to help a man or men outside the system. If we focus upon the human operator within the system, however, we see that, in some areas of technology, a fantastic change has taken place during the last few years. "Mechanical extension" has given way to replacement of men, to automation, and the men who remain are there more to help than to be helped. In some instances, particularly in large computer-centered information and control systems, the human operators are responsible mainly for functions that it proved infeasible to automate. Such systems ("humanly extended machines," North might call them) are not symbiotic systems. They are "semi-automatic" systems, systems that started out to be fully automatic but fell short of the goal.

Man-computer symbiosis is probably not the ultimate paradigm for complex technological systems. It seems entirely possible that, in due course, electronic or chemical "machines" will outdo the human brain in most of the functions we now consider exclusively within its province. Even now, Gelernter's IBM-704 program for proving theorems in plane geometry proceeds at about the same pace as Brooklyn high school students, and makes similar errors.[3] There are, in fact, several theorem-proving, problem-solving, chess-playing, and pattern-recognizing programs (too many for complete reference)[4] capable of rivaling human intellectual performance in restricted areas; and Newell, Simon, and Shaw's[5] "general problem solver" may remove some of the restrictions. In short, it seems worthwhile to avoid argument with (other) enthusiasts for artificial intelligence by conceding dominance in the distant future of cerebration to machines alone. There will nevertheless be a fairly long interim during which the main intellectual advances will be made by men and computers working together in intimate association. A multidisciplinary study group, examining future research and development problems of the Air Force, estimated that it would be 1980 before developments in artificial intelligence make it possible for machines alone to do much thinking or problem solving of military significance. That would leave, say, five years to develop man-computer symbiosis and 15 years to use it. The 15 may be 10 or 500, but those years should be intellectually the most creative and exciting in the history of mankind.

2 Aims of Man-Computer Symbiosis

Present-day computers are designed primarily to solve preformulated problems or to process data according to predetermined procedures. The course of the computation may be conditional upon results obtained during the computation, but all the alternatives must be foreseen in advance. (If an unforeseen alternative arises, the whole process comes to a halt and awaits the necessary extension of the program.) The requirement for preformulation or predetermination is sometimes no great disadvantage. It is often said that programming for a computing machine forces one to think clearly, that it disciplines the thought process. If the user can think his problem through in advance, symbiotic association with a computing machine is not necessary.

However, many problems that can be thought through in advance are very difficult to think through in advance. They would be easier to solve, and they could be solved faster, through an intuitively guided trial-and-error procedure in which the computer cooperated, turning up flaws in the reasoning or revealing unexpected turns in the solution. Other problems simply cannot be formulated without computing machine aid. Poincaré anticipated the frustration of an important group of would-be computer users when he said, "The question is not, 'What is the answer?' The question is, 'What is the question?' " One of the main aims of man-computer symbiosis is to bring the computing machine effectively into the formulative parts of technical problems.

The other main aim is closely related. It is to bring computing machines effectively into processes of thinking that must go on in "real time," time that moves too fast to permit using computers in conventional ways. Imagine trying, for example, to direct a battle with the aid of a computer on such a schedule as this. You formulate your problem today. Tomorrow you spend with a programmer. Next week the computer devotes 5 minutes to assembling your program and 47 seconds to calculating the answer to your problem. You get a sheet of paper 20 feet long, full of numbers that, instead of providing a final solution, only suggest a tactic that should be explored by simulation. Obviously, the battle would be over before the second step in its planning was begun. To think in interaction with a computer in the same way that you think with a colleague whose competence supplements your own will require much tighter coupling between man and machine than is suggested by the example and than is possible today.

3 Need for Computer Participation in Formulative and Real-Time Thinking

The preceding paragraphs tacitly made the assumption that, if they could be introduced effectively into the thought process, the functions that can be performed by data-processing machines would improve or facilitate thinking and problem solving in an important way. That assumption may require justification.

3.1 *A Preliminary and Informal Time-and-Motion Analysis of Technical Thinking*

Despite the fact that there is a voluminous literature on thinking and problem solving, including intensive case-history studies of the process of invention, I could find nothing comparable to a time-and-motion-study analysis of the mental work of a person engaged in a scientific or technical enterprise. In the spring and summer of 1957, therefore, I tried to keep track of what one moderately technical person actually did during the hours he regarded as devoted to work. Although I was aware of the inadequacy of the sampling, I served as my own subject.

It soon became apparent that the main thing I did was to keep records, and the project would have become an infinite regress if the keeping of records had been carried through in the detail envisaged in the initial plan. It was not. Nevertheless, I obtained a picture of my activities that gave me pause. Perhaps my spectrum is not typical—I hope it is not, but I fear it is.

About 85 percent of my "thinking" time was spent getting into a position to think, to make a decision, to learn something I needed to know. Much more time went into finding or obtaining information than into digesting it. Hours went into the plotting of graphs, and other hours into instructing an assistant how to plot. When the graphs were finished, the relations were obvious at once, but the plotting had to be done in order to make them so. At one point, it was necessary to compare six experimental determinations of a function relating speech-intelligibility to speech-to-noise ratio. No two experimenters had used the same definition or measure of speech-to-noise ratio. Several hours of calculating were required to get the data into comparable form. When they were in comparable form, it took only a few seconds to determine what I needed to know.

Throughout the period I examined, in short, my "thinking" time was devoted mainly to activities that were essentially clerical or mechanical: searching, calculating, plotting, transforming, determining the logical or dynamic consequences of a set of assumptions or hypotheses, preparing the way for a decision or an insight. Moreover, my choices of what to attempt and what not to attempt were determined to an embarrassingly great extent by considerations of clerical feasibility, not intellectual capability.

The main suggestion conveyed by the findings just described is that the operations that fill most of the time allegedly devoted to technical thinking are operations that can be performed more effectively by machines than by men. Severe problems are posed by the fact that these operations have to be performed upon diverse variables and in unforeseen and continually changing sequences. If those problems can be solved in such a way as to create a symbiotic relation between a man and a fast information-retrieval and data-processing machine, however, it seems evident that the cooperative interaction would greatly improve the thinking process.

It may be appropriate to acknowledge, at this point, that we are using the term "computer" to cover a wide class of calculating, data-processing, and information-storage-and-retrieval machines. The capabilities of machines in this class are increasing almost daily. It is therefore hazardous to make general statements about capabilities of the class. Perhaps it is equally hazardous to make general statements about the capabilities of men. Nevertheless, certain genotypic differences in capability between men and computers do stand out, and they have a bearing on the nature of possible man-computer symbiosis and the potential value of achieving it.

As has been said in various ways, men are noisy, narrow-band devices, but their nervous systems have very many parallel and simultaneously active channels. Relative to men, computing machines are very fast and very accurate, but they are constrained to perform only one or a few elementary operations at a time. Men are flexible, capable of "programming themselves contingently" on the basis of newly received information. Computing machines are single-minded, constrained by their "pre-programming." Men naturally speak redundant languages organized around unitary objects and coherent actions and employing 20 to 60 elementary symbols. Computers "naturally" speak nonredundant languages, usually with only two elementary symbols and no inherent appreciation either of unitary objects or of coherent actions.

To be rigorously correct, those characterizations would have to include many qualifiers. Nevertheless, the picture of dissimilarity (and therefore potential supplementation) that they present is essentially valid. Computing machines can do readily, well, and rapidly many things that are difficult or impossible for man, and men can do readily and well, though not rapidly, many things that are difficult or impossible for computers. That suggests that a symbiotic cooperation, if successful in integrating the positive characteristics of men and computers, would be of great value. The differences in speed and in language, of course, pose difficulties that must be overcome.

4 Separable Functions of Men and Computers in the Anticipated Symbiotic Association

It seems likely that the contributions of human operators and equipment will blend together so completely in many operations that it will be difficult to separate them neatly in analysis. That would be the case if, in gathering data on which to base a decision, for example, both the man and the computer came up with relevant precedents from experience and if the computer then suggested a course of action that agreed with the man's intuitive judgment. (In theorem-proving programs, computers find precedents in experience, and in the SAGE System, they suggest courses of action. The foregoing is not a far-fetched example.) In other operations, however, the contributions of men and equipment will be to some extent separable.

Men will set the goals and supply the motivations, of course, at least in the early years. They will formulate hypotheses. They will ask questions. They will think of mechanisms, procedures, and models. They will remember that such-and-such a person did some possibly relevant work on a topic of interest back in 1947, or at any rate shortly after World War II, and they will have an idea in what journals it might have been published. In general, they will make approximate and fallible, but leading, contributions, and they will define criteria and serve as evaluators, judging the contributions of the equipment and guiding the general line of thought.

In addition, men will handle the very-low-probability situations when such situations do actually arise. (In current man-machine systems, that is one of the human operator's most important functions. The sum of the probabilities of very-low-probability alternatives is often much too large to neglect.) Men will fill in the gaps, either in the problem solution or in the computer program, when the computer has no mode or routine that is applicable in a particular circumstance.

The information-processing equipment, for its part, will convert hypotheses into testable models and then test the models against data (which the human operator may designate roughly and identify as relevant when the computer presents them for his approval). The equipment will answer questions. It will simulate the mechanisms and models, carry out the procedures, and display the results to the operator. It will transform data, plot graphs ("cutting the cake" in whatever way the human operator specifies, or in several alternative ways if the human operator is not sure what he wants). The equipment will interpolate, extrapolate, and transform. It will convert static equations or logical statements into dynamic models so the human operator can examine their behavior. In general, it will carry out the routinizable, clerical operations that fill the intervals between decisions.

In addition, the computer will serve as a statistical-inference, decision-theory, or game-theory machine to make elementary evaluations of suggested courses of action whenever there is enough basis to support a formal statistical analysis. Finally, it will do as much diagnosis, pattern-matching, and relevance-recognizing as it profitably can, but it will accept a clearly secondary status in those areas. . . .

Douglas Engelbart

<< 9 >> "Augmenting Human Intellect: A Conceptual Framework" (1962)

Douglas Engelbart at the Augmentation Research Center. Courtesy of Douglas Engelbart, Bootstrap Institute.

"The term 'intelligence amplification' seems applicable to our goal of augmenting the human intellect in that the entity to be produced will exhibit more of what can be called intelligence than an unaided human could; we will have amplified the intelligence of the human by organizing his intellectual capabilities into higher levels of synergistic structuring."

<< Douglas Engelbart is one of the most influential thinkers in the history of personal computing. He is best known as the groundbreaking engineer who invented such mainstays of the personal computer as the mouse, windows, email, and the word processor. He has received less recognition for his pioneering concepts behind the interactive computer networks that are the basis for the Internet. Engelbart led one of the most important projects funded by ARPA in the 1960s: a networked environment designed to support collaborative interaction between people using computers. It was dubbed the oNLine System (NLS). This historic prototype, developed at the Augmentation Research Center of the Stanford Research Institute (SRI) and unveiled in 1968 at the Fall Joint Computer Conference in San Francisco, influenced the development of the first personal computer and the graphical user interface at Xerox PARC in the early 1970s.

Engelbart's experience as a radar specialist in wartime led him to consider the potential for a computer connected to a monitor display and an input device (such as a keyboard), which, in the fifties and sixties, transformed the way people interacted with computers. In this paper, Engelbart describes in detail how such a computer might function—a description that predicts in remarkable detail the fundamental qualities of a modern personal computer. The linking of people and computers using this approach to interactivity would result in the use of computers to "solve the world's problems," Engelbart wrote, by boosting the capacities of the mind's intellectual faculties.

Engelbart reasoned that networked computing would not only make individuals more intellectually effective; it would enable a collaborative method of sharing knowledge. At the time, the idea that computers could be operated intuitively by teams engaged in creative work was highly unorthodox. Engelbart proposed a language, process, methodology, and conceptual framework for the real-time interaction of collaborative computing. He was among the first to consider using computer networks to create collective knowledge bases among groups of professionals. This approach has since transformed many of our institutions, including the military, universities, museums, and libraries, as they have extended their reach into cyberspace. Engelbart's theories about the power of collective knowledge via a real-time electronic medium have also led to other forms of many-to-many human-computer interaction, including such collective online activities as virtual communities, on-line forums, and teleconferencing. >>

I. Introduction

A. General

By "augmenting human intellect" we mean increasing the capability of a man to approach a complex problem situation, to gain comprehension to suit his particular needs, and to derive solutions to problems. Increased capability in this respect is taken to mean a mixture of the following: more-rapid comprehension, better comprehension, the possibility of gaining a useful degree of comprehension in a situation that previously was too complex, speedier solutions, better solutions, and the possibility of finding solutions to problems that before seemed insoluble. And by "complex situations" we include the professional problems of diplomats, executives, social scientists, life scientists, physical scientists, attorneys, designers—whether the problem situation exists for twenty minutes or twenty years. We do not speak of isolated clever tricks that help in particular situations. We refer to a way of life in an integrated domain where hunches, cut-and-try, intangibles, and the human "feel for a situation" usefully co-exist with powerful concepts, streamlined terminology and notation, sophisticated methods, and high-powered electronic aids.

Man's population and gross product are increasing at a considerable rate, but the *complexity* of his problems grows still faster, and the *urgency* with which solutions must be found becomes steadily greater in response to the increased rate of activity and the increasingly global nature of that activity. Augmenting man's intellect, in the sense defined above, would warrant full pursuit by an enlightened society if there could be shown a reasonable approach and some plausible benefits.

This report covers the first phase of a program aimed at developing means to augment the human intellect. These "means" can include many things—all of which appear to be but extensions of means developed and used in the past to help man apply his native sensory, mental, and motor capabilities—and we consider the whole system of a human and his augmentation means as a proper field of search for practical possibilities. It is a very important system to our society, and like most systems its performance can best be improved by considering the whole as a set of interacting components rather than by considering the components in isolation.

This kind of system approach to human intellectual effectiveness does not find a ready-made conceptual framework such as exists for established disciplines. Before a research program can be designed to pursue such an approach intelligently, so that practical benefits might be derived within a reasonable time while also producing results of long-range significance, a conceptual framework must be searched out—a framework that provides orientation as to the important factors of the system, the relationships among these factors, the types of change among the system factors that offer likely improvements in performance, and the sort of research goals and methodology that seem promising.[1]

In the first (search) phase of our program we have developed a conceptual framework that seems satisfactory for the current needs of designing a research phase. Section II contains the essence of this framework as derived from several different ways of looking at the system made up of a human and his intellect-augmentation means.

The process of developing this conceptual framework brought out a number of significant realizations: that the intellectual effectiveness exercised today by a given human has little likelihood of being intelligence limited—that there are dozens of disciplines in engineering, mathematics, and the social, life, and physical sciences that can contribute improvements to the system of intellect-augmentation means; that any one such improvement can be expected to trigger a chain of coordinating improvements; that until every one of these disciplines comes to a standstill *and* we have exhausted all the improvement possibilities we could glean from it, we can expect to continue to develop improvements in this "human-intellect" system; that there is no particular reason not to expect gains in personal intellectual effectiveness from a concerted system-oriented approach that compare to those made in personal geographic mobility since horseback and sailboat days. . . .

To give the reader an initial orientation about what sort of thing this computer-aided working system might be, we include below a short description of a possible system of this sort. This illustrative example is not to be considered a description of the actual system that will emerge from the program. It is given only to show the general direction of the work, and is clothed in fiction only to make it easier to visualize.

Let us consider an "augmented" architect at work. He sits at a working station that has a visual display screen some three feet on a side; this is his working surface, and is controlled by a computer (his "clerk") with which he can communicate by means of a small keyboard and various other devices.

He is designing a building. He has already dreamed up several basic layouts and structural forms, and is trying them out on the screen. The surveying data for the layout he is working on now have already been entered, and he has just coaxed the "clerk" to show him a perspective view of the steep hillside building site with the roadway above, symbolic representations of the various trees that are to remain on the lot, and the service tie points for the different utilities. The view occupies the left two-thirds of the screen. With a "pointer," he indicates two points of interest, moves his left hand rapidly over the keyboard, and the distance and elevation between the points indicated appear on the right-hand third of the screen.

Now he enters a reference line with his "pointer" and the keyboard. Gradually the screen begins to show the work he is doing—a neat excavation appears in the hillside, revises itself slightly, and revises itself again. After a moment, the architect changes the scene on the screen to an overhead plan view of the site, still showing the excavation. A few minutes of study, and he enters on the keyboard a list of items, checking each one as it appears on the screen, to be studied later.

Ignoring the representation on the display, the architect next begins to enter a series of specifications and data—a six-inch slab floor, twelve-inch concrete walls eight feet high within the excavation, and so on. When he has finished, the revised scene appears on the screen. A structure is taking shape. He examines it, adjusts it, pauses long enough to ask for handbook or catalog information from the "clerk" at various points, and readjusts accordingly. He often recalls from the "clerk" his working lists of specifications and considerations to refer to them, modify them, or add to them. These lists grow into an ever-more-detailed, interlinked structure, which represents the maturing thought behind the actual design.

Prescribing different planes here and there, curved surfaces occasionally, and moving the whole structure about five feet, he finally has the rough external form of the building balanced nicely with the setting and he is assured that this form is basically compatible with the materials to be used as well as with the function of the building.

Now he begins to enter detailed information about the interior. Here the capability of the "clerk" to show him any view he wants to examine (a slice of the interior, or how the structure would look from the roadway above) is important. He enters particular fixture designs, and examines them in a particular room. He checks to make sure that sun glare from the windows will not blind a driver on the roadway, and the "clerk" computes the information that one window will reflect strongly onto the roadway between 6 and 6:30 on midsummer mornings.

Next he begins a functional analysis. He has a list of the people who will occupy this building, and the daily sequences of their activities. The "clerk" allows him to follow each in turn, examining how doors swing, where special lighting might be needed. Finally he has the "clerk" combine all of these sequences of activity to indicate spots where traffic is heavy in the building, or where congestion might occur, and to determine what the severest drain on the utilities is likely to be.

All of this information (the building design and its associated "thought structure") can be stored on a tape to represent the "design manual" for the building. Loading this tape into his own "clerk," another architect, a builder, or the client can maneuver within this "design manual" to pursue whatever details or insights are of interest to him—and can append special notes that are integrated into the "design manual" for his own or someone else's later benefit.

In such a future working relationship between human problem-solver and computer "clerk," the capability of the computer for executing mathematical processes would be used whenever it was needed. However, the computer has many other capabilities for manipulating and displaying information that can be of significant benefit to the human in nonmathematical processes of planning, organizing, studying, etc. Every person who does his thinking with symbolized concepts (whether in the form of the English language, pictographs, formal logic, or mathematics) should be able to benefit significantly.

B. *Objective of the Study*

The objective of this study is to develop a conceptual framework within which could grow a coordinated research and development program whose goals would be the following: (1) to find the factors that limit the effectiveness of the individual's basic information-handling capabilities in meeting the various needs of society for problem solving in its most general sense; and (2) to develop new techniques, procedures, and systems that will better match these basic capabilities to the needs, problems, and progress of society. We have placed the following specifications on this framework:

1. That it provide perspective for both long-range basic research and research that will yield practical results soon.
2. That it indicate what this augmentation will actually involve in the way of changes in working environment, in thinking, in skills, and in methods of working.
3. That it be a basis for evaluating the possible relevance of work and knowledge from existing fields and for assimilating whatever is relevant.

4. That it reveal areas where research is possible and ways to assess the research, be a basis for choosing starting points, and indicate how to develop appropriate methodologies for the needed research.

Two points need emphasis here. First, although a conceptual framework has been constructed, it is still rudimentary. Further search, and actual research, are needed for the evolution of the framework. Second, even if our conceptual framework did provide an accurate and complete basic analysis of the system from which stems a human's intellectual effectiveness, the explicit nature of future improved systems would be highly affected by (expected) changes in our technology or in our understanding of the human being.

II. Conceptual Framework

A. General

The conceptual framework we seek must orient us toward the real possibilities and problems associated with using modern technology to give direct aid to an individual in comprehending complex situations, isolating the significant factors, and solving problems. To gain this orientation, we examine how individuals achieve their present level of effectiveness, and expect that this examination will reveal possibilities for improvement.

The entire effect of an individual on the world stems essentially from what he can transmit to the world through his limited motor channels. This in turn is based on information received from the outside world through limited sensory channels; on information, drives, and needs generated within him; and on his processing of that information. His processing is of two kinds: that which he is generally conscious of (recognizing patterns, remembering, visualizing, abstracting, deducing, inducing, etc.), and that involving the unconscious processing and mediating of received and self-generated information, and the unconscious mediating of conscious processing itself.

The individual does not use this information and this processing to grapple directly with the sort of complex situation in which we seek to give him help. He uses his innate capabilities in a rather more indirect fashion, since the situation is generally too complex to yield directly to his motor actions, and always too complex to yield comprehensions and solutions from direct sensory inspection and use of basic cognitive capabilities. For instance, an aborigine who possesses all of our *basic* sensory-mental-motor capabilities, but does not possess our background of indirect knowledge and procedure, cannot organize the proper direct actions necessary to drive a car through traffic, request a book from the library, call a committee meeting to discuss a tentative plan, call someone on the telephone, or compose a letter on the typewriter.

Our culture has evolved means for us to organize the little things we can do with our basic capabilities so that we can derive comprehension from truly complex situations, and accomplish the processes of deriving and implementing problem solutions. The ways in which human capabilities are thus extended are here called *augmentation means,* and we define four basic classes of them:

1. *Artifacts*—physical objects designed to provide for human comfort, for the manipulation of things or materials, and for the manipulation of symbols.
2. *Language*—the way in which the individual parcels out the picture of his world into the concepts that his mind uses to model that world, and the symbols that he attaches to those concepts and uses in consciously manipulating the concepts ("thinking").
3. *Methodology*—the methods, procedures, strategies, etc., with which an individual organizes his *goal-centered* (problem-solving) activity.
4. *Training*—the conditioning needed by the human being to bring his skills in using Means 1, 2, and 3 to the point where they are operationally effective.

The system we want to improve can thus be visualized as a trained human being together with his artifacts, language, and methodology. The explicit new system we contemplate will involve as artifacts computers, and computer-controlled information-storage, information-handling, and information-display devices. The aspects of the conceptual framework that are discussed here are primarily those relating to the human being's ability to make significant use of such equipment in an integrated system.

Pervading all of the augmentation means is a particular structure or organization. While an untrained aborigine cannot drive a car through traffic, because he cannot leap the gap between his cultural background and the kind of world that contains cars and traffic, it is possible to move step by step through an organized training program that will enable him to drive effectively and safely. In other words, the human mind neither learns nor acts by large leaps, but by steps organized or structured so that each one depends upon previous steps.

Although the size of the step a human being can take in comprehension, innovation, or execution is small in comparison to the over-all size of the step needed to solve a complex problem, human beings nevertheless do solve complex problems. It is the augmentation means that serve to break down a large problem in such a way that the human being can walk through it with his little steps, and it is the structure or organization of these little steps or actions that we discuss as *process hierarchies.*

Every process of thought or action is made up of sub-processes. Let us consider such examples as making a pencil stroke, writing a letter of the alphabet, or making a plan. Quite a few discrete muscle movements are organized into the making of a pencil stroke; similarly, making particular pencil strokes and making a plan for a letter are complex processes in themselves that become sub-processes to the over-all writing of an alphabetic character.

Although every sub-process is a process in its own right, in that it consists of further sub-processes, there seems to be no point here in looking for the ultimate "bottom" of the process-hierarchical structure. There seems to be no way of telling whether or not the apparent "bottoms" (processes that cannot be further subdivided) exist in the physical world or in the limitations of human understanding.

In any case, it is not necessary to begin from the "bottom" in discussing particular process hierarchies. No person uses a process that is completely unique every time he tackles something new. Instead, he begins from a group of basic sensory-mental-motor process capabilities, and adds to these certain of the process capabilities of his artifacts. There are only a finite number of such basic human and artifact capabilities from which to draw. Furthermore, even quite different higher-order processes may have in common relatively high-order sub-processes.

When a man writes prose text (a reasonably high-order process), he makes use of many processes as sub-processes that are common to other high-order processes. For example, he makes use of planning, composing, dictating. The process of writing is utilized as a sub-process within many different processes of a still higher order, such as organizing a committee, changing a policy, and so on.

What happens, then, is that each individual develops a certain repertoire of process capabilities from which he selects and adapts those that will compose the processes that he executes. This repertoire is like a tool kit, and just as the mechanic must know what his tools can do and how to use them, so the intellectual worker must know the capabilities of his tools and have good methods, strategies, and rules of thumb for making use of them. All of the process capabilities in the individual's repertoire rest ultimately upon basic capabilities within him or his artifacts, and the entire repertoire represents an inter-knit, hierarchical structure (which we often call the *repertoire hierarchy*).

We find three general categories of process capabilities within a typical individual's repertoire. There are those that are executed completely within the human integument, which we call *explicit-human* process capabilities; there are those possessed by artifacts for executing processes without human intervention, which we call *explicit-artifact* process capabilities; and there are what we call the *composite* process capabilities, which are derived from hierarchies containing both of the other kinds.

We assume that it is our H-LAM/T system (Human using Language, Artifacts, Methodology, in which he is Trained) that has the capability and that performs the process in any instance of use of this repertoire. Let us look within the process structure for the LAM/T ingredients, to get a better "feel" for our models. Consider the process of writing an important memo. There is a particular concept associated with this process—that of putting information into a formal package and distributing it to a set of people for a certain kind of consideration—and the type of information package associated with this concept has been given the special name of *memorandum*. Already the system language shows the effect of this process—i.e., a concept and its name.

The memo-writing process may be executed by using a set of process capabilities (in intermixed or repetitive form) such as the following: planning, developing subject matter, composing text, producing hard copy, and distributing. There is a definite way in which these sub-processes will be organized that represents part of the system methodology. Each of these sub-processes represents a functional concept that must be a part of the system language if it is to be organized effectively into the human's way of doing things, and the symbolic portrayal of each concept must be such that the human can work with it and remember it.

If the memo is simple, a paragraph or so in length, then the first three processes may well be of the explicit-human type (i.e., it may be planned, developed, and composed within the mind) and the last two of the composite type. If it is a complex memo, involving a good deal of careful planning and development, then all of the sub-processes might well be of the composite type (e.g., at least including the use of pencil and paper artifacts), and there might be many different applications of some of the process capabilities within the total process (i.e., successive drafts, revised plans).

The set of sub-process capabilities discussed so far, if called upon in proper occasion and sequence, would indeed enable the execution of the memo-writing process. However, the very process of organizing and supervising the utilization of these sub-process capabilities is itself a most important sub-process of the memo-writing process. Hence, the sub-process capabilities as listed would not be complete without the addition of a seventh capability—what we call the *executive* capability. This is the capability stemming from habit, strategy, rules of thumb, prejudice, learned method, intuition, unconscious dictates, or combinations thereof, to call upon the appropriate sub-process capabilities with a particular sequence and timing. An executive process (i.e., the exercise of an executive capability) involves such sub-processes as planning, selecting, and supervising, and it is really the executive processes that embody all of the methodology in the H-LAM/T system.

To illustrate the capability-hierarchy features of our conceptual framework, let us consider an artifact innovation appearing directly within the relatively low-order capability for composing and modifying written text, and see how this can affect a (or, for instance, your) hierarchy of capabilities. Suppose you had a new writing machine—think of it as a high-speed electric typewriter with some special features. You could operate its keyboard to cause it to write text much as you could use a conventional typewriter. But the printing mechanism is more complicated; besides printing a visible character at every stroke, it adds special encoding features by means of invisible selective components in the ink and special shaping of the character.

As an auxiliary device, there is a gadget that is held like a pencil and, instead of a point, has a special sensing mechanism that you can pass over a line of the special printing from your writing machine (or one like it). The signals which this reading stylus sends through the flexible connecting wire to the writing machine are used to determine which characters are being sensed and thus to cause the automatic typing of a duplicate string of characters. An information-storage mechanism in the writing machine permits you to sweep the reading stylus over the characters much faster than the writer can type; the writer will catch up with you when you stop to think about what word or string of words should be duplicated next, or while you reposition the straight-edge guide along which you run the stylus.

This writing machine would permit you to use a new process of composing text. For instance, trial drafts could rapidly be composed from re-arranged excerpts of old drafts, together with new words or passages which you stop to type in. Your first draft could represent a free outpouring of thoughts in any order, with the inspection of foregoing thoughts continuously stimulating new considerations and ideas to be entered. If the tangle of thoughts represented by the draft became too complex, you would compile a reordered draft quickly. It would be practical for you to accommodate more complexity in the trails of thought you might build in search of the path that suits your needs.

You can integrate your new ideas more easily, and thus harness your creativity more continuously, if you can quickly and flexibly change your working record. If it is easier to update any part of your working record to accommodate new developments in thought or circumstance, you will find it easier to incorporate more complex procedures in your way of doing things. This will probably allow you to accommodate the extra burden associated with, for instance, keeping and using special files whose contents are both contributed to and utilized by any current work in a flexible manner—which in turn enables you to devise and use even-more-complex procedures to better harness your talents in your particular working situation.

The important thing to appreciate here is that a direct new innovation in one particular capability can have far-reaching effects throughout the rest of your capability hierarchy. A change can propagate *up* through the capability hierarchy; higher-order capabilities that can utilize the initially changed capability can now reorganize to take special advantage of this change and of the intermediate higher-capability changes. A change can propagate *down* through the hierarchy as a result of new capabilities at the high level and modification possibilities latent in lower levels. These latent capabilities may previously have been unusable in the hierarchy and become usable because of the new capability at the higher level.

The writing machine and its flexible copying capability would occupy you for a long time if you tried to exhaust the reverberating chain of associated possibilities for making useful innovations within your capability hierarchy. This one innovation could trigger a rather extensive redesign of this hierarchy; your way of accomplishing many of your tasks would change considerably. Indeed, this process characterizes the sort of evolution that our intellect-augmentation means have been undergoing since the first human brain appeared.

To our objective of deriving orientation about possibilities for actively pursuing an increase in human intellectual effectiveness, it is important to realize that we must be prepared to pursue such new-possibility chains throughout the *entire* capability hierarchy (calling for a "system" approach). It is also important to realize that we must be oriented to the *synthesis* of new capabilities from reorganization of other capabilities, both old and new, that exist throughout the hierarchy (calling for a "system-engineering" approach).

B. *The Basic Perspective*

Individuals who operate effectively in our culture have already been considerably "augmented." Basic human capabilities for sensing stimuli, performing numerous mental operations, and for communicating with the outside world, are put to work in our society within a system—an H-LAM/T system—the individual augmented by the language, artifacts, and methodology in which he is trained. Furthermore, we suspect that improving the effectiveness of the individual as he operates in our society should be approached as a system-engineering problem—that is, the H-LAM/T system should be studied as an interacting whole from a synthesis-oriented approach.

This view of the system as an interacting whole is strongly bolstered by considering the repertoire hierarchy of process capabilities that is structured from the basic ingredients within the H-LAM/T system. The realization that any potential change in language, artifact, or methodology has importance only relative to its use within a process, and that a new process capability appearing anywhere within that hierarchy can make practical a new consideration of latent change possibilities in many other parts of the hierarchy—possibilities in either language, artifacts, or methodology—brings out the strong interrelationship of these three augmentation means.

Increasing the effectiveness of the individual's use of his basic capabilities is a problem in redesigning the changeable parts of a system. The system is actively engaged in the continuous processes (among others) of developing comprehension within the individual and of solving problems; both processes are subject to human motivation, purpose, and will. To redesign the system's capability for performing these processes means redesigning all or part of the repertoire hierarchy. To redesign a structure, we must learn as much as we can of what is known about the basic materials and components as they are utilized within the structure; beyond that, we must learn how to view, to measure, to analyze, and to evaluate in terms of the functional whole and its purpose. In this particular case, no existing analytic theory is by itself adequate for the purpose of analyzing and evaluating over-all system performance; pursuit of an improved system thus demands the use of *experimental* methods.

It need not be just the very sophisticated or formal process capabilities that are added or modified in this redesign. Essentially any of the processes utilized by a representative human today—the processes that he thinks of when he looks ahead to his day's work—are composite processes of the sort that involve external composing and manipulating of symbols (text, sketches, diagrams, lists, etc.).

Many of the external composing and manipulating (modifying, rearranging) processes serve such characteristically "human" activities as playing with forms and relationships to see what develops, cut-and-try multiple-pass development of an idea, or listing items to reflect on and then rearranging and extending them as thoughts develop.

Existing, or near-future, technology could certainly provide our professional problem-solvers with the artifacts they need to have for duplicating and rearranging text before their eyes, quickly and with a minimum of human effort. Even so apparently minor an advance could yield total changes in an individual's repertoire hierarchy that would represent a great increase in over-all effectiveness. Normally the necessary equipment would enter the market slowly; changes from the expected would be small, people would change their ways of doing things a little at a time, and only gradually would their accumulated changes create markets for more radical versions of the equipment. Such an evolutionary process has been typical of the way our repertoire hierarchies have grown and formed.

But an active research effort, aimed at exploring and evaluating possible integrated changes throughout the repertoire hierarchy, could greatly accelerate this evolutionary process. The research effort could guide the product development of new artifacts toward taking long-range meaningful steps; simultaneously, competitively minded individuals who would respond to demonstrated methods for achieving greater personal effectiveness would create a market for the more radical equipment innovations. The guided evolutionary process could be expected to be considerably more rapid than the traditional one.

The category of "more radical innovations" includes the digital computer as a tool for the personal use of an individual. Here there is not only promise of great flexibility in the composing and rearranging of text and diagrams before the individual's eyes, but also promise of many other process capabilities that can be integrated into the H-LAM/T system's repertoire hierarchy.

C. Detailed Discussion of the H-LAM/T System

1. THE SOURCE OF INTELLIGENCE When one looks at a computer system that is doing a very complex job, he sees on the surface a machine that can execute some extremely sophisticated processes. If he is a layman, his concept of what provides this sophisticated capability may endow the machine with a mysterious power to sweep information through perceptive and intelligent synthetic thinking devices. Actually, this sophisticated capability results from a very clever

organizational hierarchy, so that pursuit of the source of intelligence within this system would take one down through layers of functional and physical organization that become successively more primitive.

To be more specific, we can begin at the top and list the major levels down through which we would pass if we successively decomposed the functional elements of each level, in search of the "source of intelligence." A programmer could take us down through perhaps three levels (depending upon the sophistication of the total process being executed by the computer) perhaps depicting the organization at each level with a flow chart. The first level down would organize functions corresponding to statements in a problem-oriented language (e.g., ALGOL or COBOL), to achieve the desired over-all process. The second level down would organize lesser functions into the processes represented by first-level statements. The third level would perhaps show how the basic machine commands (or rather the processes which they represent) were organized to achieve each of the functions of the second level.

Then a machine designer could take over, and with a block diagram of the computer's organization he could show us (Level 4) how the different hardware units (e.g., random-access storage, arithmetic registers, adder, arithmetic control) are organized to provide the capability of executing sequences of the commands used in Level 3. The logic designer could then give us a tour of Level 5, also using block diagrams, to show us how such hardware elements as pulse gates, flip-flops, and AND, OR, and NOT circuits can be organized into networks giving the functions utilized at Level 4. For Level 6 a circuit engineer could show us diagrams revealing how components such as transistors, resistors, capacitors, and diodes can be organized into modular networks that provide the functions needed for the elements of Level 5.

Device engineers and physicists of different kinds could take us down through more layers. But rather soon we have crossed the boundary between what is man-organized and what is nature-organized, and are ultimately discussing the way in which a given physical phenomenon is derived from the intrinsic organization of sub-atomic particles, with our ability to explain succeeding layers blocked by the exhaustion of our present human comprehension.

If we then ask ourselves where that intelligence is embodied, we are forced to concede that it is elusively distributed throughout a hierarchy of functional processes—a hierarchy whose foundation extends down into natural processes below the depth of our comprehension. If there is any one thing upon which this "intelligence" depends, it would seem to be *organization.* The biologists and physiologists use a term "synergism" to designate (from *Webster's Unabridged*

Dictionary, Second Edition) the "... cooperative action of discrete agencies such that the total effect is greater than the sum of the two effects taken independently ..." This term seems directly applicable here, where we could say that synergism is our most likely candidate for representing the actual source of intelligence.

Actually, each of the social, life, or physical phenomena we observe about us would seem to derive from a supporting hierarchy of organized functions (or processes), in which the synergistic principle gives increased phenomenological sophistication to each succeedingly higher level of organization. In particular, the intelligence of a human being, derived ultimately from the characteristics of individual nerve cells, undoubtedly results from synergism.

2. INTELLIGENCE AMPLIFICATION It has been jokingly suggested several times during the course of this study that what we are seeking is an "intelligence amplifier." (The term is attributed originally to W. Ross Ashby.[2] At first this term was rejected on the grounds that in our view one's only hope was to make a better match between existing human intelligence and the problems to be tackled, rather than in making man more intelligent. But deriving the concepts brought out in the preceding section has shown us that indeed this term does seem applicable to our objective.)

Accepting the term "intelligence amplification" does not imply any attempt to increase native human intelligence. The term "intelligence amplification" seems applicable to our goal of augmenting the human intellect in that the entity to be produced will exhibit more of what can be called intelligence than an unaided human could; we will have amplified the intelligence of the human by organizing his intellectual capabilities into higher levels of synergistic structuring. What possesses the amplified intelligence is the resulting H-LAM/T system, in which the LAM/T augmentation means represent the amplifier of the human's intelligence.

In amplifying our intelligence, we are applying the principle of synergistic structuring that was followed by natural evolution in developing the basic human capabilities. What we have done in the development of our augmentation means is to construct a superstructure that is a synthetic extension of the natural structure upon which it is built. In a very real sense, as represented by the steady evolution of our augmentation means, the development of "artificial intelligence" has been going on for centuries.

3. TWO-DOMAIN SYSTEM The human and the artifacts are the only physical components in the H-LAM/T system. It is upon their capabilities that the ulti-

mate capability of the system will depend. This was implied in the earlier statement that every composite process of the system decomposes ultimately into explicit-human and explicit-artifact processes. There are thus two separate domains of activity within the H-LAM/T system: that represented by the human, in which all explicit-human processes occur; and that represented by the artifacts, in which all explicit-artifact processes occur. In any composite process, there is cooperative interaction between the two domains, requiring interchange of energy (much of it for information exchange purposes only). Figure 1 depicts this two domain concept and embodies other concepts discussed below.

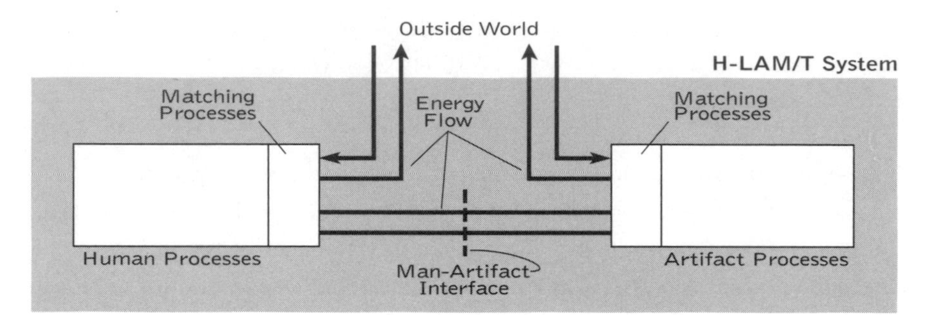

FIGURE 1. *Portrayal of the Two Actove Domains within the H-LMA/T System.*

Where a complex machine represents the principal artifact with which a human being cooperates, the term "man-machine interface" has been used for some years to represent the boundary across which energy is exchanged between the two domains. However, the "man-artifact interface" has existed for centuries, ever since humans began using artifacts and executing composite processes.

Exchange across this "interface" occurs when an explicit-human process is coupled to an explicit-artifact process. Quite often these coupled processes are designed for just this exchange purpose, to provide a functional match between other explicit-human and explicit-artifact processes buried within their respective domains that do the more significant things. For instance, the finger and hand motions (explicit-human processes) activate key-linkage motions in the typewriter (couple to explicit-artifact processes). But these are only part of the matching processes between the deeper human processes that direct a given word to be typed and the deeper artifact processes that actually imprint the ink marks on the paper.

The outside world interacts with our H-LAM/T system by the exchange of energy with either the individual or his artifact. Again, special processes are often designed to accommodate this exchange. However, the direct concern of our present study lies within the system, with the internal processes that are and can be significantly involved in the effectiveness of the *system* in developing the human's comprehension and pursuing the human's goals. . . .

5. Capability Repertoire Hierarchy The concept of our H-LAM/T system possessing a repertoire of capabilities that is structured in the form of a hierarchy is most useful in our study. We shall use it in the following to tie together a number of considerations and concepts.

There are two points of focus in considering the design of new repertoire hierarchies: the materials with which we have to work, and the principles by which new capability is constructed from these basic materials.

a. Basic Capabilities

Materials in this context are those capabilities in the human and in the artifact domains from which all other capabilities in the repertoire hierarchy must be constructed. Each such basic capability represents a type of functional component with which the system can be built, and a thorough job of redesigning the system calls for making an inventory of the basic capabilities available. Because we are exploring for perspective, and not yet recommending research activities, we are free to discuss and define in more detail what we mean by "basic capability" without regard to the amount of research involved in making an actual inventory.

The two domains, human and artifact, can be explored separately for their basic capabilities. In each we can isolate two classes of basic capability; these classes are distinguished according to whether or not the capability has been put to use within out augmentation means. The first class (those in use) can be found in a methodical manner by analyzing present capability hierarchies. For example, select a given capability, at any level in the hierarchy, and ask yourself if it can be usefully changed by any means that can be given consideration in the augmentation research contemplated. If it can, then it is not basic but it can be decomposed into an eventual set of basic capabilities. As you proceed down through the hierarchy, you will begin to encounter capabilities that cannot be usefully changed, and these will make up your inventory of basic capabilities. Ultimately, every such recursive decomposition of a given capability in the hierarchy will find every one of its branching paths terminated by basic capabilities. Beginning such decomposition search with different capabilities in the hierarchy will eventually uncover all of those basic capabilities used within that hierarchy or augmentation system. Many of the branching paths in the decomposition of a given higher-order capability will terminate in the same basic capability, since a given basic capability will often be used within many different higher-order capabilities.

Determining the class of basic capabilities not already utilized within existing augmentation systems requires a different exploration method. Examples of this method occur in technological research, where analytically oriented researchers search for new understandings of phenomena that can add to the research engineer's list of things to be used in the synthesis of better artifacts.

Before this inventorying task can be pursued in any specific instance, some criteria must be established as to what possible changes within the H-LAM/T system can be given serious consideration. For instance, some research situations might have to disallow changes which require extensive retraining, or which require undignified behavior by the human. Other situations might admit changes requiring years of special training, very expensive equipment, or the use of special drugs.

The capability for performing a certain finger action, for example, may not be basic in our sense of the word. Being able to extend the finger a certain distance would be basic, but the strength and speed of a particular finger motion and its coordination with higher actions generally are usefully changeable and therefore do not represent basic capabilities. What would be basic in this case would perhaps be the processes whereby strength could be increased and coordinated movement patterns learned, as well as the basic movement range established by the mechanical-limit loci of the muscle-tendon-bone system. Similar capability breakdowns will occur for sensory and cognitive capabilities.

b. Structure Types

1) General The fundamental principle used in building sophisticated capabilities from the basic capabilities is structuring—the special type of structuring (which we have termed synergetic) in which the organization of a group of elements produces an effect greater than the mere addition of their individual effects. Perhaps "purposeful" structuring (or organization) would serve us as well, but since we aren't sure yet how the structuring concept must mature for our needs, we shall tentatively stick with the special modifier, "synergetic." We are developing a growing awareness of the significant and pervasive nature of such structure within every physical and conceptual thing we inspect, where the hierarchical form seems almost universally present as stemming from successive levels of such organization.

The fundamental entities that are being structured in each and every case seems to be what we could call processes, where the most basic of physical processes (involving fields, charges, and momenta associated with the dynamics of fundamental particles) appear to be the hierarchical base. There are dynamic electro-optical-mechanical processes associated with the function of our artifacts, as well as metabolic, sensory, motor, and cognitive processes of the human, which we find to be relatively fundamental components within the structure of our H-LAM/T system—and each of these seems truly to be ultimately based (to our degree of understanding) upon the above mentioned basic physical processes. The elements that are organized to give fixed structural form to our physical objects—e.g., the "element" of tensile strength of a material—are also derived from what we could call synergetic structuring of the most basic physical processes.

But at the level of the capability hierarchy where we wish to work, it seems useful to us to distinguish several different types of structuring—even though each type is fundamentally a structuring of the basic physical processes. Tentatively we have isolated five such types—although we are not sure how many we shall ultimately want to use in considering the problem of augmenting the human intellect, nor how we might divide and subdivide these different manifestations of physical-process structuring. We use the terms "mental structuring," "concept structuring," "symbol structuring," "process structuring," and "physical structuring."

2) Mental Structuring Mental structuring is what we call the internal organization of conscious and unconscious mental images, associations, or concepts (or whatever it is that is organized within the human mind) that somehow manages to provide the human with understanding and the basis for such as judgment, intuition, inference, and meaningful action with respect to his environment. There is a term used in psychology, "cognitive structure," which so far seems to represent just what we want for our concept of mental structure, but we will not adopt it until we become more sure of what the accepted psychological meaning is and of what we want for our conceptual framework.

For our present purpose, it is irrelevant to worry over what the fundamental mental "things" being structured are, or what mechanisms are accomplishing the structuring or making use of what has been structured. We feel reasonably safe in assuming that learning involves some kind of meaningful organization within the brain, and that whatever is so organized or structured represents the operating model of the individual's universe to the mental mechanisms that derive his behavior. And further, our assumption is that when the human in our H-LAM/T system makes the key decision or action that leads to the solution of a complex problem, it will stem from the state of his mental structure at that time. In this view, then, the basic purpose of the system's activity on that problem up to that point has been to develop his mental structure to the state from which the mental mechanisms could derive the key action.

Our school systems attest that there are specific experiences that can be given to a human that will result in development of his mental structure to the point where the behavior derived therefrom by his mental mechanisms shows us that he has gained new comprehension—in other words, we can do a certain amount from outside the human toward developing his mental structure. Independent students and researchers also attest that internally directed behavior on the part of an individual can directly aid his structure-building process.

We don't know whether a mental structure is developed in a manner analogous to (a) development of a garden, where one provides a good environment, plants the seeds, keeps competing weeds and injurious pests out, but otherwise has to let natural processes take their course, or to (b) development of a basketball team, where much exercise of skills, patterns, and strategies must be provided so that natural processes can slowly knit together an integration, or to (c) development of a machine, where carefully formed elements are assembled in a precise, planned manner so that natural phenomena can immediately yield planned function. We don't know the processes, but we can and have developed empirical relationships between the experiences given a human and the associated manifestations of developing comprehension and capability, and we see the near-future course of the research toward augmenting the human's intellect as depending entirely upon empirical findings (past and future) for the development of better means to serve the development and use of mental structuring in the human.

We don't mean to imply by this that we renounce theories of mental processes. What we mean to emphasize is that pursuit of our objective need not wait upon the understanding of the mental processes that accomplish (what we call) mental structuring and that derive behavior therefrom. It would be to ignore the emphases of our own conceptual framework not to make fullest use of any theory that provided a working explanation for a group of empirical data. What's more, our entire conceptual framework represents the first pass at a "theoretical" model with which to organize our thinking and action.

3) Concept Structuring Within our framework we have developed the working assumption that the manner in which we seem to be able to provide experiences that favor the development of our mental structures is based upon concepts as a "medium of exchange." We view a concept as a tool that can be grasped and used by the mental mechanisms, that can be composed, interpreted, and used by the natural mental substances and processes. The grasping and handling done by these mechanisms can often be facilitated if the concept is given an explicit "handle" in the form of a representative symbol. Somehow the mental mechanisms can learn to manipulate images (or something) of symbols in a meaningful way and remain calmly confident that the associated conceptual manipulations are within call.

Concepts seem to be structurable, in that a new concept can be composed of an organization of established concepts. For present purposes, we can view a *concept structure* as something which we might try to develop on paper for ourselves or work with by conscious thought processes, or as something which we try to communicate to one another in serious discussion. We assume that, for a given unit of comprehension to be imparted, there is a concept structure (which can be consciously developed and displayed) that can be presented to an individual in such a way that it is mapped into a corresponding mental structure which provides the basis for that individual's "comprehending" behavior. Our working assumption also considers that some concept structures would be better for this purpose than others, in that they would be more easily mapped by the individual into workable mental structures, or in that the resulting mental structures enable a higher degree of comprehension and better solutions to problems, or both.

A concept structure often grows as part of a cultural evolution—either on a large scale within a large segment of society, or on a small scale within the activity domain of an individual. But it is also something that can be directly designed or modified, and a basic hypothesis of our study is that better concept structures can be developed—structures that when mapped into a human's mental structure will significantly improve his capability to comprehend and to find solutions within his complex-problem situations.

A natural language provides its user with a ready-made structure of concepts that establishes a basic mental structure, and that allows relatively flexible, general-purpose concept structuring. Our concept of "language" as one of the basic means for augmenting the human intellect embraces all of the concept structuring which the human may make use of.

4) Symbol Structuring The other important part of our "language" is the way in which concepts are represented—the symbols and *symbol structures.* Words structured into phrases, sentences, paragraphs, monographs—charts, lists, diagrams, tables, etc. A given structure of concepts can be represented by any of an infinite number of different symbol structures, some of which would be much better than others for enabling the human perceptual and cognitive apparatus to search out and comprehend the conceptual matter of significance and/or interest to the human. For instance, a concept structure involving many numerical data would generally be much better represented with Arabic rather than Roman numerals and quite likely a graphic structure would be better than a tabular structure.

But it is not only the *form* of a symbol structure that is important. A problem solver is involved in a stream of conceptual activity whose course serves his mental needs of the moment. The sequence and nature of these needs are quite variable, and yet for each need he may benefit significantly from a form of symbol structuring that is uniquely efficient for that need.

Therefore, besides the forms of symbol structures that can be constructed and portrayed, we are very much concerned with the speed and flexibility with which one form can be transformed into another, and with which new material can be located and portrayed.

We are generally used to thinking of our symbol structures as a pattern of marks on a sheet of paper. When we want a different symbol-structure view, we think of shifting our point of attention on the sheet, or moving a new sheet into position. But another kind of view might be obtained by extracting and ordering all statements in the local text that bear upon Consideration A of the argument—or by replacing all occurrences of specified esoteric words by one's own definitions. This sort of "view generation" becomes quite feasible with a computer-controlled display system, and represents a very significant capability to build upon.

With a computer manipulating our symbols and generating their portrayals to us on a display, we no longer need think of our looking at *the* symbol structure which is stored—as we think of looking at *the* symbol structures stored in notebooks, memos, and books. What the computer actually stores need be none of our concern, assuming that it can portray symbol structures to us that are consistent with the form in which we think our information is structured.

A given concept structure can be represented with a symbol structure that is completely compatible with the computer's internal way of handling symbols, with all sorts of characteristics and relationships given explicit identifications that the user may never directly see. In fact, this structuring has immensely greater potential for accurately mapping a complex concept structure than does a structure an individual would find it practical to construct or use on paper.

The computer can transform back and forth between the two-dimensional portrayal on the screen, of some limited view of the total structure, and the aspect of the n-dimensional internal image that represents this "view." If the human adds to or modifies such a "view," the computer integrates the change into the internal-image symbol structure (in terms of the computer's favored symbols and structuring) and thereby automatically detects a certain proportion of his possible conceptual inconsistencies.

Thus, inside this instrument (the computer) there is an internal-image, computer-symbol structure whose convolutions and multi-dimensionality we can learn to shape to represent to hitherto unattainable accuracy the concept structure we might be building or working with. This internal structure may have a form that is nearly incomprehensible to the direct inspection of a human (except in minute chunks).

But let the human specify to the instrument his particular conceptual need of the moment, relative to this internal image. Without disrupting its own internal reference structure in the slightest, the computer will effectively stretch, bend, fold, extract, and cut as it may need in order to assemble an internal substructure that is its response, structured in its own internal way. With the set of standard translation rules appropriate to the situation, it portrays to the human via its display a symbol structure designed for *his* quick and accurate perception and comprehension of the conceptual matter pertinent to this internally composed substructure.

No longer does the human work on stiff and limited symbol structures, where much of the conceptual content can only be implicitly designated in an indirect and distributed fashion. These new ways of working are basically available with today's technology—we have but to free ourselves from some of our limiting views and begin experimenting with compatible sets of structure forms and processes for human concepts, human symbols, and machine symbols.

5) Process Structuring Essentially everything that goes on within the H-LAM/T system and that is of direct interest here involves the manipulation of concept and symbol structures in service to the mental structure. Therefore, the processes within the H-LAM/T system that we are most interested in developing are those that provide for the manipulation of all three types of structure. This brings us to the fourth category of structuring, *process* structuring.

As we are currently using it, the term *process structuring* includes the organization, study, modification, and execution of processes and process structures. Whereas concept structuring and symbol structuring together represent the language component of our augmentation means, process structuring represents the methodology component (plus a little more, actually). There has been enough previous discussion of process structures that we need not describe the notion here, beyond perhaps an example or two. The individual processes (or actions) of my hands and fingers have to be cooperatively organized if the typewriter is to do my bidding. My successive actions throughout my working day are meant to cooperate toward a certain over-all professional goal.

Many of the process structures are applied to the task of organizing, executing, supervising, and evaluating other process structures. Many of them are applied to the formation and manipulation of symbol structures (the purpose of which will often be to support the conceptual labor involved in process structuring).

6) Physical Structuring Physical structuring, the last of the five types which we currently use in our conceptual framework, is nearly self-explanatory. It pretty well represents the artifact component of our augmentation means, insofar as their actual physical construction is concerned. . . .

e. Flexibility in the Executive Role

[T]here is finite human capability which must be divided between executive and direct-contributive activities [in the H-LAM/T system]. An important aspect of the multi-role activity of the human in the system is the development and manipulation of the symbol structures associated with *both* his direct-contributive roles and his executive roles.

When the system encounters a complex situation in which comprehension and problem solutions are being pursued, the direct-contributive roles require the development of symbol structures that portray the concepts involved within the situation. But executive roles in a complex problem situation also require conceptual activity—e.g., comprehension, selection, supervision—that can benefit from well-designed symbol structures and fast, flexible means for manipulating and displaying them. For complex processes, the executive problem posed to the human (of gaining the necessary comprehension and making a good plan) may be tougher than the problem he faced in the role of direct-contributive worker. If the flexibility desired for the process hierarchies (to make room for human cut-and-try methods) is not to be degraded or abandoned, the executive activity will have to be provided with fast and flexible symbol-structuring techniques.

The means available to humans today for developing and manipulating these symbol structures are both laborious and inflexible. It is hard enough to develop an initial structure of diagrams and text, but the amount of effort required to make changes is often prohibitively great; one settles for inflexibility. Also, the kind of generous flexibility that would be truly helpful calls for added symbol structuring just to keep track of the trials, branches, and reasoning thereto that are involved in the development of the subject structure; our present symbol-manipulation means would very soon bog down completely among the complexities that are involved in being more than just a little bit flexible.

We find that the humans in our H-LAM/T systems are essentially working continuously within a symbol structure of some sort, shifting their attention from one structure to another as they guide and execute the processes that ultimately provide them with the comprehension and the problem solutions that they seek. This view increases our respect for the essential importance of the basic capability of composing and modifying efficient symbol structures. Such a capability depends heavily upon the particular concepts that are isolated and manipulated as entities, upon the symbology used to represent them, upon the artifacts that help to manipulate and display the symbols, and upon the methodology for developing and using symbol structures. In other words, this capability depends heavily upon proper language, artifacts, and methodology, our basic augmentation means.

When the course of action must respond to new comprehension, new insights and new intuitive flashes of possible explanations or solutions, it will not be an orderly process. Existing means of composing and working with symbol structures penalize disorderly processes very heavily, and it is part of the real promise in the automated H-LAM/T systems of tomorrow that the human can have the freedom and power of disorderly processes. . . .

VI. Conclusions

Three principal conclusions may be drawn concerning the significance and implications of the ideas that have been presented.

First, any possibility for improving the effective utilization of the intellectual power of society's problem solvers warrants the most serious consideration. This is because man's problem-solving capability represents possibly the most important resource possessed by a society. The other contenders for first importance are all critically dependent for their development and use upon this resource. Any possibility for evolving an art or science that can couple directly and significantly to the continued development of that resource should warrant doubly serious consideration.

Second, the ideas presented are to be considered in both of the above senses: the direct-development sense and the "art of development" sense. To be sure, the possibilities have long-term implications, but their pursuit and initial rewards await us now. By our view, we do not have to wait until we learn how the human mental processes work, we do not have to wait until we learn how to make computers more intelligent or bigger or faster; we can begin developing powerful and economically feasible augmentation systems on the basis of what we now know and have. Pursuit of further basic knowledge and improved machines will continue into the unlimited future, and will want to be integrated into the "art" and its improved augmentation systems—but getting started now will provide not only orientation and stimulation for these pursuits, but will give us improved problem-solving effectiveness with which to carry out the pursuits.

Third, it becomes increasingly clear that there should be action now—the sooner the better—action in a number of research communities and on an aggressive scale. We offer a conceptual framework and a plan for action, and we recommend that these be considered carefully as a basis for action. If they be considered but found unacceptable, then at least serious and continued effort should be made toward developing a more acceptable conceptual framework within which to view the overall approach, toward developing a more acceptable plan of action, or both.

This is an open plea to researchers and to those who ultimately motivate, finance, or direct them, to turn serious attention toward the possibility of evolving a dynamic discipline that can treat the problem of improving intellectual effectiveness in a total sense. This discipline should aim at producing a continuous cycle of improvements—increased understanding of the problem, improved means for developing new augmentation systems, and improved augmentation systems that can serve the world's problem solvers in general and this discipline's workers in particular. After all, we spend great sums for disciplines aimed at understanding and harnessing nuclear power. Why not consider developing a discipline aimed at understanding and harnessing "neural power"? In the long run, the power of the human intellect is really much the more important of the two.

John Cage

"Diary: Audience 1966," *A Year From Monday* (1966)

<< 10 >>

John Cage, Variations V. *Photo by Herve Gloaguen. Courtesy of Merce Cunningham Dance Foundation.*

"What'll art become? A family reunion? If so, let's have it with people in the round, each individual free to lend his attention wherever he will."

<< As a musician, composer, artist, poet, and philosopher, John Cage's work rarely fit within the traditional boundaries of artistic practice. Originally from Los Angeles, where he was a student of the Viennese composer Arnold Schoenberg, Cage's interdisciplinary interests led him far from the conventions of his musical training. In the late 1940s, during a residency at Black Mountain College, he developed his provocative "theater of mixed-means" in collaborations with the artists Robert Rauschenberg and Jasper Johns and the choreographer Merce Cunningham. These experiments gave birth to an explosion of performance art in the 1950s and 1960s that introduced all types of actions, artifacts, noises, images, and movement into the performance space, which culminated in the Happenings.

Cage embraced indeterminacy as an integral part of his process of composition; this technique led him to include the participation of the audience in the creation of his work. Inspired by Zen Buddhism, he reveled in an anarchy that dethrones the artist as the heroic, all-powerful arbiter of creative expression. He proposed instead a shift to an inclusive, participatory art that encourages interaction between artist, performer and audience, one in which the latter "can sit quietly or make noises . . . whisper, talk and even shout." This is best illustrated by his infamous piece from 1953, *4′33″*, in which the pianist David Tudor remained silent at his instrument for the prescribed duration of time. While the piece was initially met by critics as a scandal, Cage's intent was to encourage the listener to contemplate the passing of time freely while listening attentively to the random sounds of the concert hall.

In this essay, Cage links the notion of an interactive listener to the concept of the computer as an agent of participation rather than as a servile, "labor-saving" device. His insightful analysis of the changing relationship between the artwork and the viewer has set the stage for much that has since become widespread in human-computer interaction, and has profoundly influenced later generations of media artists exploring interactive strategies. >>

I. Are we an audience for computer art? The answer's not No; it's Yes. What we need is a computer that isn't labor-saving but which increases the work for us to do, that puns (this is McLuhan's idea) as well as Joyce revealing bridges (this is Brown's idea) where we thought there weren't any, turns us (my idea) not "on" but into artists. *Orthodox seating arrangement in synagogues.* Indians have known it for ages: life's a dance, a play, illusion. Lila. Maya. Twentieth-century art's opened our eyes. Now music's opened our ears. Theatre? Just notice what's around. (If what you want in India is an audience, Gita Sarabhai told me, all you need is one or two people.) II. He said: Listening to your music I find it provokes me. What should I do to enjoy it? Answer: There're many ways to help you. I'd give you a lift, for instance, if you were going in my direction, but the last thing I'd do would be to tell you how to use your own aesthetic faculties. (You see? We're unemployed. If not yet, "soon again 'twill be." We have nothing to do. So what shall we do? Sit in an audience? Write criticism? Be creative?) We used to have the artist up on a pedestal. Now he's no more extraordinary than we are. III. Notice audiences at high altitudes and audiences in northern countries tend to be attentive during performances while audiences at sea level or in warm countries voice their feelings whenever they have them. Are we, so to speak, going south in the way we experience art? Audience participation? (Having nothing to do, we do it nonetheless; our biggest problem is finding scraps of time in which to get it done. Discovery. Awareness.) "Leave the beaten track. You'll see something never seen before." *After the first performance of my piece for twelve radios, Virgil Thomson said, "You can't do that sort of thing and expect people to pay for it." Separation.* IV. When our time was given to physical labor, we needed a stiff upper lip and backbone. Now that we're changing our minds, intent on things invisible, inaudible, we have other spineless virtues: flexibility, fluency. Dreams, daily events, everything gets to and through us. (Art, if you want a definition of it, is criminal action. It conforms to no rules. Not even its own. Anyone who experiences a work of art is as guilty as the artist. It is not a question of sharing the guilt. Each one of us gets all of it.) They asked me about theatres in New York. I said we could use them. They should be small for the audiences, the performing areas large and spacious, equipped for television broadcast for those who prefer staying at home. There should be a café in connection having food and drink, no music, facilities for playing chess. V. What happened at Rochester? We'd no sooner begun playing than the audience began. Began what? Costumes. Food. Rolls of toilet paper projected in streamers from the balcony through the air. Programs, too, folded, then flown. Music, perambulations, conversations. Began festivities. *An audience can sit quietly or make noises. People can whisper, talk and even shout. An audience can sit still or it can get up and move around. People are people, not plants.* "Do you love the audience?" Certainly we do. We show it by getting out of their way. (Art and money are in this world together, and they need each other to keep on going. Perhaps they're both on their way out. Money'll become a credit card without a monthly bill. What'll art become? A family reunion?

If so, let's have it with people in the round, each individual free to lend his attention wherever he will. Meeting house.) VI. After an Oriental decade, a Tibetan Bikku returned to Toronto to teach. He told me that were he to speak the truth his audience would drop to six. Instead he gives lectures transmitting not the spirit but the understandable letter. Two hundred people listen on each occasion, all of them deeply moved. (Art's a way we have for throwing out ideas—ones we've picked up in or out of our heads. What's marvelous is that as we throw them out—these ideas—they generate others, ones that weren't even in our heads to begin with.) Charles Ives had this idea: the audience is any one of us, just a human being. He sits in a rocking chair on a verandah. Looking out toward the mountains, he sees the setting sun and hears his own symphony: it's nothing but the sounds happening in the air around him.

Roy Ascott

"Behaviourist Art and the Cybernetic Vision" (1966–1967)

<< 11 >>

Roy Ascott. Photo by Josephine Coy. Courtesy of Roy Ascott.

"If the cybernetic spirit constitutes the predominant attitude of the modern era, the computer is the supreme tool that its technology has produced. Used in conjunction with synthetic materials it can be expected to open up paths of radical change and invention in art. . . . The interaction of man and computer in some creative endeavor, involving the heightening of imaginative thought, is to be expected."

<< Since the 1960s, the British educator, artist, and theoretician Roy Ascott has been one of Europe's most active and outspoken practitioners of interactive computer art. Ten years before the personal computer came into existence, Ascott saw that interactivity in computer-based forms of expression would be an emerging issue in the arts. Intrigued by the possibilities, he built a theoretical framework for approaching interactive artworks, which brought together certain characteristics of the avant-garde (Dada, surrealism, Fluxus, Happenings, and pop art, in particular) with the science of cybernetics championed by Norbert Wiener.

Ascott's thesis on cybernetic vision in the arts begins with the premise that interactive art must free itself from the modernist ideal of the "perfect object." Like John Cage, he proposes that the artwork be responsive to the viewer, rather than fixed and static. But Ascott expands on Cage's premise in the realm of computer-based art, suggesting that the "spirit of cybernetics" offers the most effective means for achieving a two-way exchange between the artwork and its audience. Ascott challenges artists to acknowledge information technology as the most significant tool of the age, and insists that it is the artist's obligation to use this technology. Yet, unlike Nam June Paik's vision, Ascott's is not ironic; rather, it is utopian in its embrace of a new medium, excited by the potential of a thriving, dynamic exchange between technology and art to empower the spectator and deepen his or her experience. >>

The Behavioural Tendency in Modern Art

...By "Modern Art" we mean that cultural continuum of ideas, forms and human activity which differs radically from any previous era and is both expressive and formative of the attitudes and conditions of our time. To describe it as a continuum may seem contradictory to its accepted identity. It is seen popularly as an anarchic, highly diversified and chaotic situation which loses as much in coherence and continuity as it gains in novelty and imagination.

Now, undoubtedly it is anarchic, but in the good sense that interaction between artists is free and not constrained by aesthetic canons or political directives. The diversity of images, structures and ideas which it engenders is far greater than at any other period in history. And it may well seem chaotic; a common cultural consciousness is not readily apparent today. But it is our purpose to demonstrate that Modern Art is fundamentally of a piece, that there is unity in its diversity, and that the quality which unifies it is in distinct contrast to the essential nature of the art which went before it. We shall describe this quality as "behavioural" and we shall show how it evidences our present transition from the old deterministic culture to a future shaped by a Cybernetic Vision.

The analysis of this behavioural tendency will be largely confined to one broad area, that of the visual/plastic arts, since there it seems to be most marked, but in a more general sense we shall discuss the arts as a whole, illustrating their convergence and interaction in this context. We shall demonstrate how this unity of approach may be potentially part of a larger unity, an integral culture, embracing modern science and technology. And we shall warn how this unity, and the incipient cybernetic vision in art, may be inhibited by artistic attitudes which, out of ignorance and fear, are opposed to radical creative change, and view a cybernated society with indifference or hostility.

The General Characteristics of Modern Art

The dominant feature of art of the past was the wish to transmit a clearly defined message to the spectator as a more or less passive receptor, from the artist as a unique and highly individualised source. This deterministic aesthetic was centred upon the structuring, or "composition," of *facts*, of concepts of the *essence* of things, encapsulated in a factually correct visual field. Modern Art, by contrast, is concerned to initiate *events* and with the forming of concepts of *existence*. The vision of art has shifted from the field of objects to the field of behaviour and its function has become less descriptive and more purposive.

Although in Painting and Sculpture the channel of communication remains largely visual, other modalities are increasingly employed—tactile, postural, aural; so that a more inclusive term than "visual" art must be found, and the one we propose is "behavioural." This behavioural tendency dominates art now in all its aspects. The artist, the artifact and the spectator are all involved in a more behavioural context. We find an insistence on polemic, formal ambiguity and instability, uncertainty and room for change in the images and forms of Modern Art. And these factors predominate not for esoteric or obscurantist reasons but to draw the spectator into active participation in the act of creation; to extend him, via the artifact, the opportunity to become involved in creative behaviour on all levels of experience—physical, emotional and conceptual. A feedback loop is established so that the evolution of the artwork/experience is governed by the intimate involvement of the spectator. As the process is open-ended the spectator now engages in decision-making play.

Creative Participation

We may say that the boundaries between making art, the artifact itself, and the experience of the work are no longer clearly defined. Or, more precisely, that the tendency for this to be so is evident. There are still in this transitional period many artists who contrive to force the new sensibility into old moulds, just as in technology there are many industrialists who attempt to squeeze cybernation into a nineteenth-century structure of operations.

The participational, inclusive form of art has as its basic principle "feedback," and it is this loop which makes of the triad artist/artwork/observer an integral whole. For art to switch its role from the private, exclusive arena of a rarefied elite to the public, open field of general consciousness, the artist has had to create more flexible structures and images offering a greater variety of readings than were needed in art formerly. This situation, in which the artwork exists in a perpetual state of transition where the effort to establish a final resolution must come from the observer, may be seen in the context of games. We can say that in the past the artist played to win, and so set the conditions that he always dominated the play. The spectator was positioned to lose, in the sense that his moves were predetermined and he could form no strategy of his own. Nowadays we are moving towards a situation in which the game is never won but remains perpetually in a state of play. While the general context of the art-experience is set by the artist, its evolution in any specific sense is unpredictable and dependent on the total involvement of the spectator.

Where once the function of art was to create an equilibrium, establish a harmony on the public level of relatively passive reception, we now find art as a more strident agent of change, effecting a jolt to the whole human organism, a catalyst which sets up patterns of behaviour, of thought and emotion, which are unpredictable in any fine sense. We observe in the painting of Poussin, for example, the wish to fix a set of relationships in the spectator's consciousness, to reinforce these absolutes by the stability of the formal composition; he communicates but by a one-way channel. The modern artist, on the other hand, is primarily motivated to initiate a *dialogue,* to set feelings and ideas in motion, to enrich the artistic experience with feedback from the spectator's response.

This cybernetic process of retraction generates a constant stream of new and unfamiliar relationships, associative links and concepts. Each artwork becomes a sort of behavioural Tarot pack, presenting coordinates which can be endlessly reshuffled by the spectator, always to produce meaning. This is achieved principally in one of two ways: either the artifact has a definitive form but contains only a small amount of low-definition information; or its physical structure is such that its individual constituent parts can change their relationships, either by the direct manipulation of the spectator, or by his shifting viewpoint, or by the agency of electrical or other natural power. The active involvement of the spectator can be thought of as removing uncertainty about a set of possibilities. Deep involvement and interplay produces information. The "set" of the artwork has variety only in so far as the observer participates. The variety of the set is a measure of the uncertainty involved. An important characteristic of Modern Art, then, is that it offers a high degree of uncertainty and permits a great intensity of participation.

As to the artist's role, it can be said to function on two levels simultaneously, the private and the social. In the first case, the primacy of a total behavioural involvement in the activity or *process* of making art is apparent. The artist is not goal-directed in the sense of working towards a predetermined art object. The artifact is essentially the result of his creative behaviour, rather than the reason for it. The growth of a painting or sculpture or environment is of more importance than the achievement of its final form. Indeed, unlike Classical Art, there is no point at which it can be said to have reached a final form. From the social point of view the artist's behaviour is a Ritual in which he acts out the role of the Free Man controlling his world by taking endless risks as he plunges into the unknown territories of Form and Idea. It is a paradigm of a condition to which the human being constantly aspires, where freedom and responsibility combine to reduce our anxiety of the unknown and unpredictable while enlarging our experience of the unfamiliar and irresistible.

At this early stage of a radically new culture the artist is doing little more than exploring his new relationship to the spectator. He is searching for new ways of handling ideas, for more flexible and adaptive structures to contain them; he is attempting to generate new carrier waves for the modulations of contemporary experience; and he is searching the resources of technology to expand his repertoire of skills. His concern is to affirm that dialogue is possible—that is the content and the message of art now; and that is why, seen from the deterministic point of view, art may seem devoid of content and the artist to have nothing to say. The modern means of communication, of feedback and viable interplay—these are the content of art. The artist's message is that the extension of creative behaviour into everyday experience is possible.

The message is timely and apposite at a period in which we can anticipate the reduction of labour to a minimum and expect the creative use of leisure time to be the main preoccupation of our lives. And even if the artist were to have fully explored the new channels of communication and thoroughly exploited the media and techniques of modern technology, it is unlikely that his attitude would change. He would continue to avoid the limitations of an aesthetic geared to the transmission of finite messages or the formulation of fixed attitudes and absolute values. He will continue, instead, to provide a *matrix* for ideas and feelings from which the participants in his work may construct for themselves new experiences and unfamiliar patterns of behaviour. . . .

The Cybernetic Vision in Art

By this term we do not mean "the Art of Cybernetics" nor do we refer to an art concerned to illustrate Cybernetics, nor yet an art embodying cybernetic machines or Robot Art—although any one of these things might be involved at some point, and again, they might not.

We are referring to the spirit of Cybernetics which may inform art and in turn be enriched by it. We contrast the Cybernetic Vision in Art to the Deterministic Vision of the past which has already been outlined. We say of Cybernetics that, before it is a method or an applied science, it is a field of knowledge which shapes our philosophy, influences our behaviour and extends our thought.

We are moving towards a fully cybernated society[1] where processes of retraction, instant communication, autonomic flexibility will inform every aspect of our environment. In that forming society, of which we are a part, the cybernetic spirit finds its expression in the Human Science[2] and in Environmental Technology; the two poles between which we act out our existence. It is the spirit of our understanding of life at its simplest and most complex levels, and a large measure of our ability to control it.

The economic and social effects of automation in the cybernated society will be profound.[3] The effects of our transition of that future state are already felt, particularly in the United States. Matters of leisure, class formation, political and economic power have already called for revision and new thinking. Cybernetics already dominates our more advanced concepts of transport, shelters, storage and other day to day matters of control and communication and has caused the radical transformation of many industrial and commercial procedures. The effect of the computer on human thought is currently the subject of vigorous discussion in academic circles; the man/computer relationship is seen to be as much a question of identity as of methodology.

Fundamentally Cybernetics concerns the idea of the perfectibility of systems; it is concerned in practice with the procurement of effective action by means of self-organising systems. It recognises the idea of the perfectibility of Man, of the possibility of further evolution in the biological and social sphere. In this it shares its optimism with Molecular Biology. Bio-cybernetics, the simulation of living processes, genetic manipulation, the behavioural sciences, automatic environments, together constitute an understanding of the human being which calls for and will in time produce new human values and a new morality.

How does the artist stand in relation to these radical changes? On the level of opinions or concepts he is and will be free to accept or reject them. But on the level of deep human experience they will "alter sense ratios or patterns of perception steadily and without resistance."[4] The artist is faced with two possibilities; either to be carried along in the stream of events, mindlessly, half aware and perhaps bitter and hostile as a result; or he can come to terms with his world, shape it and develop it by understanding its underlying cybernetic characteristics. Awareness of these underlying forces will sharpen his perception; the utilisation of new techniques will enlarge his powers of thought and creative action; he will be empowered to construct a vision in art which will enhance the cybernated society as much as it will be enriched by it. Understanding and awareness, in short, are the conditions for optimism in art.

There is reason to suppose that a unity of art, science and human values is possible; there is no doubt that it is desirable.[5] More specifically we propose that an essentially *cybernetic* vision could unify and feed such culture. The grounds for supposing that Art has anticipated this integral situation and is prepared for it can be found in the emphatically *behavioural* tendency which it displays. Cybernetics is consistent with Behaviourist Art; it can assist in its evolution just as, in turn, a behavioural synthesis can embody a Cybernetic Vision.

Cybernetics and Behaviourist Art

It is necessary to differentiate between *"l'esprit cybernétique"*[6] as we have tried to describe it above, and Cybernetics as a descriptive method. Now, art like any process or system can be examined from the cybernetic point of view; it can also derive technical and theoretical support from this science—as in the past it has done from Optics or Geometry. This is not unimportant since the artist's range can be extended considerably, as we briefly indicate below. But it is important to remember that the Cybernetic Vision in Art, which will unify art with a cybernated society, is a matter of "stance," a fundamental attitude to events and human relationships, before it is in any sense a technical or procedural matter.

Behaviourist Art constitutes a retroactive process of human involvement, in which the artifact functions as both matrix and catalyst. As matrix, it is the substance between two sets of behaviours; it neither exists for itself nor by itself. As a catalyst, it triggers changes in the spectator's total behaviour. Its structure must be adaptive implicitly or physically, to accommodate the spectator's responses, in order that the creative evolution of form and idea may take place. The basic principle is *feedback.* The system Artifact/Observer furnishes its own controlling energy; a function of an output variable (observer response) is to act as an input variable, which introduces more variety into the system and leads to more variety in the output (observer's experience). This rich interplay derives from what is a self-organising system in which there are two controlling factors; one, the spectator is a self-organising sub-system; the other, the artwork is not usually at present homeostatic.

There is no a priori reason why the artifact should not be a self-organising system; an organism, as it were, which derives its initial programme or code from the artist's creative activity, and then evolves its specific artistic identity and function in response to the environments which it encounters. The artist's creative activity is also dependent on feedback; the changes which he effects in his immediate environment (or "arena") by means of tools and media set up configurations which feed back to affect his subsequent decisions and actions. Thus Modern Art, with its fundamental behavioural quality, is the art of *the organisation of effects.* And when all the control factors, including the artwork itself, are effectively homeostatic, art will be concerned with the automatic control of effects. Cybernetics, of course, is the science of the organisation of effects, and of the automatic control of effects.[7]

Equally, there is no a priori reason why the artwork *should* become a self-organising system; the basic feedback process of behaviourist art operates within the conventions of painting and sculpture, provided that they display low definition, multiple associations and indeterminate content, within parameters which are, at least implicitly, flexible. And, as we have suggested already, this is nowadays the case—even to the extent of providing a more or less empty receptacle (the canvas) into which the spectator can project his own imaginative world, e.g., Yves Klein, Ad Reinhard.

The Computer and Growth Systems

However, historically it has been a characteristic of the artist to reach out to the tools and materials which the technology of his time produces, just as his perception and patterns of thought have tended to identify with scientific and philosophical attitudes of the period. If the cybernetics spirit constitutes the predominant *attitude* of the modern era, the computer is the supreme *tool* that its technology has produced. Used in conjunction with synthetic materials it can be expected to open up paths of radical change and invention in art. For it is not simply a physical tool in the sense that an aluminium casting plant or CO_2 welding gear are tools, i.e., extensions of physical power. It is a tool for the mind, an instrument for the magnification of thought, potentially an intelligence amplifier.[8] The interaction of man and computer in some creative endeavour, involving the heightening of imaginative thought, is to be expected. Moreover the interaction of Artifact and computer, in the context of the behavioural structure, is equally foreseeable.

Experiments are already taking place. We have cited Schoffer's use of a computer in some of his structures. In music Iannis Xenakis has made extensive use of an IBM 7090—a process in which he "specifies the duration and density of sound events, leaving the parameters of pitch, velocity and dynamics to the computer." The "Light-Harp" project of Haukeland and Nordheim, an environmental sculpture emitting sound in relation to the quality of local light, with sound sources changing position within the structure, calls for a highly sophisticated control and communications system within it.

The computer may be linked to an artwork and the artwork may in some sense *be* a computer. The necessary conditions of behaviourist art are that the spectator is involved and that the artwork in some way *behaves.* Now, it seems likely that in the artist's attempt to create structures which are probabilistic, the artifact may result from biological modelling. In short, it may be developed with the properties of growth. Cybernetics already furnishes models which could assist in this development, e.g., Beer's Fungoid Systems and research into chemical and chemical-colloidal computers.[9] The potential for the future is enormous.[10]

The cybernetic vision not only shapes modern science and technology, integrating and bridging disparate fields of knowledge and improving artificial control and communication systems by the understanding of complex natural processes, but it can be expected to find expression and enlargement in Art. It can assist in the evolution of art, serving to increase its variety and vigour. . . .

Myron Krueger

<< 12 >> "Responsive Environments" (1977)

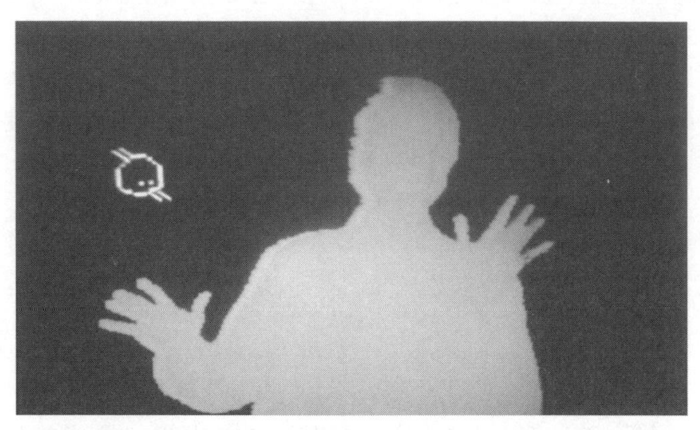

Myron Krueger, Videoplace. © *Myron Krueger. Courtesy of Myron Krueger.*

"The responsive environment has been presented as the basis for a new aesthetic medium based on real-time interaction between men and machines. In the long range it augurs a new realm of human experience, artificial realities which seek not to stimulate the physical world but to define arbitrary, abstract and otherwise impossible relationships between action and result."

<< Myron Krueger was among the first artists to explore the computer as a central component in interactive art. Originally trained as a computer scientist, Krueger, under the influence of John Cage's experiments in indeterminacy and audience participation, pioneered human-computer interaction in the context of physical environments. Beginning in 1969, he collaborated with artist and engineer colleagues at the University of Wisconsin to create artworks that responded to the movement and gesture of the viewer through an elaborate system of sensing floors, graphic tables, and video cameras. In this essay, which reflects the influence of cybernetics on the arts and sciences, Krueger discusses human-computer interaction as the basis for an emerging art form.

At the heart of Krueger's contribution to interactive computer art was the notion of the artist as a "composer" of intelligent, real-time computer-mediated spaces, or "responsive environments," as he called them. Krueger "composed" environments in which the computer responded to the gestures of the audience by interpreting, and even anticipating, their actions. In such works as *Metaplay* and *Videoplace,* both from the 1970s, images of audience members would be projected on an eight-by-ten-foot screen; superimposed on the screen would be animated figures and special effects designed by the artist. Audience members could "touch" each other's video-generated silhouettes, as well as manipulate the odd, playful assortment of graphical objects and animated organisms that appeared on the screen, imbued with the presence of artificial life.

In Krueger's artworks the images of spectators video-broadcast from separate locations would be projected into a single space. This collage effect anticipated subsequent developments in telepresence, which can be described as the state of being present simultaneously in more than one location. In these early explorations of telematic art, audience members experienced a sensation of virtual proximity, as if they were standing next to each other in the same space, though they were separated by a significant distance. This aspect of Krueger's work makes it a precursor to today's global communications technology, including video-conferencing, networked virtual worlds, and live on-line chat. >>

Introduction

Man-machine interaction is usually limited to a seated man poking at a machine with his fingers or perhaps waving a wand over a data tablet. Seven years ago, I was dissatisfied with such a restricted dialogue and embarked on research exploring more interesting ways for men and machines to relate. The result was the concept of a responsive environment in which a computer perceives the actions of those who enter and responds intelligently through complex visual and auditory displays.

Over a period of time the computer's displays establish a context within which the interaction occurs. It is within this context that the participant chooses his next action and anticipates the environment's response. If the response is unexpected, the environment has changed the context and the participant must reexamine his expectations. The experience is controlled by a composition which anticipates the participant's actions and flirts with his expectations.

This paper describes the evolution of these concepts from their primitive beginnings to my current project, *Videoplace,* which provides a general tool for devising many interactions. Based on these examples an interactive art form is defined and its promise identified. While the environments described were presented with aesthetic intent, their implications go beyond art. In the final section, applications in education, psychology and psychotherapy are suggested.

Glowflow

In 1969, I became involved in the development of *Glowflow,* a computer art project conceived by Dan Sandin, Jerry Erdman and Richard Venezsky at the University of Wisconsin. It was designed in an atmosphere of encounter between art and technology. The viewer entered a darkened room in which glowing lines of light defined an illusory space. The display was accomplished by pumping phosphorescent particles through transparent tubes attached to the gallery walls. These tubes passed through opaque columns concealing lights which excited the phosphors. A pressure sensitive pad in front of each of the six columns enabled the computer to respond to footsteps by lighting different tubes or changing the sounds generated by a Moog synthesizer or the origin of these sounds. However, the artists' attitude toward the capacity for response was ambivalent. They felt that it was important that the environment respond, but not that the audience be aware of it. Delays were introduced between the detection of a participant and the computer's response so that the contemplative mood of the environment would not be destroyed by frantic attempts to elicit more responses.

While *Glowflow* was quite successful visually, it succeeded more as a kinetic sculpture than as a responsive environment. However, the *Glowflow* experience led me to a number of decisions:

1. Interactive art is potentially a richly composable medium quite distinct from the concerns of sculpture, graphic art or music.
2. In order to respond intelligently the computer should perceive as much as possible about the participant's behavior.
3. In order to focus on the relationships between the environment and the participants, rather than among participants, only a small number of people should be involved at a time.
4. The participants should be aware of how the environment is responding to them.
5. The choice of sound and visual response systems should be dictated by their ability to convey a wide variety of conceptual relationships.
6. The visual responses should not be judged as art nor the sounds as music. The only aesthetic concern is the quality of the interaction.

METAPLAY

Following the *Glowflow* experience, I conceived and directed *Metaplay,* which was exhibited in the Memorial Union Gallery of the University of Wisconsin for a month in 1970. It was supported by the National Science Foundation, the Computer Science Department, the Graduate School and the loan of a PDP-12 by Digital Equipment Corporation.

Metaplay's focus reflected my reactions to *Glowflow.* Interaction between the participants and the environment was emphasized; the computer was used to facilitate a unique real-time relationship between the artist and the participant. An 8′ × 10′ rear-projection video screen dominated the gallery. The live video image of the viewer and a computer graphic image drawn by an artist, who was in another building, were superimposed on this screen. Both the viewer and the artist could respond to the resulting image.

Hardware

The image communications started with an analogue data tablet which enabled the artist to draw or write on the computer screen. The person doing the drawing did not have to be an artist, but the term is used for convenience. One video camera, in the Computer Center, was aimed at the display screen of the Adage Graphic Display Computer. A second camera, a mile away in the gallery, picked up the live image of people in the room. A television cable transmitted the video computer image from the Computer Center to the gallery and the two signals were mixed so that the computer image overlayed the live image. The composite image was projected on the 8′ × 10′ screen in the gallery and was simultaneously transmitted back to the Computer Center where it was displayed on a video monitor providing feedback for the artist.

The artist could draw on the Adage screen using a data tablet. By using function switches, potentiometers and the teletype keyboard the pictures could be rapidly modified or the mode of drawing itself altered. In addition to the effects of simple drawings, the image could be moved around the screen, image size could be controlled and the picture could be repeated up to ten times on the screen displaced by variable X, Y and size increments. A tail of a fixed number of line segments could be drawn allowing the removal of a segment at one end while another was added at the opposite end. An image could be rotated in 3-space under control of the pen. Although this was not true rotation, the visual effect was similar. A simple set of transformations under potentiometer and tablet control yielded apparent animation of people's outlines. Finally, previously defined images could be recalled or exploded. While it might seem that the drawing could be done without a computer, the ability to rapidly erase, recall and transform images required considerable processing and created a far more powerful means of expression than pencil and paper could provide.

Interaction

These facilities provided a rich repertoire for an unusual dialogue. The artist could draw pictures on the participants' images or communicate directly by writing words on the screen. He could induce people to play a game like Tic-Tac-Toe or play with the act of drawing, starting to draw one kind of picture only to have it transformed into another by interpolation.

Live Graffiti

One interaction derived from the artist's ability to draw on the image of the audience. He could add graffiti-like features or animate a drawn outline of a person so that it appeared to dance to the music in the gallery. The artist tried various approaches to involve people in the interaction. Failing to engage one person, he would seek someone more responsive.

It was important to involve the participants in the act of drawing. However, the electronic wand designed for this purpose did not work reliably. What evolved was a serendipitous solution. One day as I was trying to draw on a student's hand, he became confused and moved it. When I erased my scribblings and started over, he moved his hand again. He did this repeatedly until it became a game. Finally, it degenerated to the point where I was simply tracking the image of his hand with the computer line. In effect, by moving his hand he could draw on the screen before him.

The relationship established with this participant was developed as one of the major themes of *Metaplay*. It was repeated and varied until it became an aesthetic medium in itself. With each person we involved in this way, we tried to preserve the pleasure of the original discovery. After playing some graffiti games with each group that entered, we would focus on a single individual and draw around the image of his hand. After an initial reaction of blank bewilderment, the self-conscious person would make a nervous gesture. The computer line traced the gesture. A second gesture, followed by the line, was the key to discovery. One could draw on the video screen with his finger! Others in the group, observing this phenomenon, would want to try it too. The line could be passed from one person's finger to another's. Literally hundreds of interactive vignettes developed within this simple communication channel.

Drawing by this method was a rough process. Pictures of any but the simplest shapes were unattainable. This was mainly because of the difficulty of tracking a person's finger. Happily, neither the artist nor the audience was concerned about the quality of the drawings. What was exciting was interacting in this novel way through a man-computer-video link spanning a mile.

Psychic Space

The next step in the evolution of the responsive environment was *Psychic Space,* which I designed and exhibited in the Memorial Union Gallery during May and June of 1971. It was implemented with the help of my students, the Computer Science Department and a National Science Foundation grant in Complex Information Processing.

Psychic Space was both an instrument for musical expression and a richly composed, interactive, visual experience. Participants could become involved in a softshoe duet with the environment, or they could attempt to match wits with the computer by walking an unpredictable maze projected on an 8′ × 10′ video screen.

Hardware

A PDP-11 had direct control of all sensing and sound in the gallery. In addition, it communicated with the Adage AGT-10 Graphic Display Computer at the Computer Center. The Adage image was transmitted over video cable to the gallery where it was rear-projected on the 8′ × 10′ screen. The participant's position on the floor was the basis for each of the interactions. The sensing was done by a 16′ × 24′ grid of pressure switches, constructed in 2′ × 4′ modules, each containing 8 switches. Since they were electronically independent, the system was able to discriminate among individuals if several were present. This independence made it easy for the programming to ignore a faulty switch until its module was replaced or repaired. Since there were 16 bits in the input words of the PDP-11, it was natural to read the 16 switches in each row across the room in parallel. Digital circuitry was then used to scan the 24 rows under computer control.

Input and Interaction

Since the goal was to encourage the participants to express themselves through the environment, the program automatically responded to the footsteps of people entering the room with electronic sound. We experimented with a number of different schemes for actually generating the sounds based on an analysis of peoples' footsteps. In sampling the floor 60 times per second we discovered that a single footstep consisted of as many as four discrete events: lifting the heel, lifting the toe, putting the heel down and putting the ball of the foot down. The first two were dubbed the "unfootstep." We could respond to each footstep or unfootstep as it occurred, or we could respond to the person's average position. A number of response schemes were tried, but the most pleasing was to start each tone only when a new switch was stepped on and then to terminate it on the next "unfootstep." Thus it was possible to get silence by jumping, or by lifting one foot, or by putting both feet on the same switch.

Typical reaction to the sounds was instant understanding, followed by a rapid-fire sequence of steps, jumps and rolls. This phase was followed by a slower, more thoughtful exploration of the environment in which more subtle and interesting relationships could be developed. In the second phase, the participant would discover that the room was organized with high notes at one end and low notes at the other. After a while, the keyboard was abruptly rotated by 90 degrees.

After a longer period of time an additional feature came into play. If the computer discovered that a person's behavior was characterized by a short series of steps punctuated by relatively long pauses, it would use the pause to establish a new kind of relationship. The sequence of steps was responded to with a series of notes as before; however, during the pause the computer would repeat these notes again. If the person remained still during the pause, the computer assumed that the relationship was understood. The next sequence of steps was echoed at a noticeably higher pitch. Subsequent sequences were repeated several times with variations each time. This interaction was experimental and extremely difficult to introduce clearly with feedback alone, i.e., without explicit instructions. The desire was for a man-machine dialogue resembling the guitar duel in the film *Deliverance.*

Maze—A Composed Environment

The maze program focused on the interaction between one individual and the environment. The participant was lured into attempting to navigate a projected maze. The intrigue derived from the maze's responses, a carefully composed sequence of relations designed to constitute a unique and coherent experience.

Hardware

The maze itself was not programmed on the PDP-11, but on the Adage located a mile away in the Computer Center. The PDP-11 transmitted the participant's floor coordinates across an audio cable to the Adage. The data was transmitted asynchronously as a serial bit stream of varying pulse widths. The Adage generated the maze image which was picked up by a TV camera and transmitted via a video cable back to the Union where it was rear-screen-projected to a size of 8′ × 10′.

Interaction

The first problem was simply to educate the person to the relationships between the floor and the screen. Initially, a diamond with a cross in it representing the person's position appeared on the screen. Physical movement in the room caused the symbol to move correspondingly on the screen. As the participant approached the screen, the symbol moved up. As he moved away, it moved down. The next step was to induce the person to move to the starting point of the maze, which had not yet appeared on the screen. To this end, another object was placed on the screen at the position which would be the starting point of the maze. The viewer unavoidably wondered what would happen if he walked his symbol to the object. The arrival of his symbol at the starting point caused the object to vanish and the maze to appear. Thus confronted with the maze, no one questioned the inevitability of walking it.

Software Boundaries

Since there were no physical constraints in the gallery, the boundaries of the maze had to be enforced by the computer. Each attempt to violate a boundary was foiled by one of many responses in the computer's repertoire. The computer could move the line, stretch it elastically, or move the whole maze. The line could disappear, seemingly removing the barrier, except that the rest of the maze would change simultaneously so no advantage was gained. In addition, the symbol representing the person could split in half at the violated boundary, with one half held stationary while the other half, the alter ego, continued to track movement. However, no progress could be made until the halves of the symbol were reunited at the violated boundary.

Even when the participant was moving legally, there were changes in the program contingent upon his position. Several times, as the goal was approached, the maze changed to thwart immediate success. Or, the relationship between the floor and the maze was altered so that movements that once resulted in vertical motion now resulted in horizontal motion. Alternatively, the symbol representing the participant could remain stationary while the maze moved.

Ultimately, success was not allowed. When reaching the goal seemed imminent, additional boundaries appeared in front of and behind the symbol, boxing it in. At this point, the maze slowly shrank to nothing. While the goal could not be reached, the composed frustration made the route interesting.

Experience

The maze experience conveyed a unique set of feelings. The video display space created a sense of detachment enhanced by the displaced feedback: movement on the horizontal plane of the floor translated onto the vertical plane of the screen. The popular stereotype of dehumanizing technology seemed fulfilled. However, the maze idea was engaging and people became involved willingly. The lack of any other sensation focused attention completely on this interaction. As the experience progressed, their perception of the maze changed. From the initial impression that it was a problem to solve, they moved to the realization that the maze was a vehicle for whimsy, playing with the concept of a maze and poking fun at their compulsion to walk it.

VIDEOPLACE

For the past two years I have been working on a project called *Videoplace,* under the aegis of the Space Science and Engineering Center of the University of Wisconsin. This work is funded by the National Endowment for the Arts and the Wisconsin Arts Board. A preliminary version was exhibited at the Milwaukee Art Center for six weeks beginning in October 1975. The development of *Videoplace* is still under way and several more years will be required before its potential is fully realized both in terms of implementing the enabling hardware and exploring its compositional possibilities.

Videoplace is a conceptual environment with no physical existence. It unites people in separate locations in a common visual experience, allowing them to interact in unexpected ways through the video medium. The term *Videoplace* is based on the premise that the act of communication creates a place that consists of all the information that the participants share at that moment. When people are in the same room, the physical and communication places are the same. When the communicants are separated by distance, as in a telephone conversation, there is still a sense of being together although sight and touch are not possible. By using television instead of telephone, *Videoplace* seeks to augment this sense of place by including vision, physical dimension and a new interpretation of touch.

Videoplace consists of two or more identical environments which can be adjacent or hundreds of miles apart. In each environment, a single person walks into a darkened room where he finds himself confronted by an 8′ × 10′ rear-view projection screen. On the screen he sees his own life-size image and the image of one or more other people. This is surprising in itself, since he is alone in the room. The other images are of people in the other environments. They see the same composite image on their screens. The visual effect is of several people in the same room. By moving around their respective rooms, thus moving their images, the participants can interact within the limitations of the video medium.

It is these apparent limitations that I am currently working to overcome. When people are physically together, they can talk, move around the same space, manipulate the same objects and touch each other. All of these actions would appear to be impossible within the *Videoplace.* However, the opposite is true. The video medium has the potential of being more rich and variable in some ways than reality itself.

It would be easy to allow the participants to talk, although I usually preclude this, to force people to focus on the less familiar kinds of interaction that the video medium provides. A sense of dimension can be created with the help of computer graphics, which can define a room or another spatial context within which the participants appear to move around. Graphics can also furnish this space with artificial objects and inhabit it with imaginary organisms. The sense of touch would seem to be impossible to duplicate. However, since the cameras see each person's image in contrast to a neutral background, it is easy to digitize the outline and to determine its orientation on the screen. It is also easy to tell if one person's image touches another's, or if someone touches a computer graphic object. Given this information the computer can make the sense of touch effective. It can currently respond with sounds when two images touch and will ultimately allow a person's image to pick up a graphic object and move it about the screen.

While the participants' bodies are bound by physical laws such as gravity, their images could be moved around the screen, shrunk, rotated, colorized and keyed together in arbitrary ways. Thus, the full power of video processing could be used to mediate the interaction and the usual laws of cause and effect replaced with alternatives composed by the artist.

The impact of the experience will derive from the fact that each person has a very proprietary feeling towards his own image. What happens to his image happens to him. In fact, when one person's image overlaps another's, there is a psychological sensation akin to touch. In *Videoplace,* this sensation can be enhanced in a number of ways. One image can occlude the other. Both images can disappear where they intersect. Both images can disappear except where they intersect. The intersection of two images can be used to form a window into another scene so two participants have to cooperate to see a third.

Videoplace need not involve more than one participant. It is quite possible to create a compelling experience for one person by projecting him into this imaginary domain alone. In fact the hardware/software system underlying *Videoplace* is not conceived as a single work but as a general facility for exploring all the possibilities of the medium to be described next.

Response Is the Medium

The environments described suggest a new art medium based on a commitment to real-time interaction between men and machines. The medium is comprised of sensing, display and control systems. It accepts inputs from or about the participant and then outputs in a way he can recognize as corresponding to his behavior. The relationship between inputs and outputs is arbitrary and variable, allowing the artist to intervene between the participant's action and the results perceived. Thus, for example, the participant's physical movement can cause sounds or his voice can be used to navigate a computer-defined visual space. It is the composition of these relationships between action and response that is important. The beauty of the visual and aural response is secondary. Response is the medium!

The distinguishing aspect of the medium is, of course, the fact that it responds to the viewer in an interesting way. In order to do this, it must know as much as possible about what the participant is doing. It cannot respond intelligently if it is unable to distinguish various kinds of behavior as they occur.

The environment might be able to respond to the participant's position, voice volume or pitch, position relative to prior position or the time elapsed since the last movement. It could also respond to every third movement, the rate of movement, posture, height, colors of clothing or time elapsed since the person entered the room. If there were several people in the room, it might respond to the distance separating them, the average of their positions or the computer's ability to resolve them, i.e., respond differently when they are very close together.

In more complex interactions like the maze, the computer can create a context within which the interaction occurs. This context is an artificial reality within which the artist has complete control of the laws of cause and effect. Thus the actions perceived by the hardware sensors are tested for significance within the current context. The computer asks if the person has crossed the boundary in the maze or has touched the image of a particular object. At a higher level the machine can learn about the individual and judge from its past experience with similar individuals just which responses would be most effective.

Currently, these systems are constrained by the total inability of the computer to make certain very useful and for the human very simple perceptual judgments, such as whether a given individual is a man or a woman or is young or old.

The perceptual system will define the limits of meaningful interaction, for the environment cannot respond to what it cannot perceive. To date the sensing systems have included pressure pads, ultrasonics and video digitizing.

As mentioned before, the actual means of output are not as important in this medium as they would be if the form were conceived as solely visual or auditory. In fact, it may be desirable that the output not qualify as beautiful in any sense, for that would distract from the central theme: the relationship established between the observer and the environment. Artists are fully capable of producing effective displays in a number of media. This fact is well known and to duplicate it produces nothing new. What is not known and remains to be tested is the validity of a responsive aesthetic.

It is necessary that the output media be capable of displaying intelligent, or at least composed, reactions, so that the participant knows which of his actions provoked it and what the relationship of the response is to his action. The purpose of the displays is to communicate the relationships that the environment is trying to establish. They must be capable of great variation and fine control. The response can be expressed in light, sound, mechanical movement, or through any means that can be perceived. So far computer graphics, video generators, light arrays and sound synthesizers have been used.

Control and Composition

The control system includes hardware and software control of all inputs and outputs as well as processing for decisions that are programmed by the artist. He must balance his desire for interesting relationships against the commitment to respond in real-time. The simplest responses are little more than direct feedback of the participant's behavior, allowing the environment to show off its perceptual system. But far more sophisticated results are possible. In fact, a given aggregation of hardware sensors, displays and processors can be viewed as an instrument which can be programmed by artists with differing sensitivities to create completely different experiences. The environment can be thought of in the following ways:

1. An entity which engages the participant in a dialogue. The environment expresses itself through light and sound while the participant communicates with physical motion. Since the experience is an encounter between individuals, it might legitimately include greetings, introductions and

farewells—all in an abstract rather than literal way. The problem is to provide an interesting personality for the environment.

2. A personal amplifier. One individual uses the environment to enhance his ability to interact with those within it. To the participants the interaction might appear similar to that described above. The result would be limited by the speed of the artist's response but improved by his sensitivity to the participants' moods. The live drawing interaction in *Metaplay* could be considered an example of this approach.
3. An environment which has sub-environments with different response relationships. This space could be inhabited by artificial organisms defined either visually or with sound. These creatures can interact with the participants as they move about the room.
4. An amplifier of physical position in a real or artificially generated space. Movements around the environment would result in much larger apparent movements in the visually represented space. A graphic display computer can be used to generate a perspective view of a modelled space as it would appear if the participant were within it. Movements in the room would result in changes in the display, so that by moving only five feet within the environment, the participant would appear to have moved fifty feet in the display. The rules of the modelled space can be totally arbitrary and physically impossible, e.g., a space where objects recede when you approach them.
5. An instrument which the participants play by moving about the space. In *Psychic Space* the floor was used as a keyboard for a simple musical instrument.
6. A means of turning the participant's body into an instrument. His physical posture would be determined from a digitized video image and the orientation of the limbs would be used to control lights and sounds.
7. A game between the computer and the participant. This variation is really a far more involving extension of the pinball machine, already the most commercially successful interactive environment.
8. An experimental parable where the theme is illustrated by the things that happen to the protagonist—the participant. Viewed from this perspective, the maze in *Psychic Space* becomes pregnant with meaning. It was impossible to succeed, to solve the maze. This could be a frustrating experience if one were trying to reach the goal. If, on the other hand, the participant maintained an active curiosity about how the maze would thwart him next, the experience was entertaining. Such poetic composition of experience is

one of the most promising lines of development to be pursued with the environments.

Implications of the Art Form

For the artist the environment augurs new relationships with his audience and his art. He operates at a metalevel. The participant provides the direct performance of the experience. The environmental hardware is the instrument. The computer acts much as an orchestra conductor controlling the broad relationships while the artist provides the score to which both performer and conductor are bound. This relationship may be a familiar one for the musical composer, although even he is accustomed to being able to recognize one of his pieces, no matter who is interpreting it. But the artist's responsibilities here become even broader than those of a composer who typically defines a detailed sequence of events. He is composing a sequence of possibilities, many of which will not be realized for any given participant who fails to take the particular path along which they lie.

Since the artist is not dedicated to the idea that his entire piece be experienced he can deal with contingencies. He can try different approaches, different ways of trying to elicit participation. He can take into account the differences among people. In the past, art has often been a one-shot, hit-or-miss proposition. A painting could accept any attention paid it, but could do little to maintain interest once it had started to wane. In an environment the loss of attention can be sensed as a person walks away. The medium can try to regain attention and upon failure, try again. The piece has a second strike capability. In fact it can learn to improve its performance, responding not only to the moment but also to the entire history of its experience.

In the environment, the participant is confronted with a completely new kind of experience. He is stripped of his informed expectations and forced to deal with the moment in its own terms. He is actively involved, discovering that his limbs have been given new meaning and that he can express himself in new ways. He does not simply admire the work of the artist; he shares in its creation. The experience he achieves will be unique to his movements and may go beyond the intentions of the artist or his understanding of the possibilities of the piece.

Finally, in an exciting and frightening way, the environments dramatize the extent to which we are savages in a world of our own creation. The layman has extremely little ability to define the limits of what is possible with current technology and so will accept all sorts of cues as representing relationships which in fact do not exist. The constant birth of such superstitions indicates how much we have already accomplished in mastering our natural environment and how difficult the initial discoveries must have been.

Applications

The responsive environment is not limited to aesthetic expression. It is a potent tool with applications in many fields. *Videoplace* clearly generalizes the act of telecommunication. It creates a form of communication so powerful that two people might choose to meet visually, even if it were possible for them to meet physically. While it is not immediately obvious that *Videoplace* is the optimum means of telecommunication, it is reasonably fair to say that it provides an infinitely richer interaction than Picturephone allows. It broadens the range of possibilities beyond current efforts at teleconferencing. Even in its fetal stage, *Videoplace* is far more flexible than the telephone is after one hundred years of development. At a time when the cost of transportation is increasing and fiber optics promise to reduce the cost of communication, it seems appropriate to research the act of communication in an intuitive sense as well as in the strictly scientific and problem-solving approaches that prevail today. . . .

Conclusion

The responsive environment has been presented as the basis for a new aesthetic medium based on real-time interaction between men and machines. In the long range it augurs a new realm of human experience, artificial realities which seek not to simulate the physical world but to define arbitrary, abstract and otherwise impossible relationships between action and result. In addition, it has been suggested that the concepts and tools of the responsive environments can be fruitfully applied in a number of fields.

What perhaps has been obscured is that these concepts are the result of a personal need to understand and express the essence of the computer in humanistic terms. An earlier project to teach people how to use the computer was abandoned in favor of exhibits which taught people about the computer by letting them experience it. *Metaplay, Psychic Space* and *Videoplace* were designed to communicate an affirmative vision of technology to the lay public. This level of education is important, for our culture cannot continue if a large proportion of our population is hostile to the tools that define it.

We are incredibly attuned to the idea that the sole purpose of our technology is to solve problems. It also creates concepts and philosophy. We must more fully explore these aspects of our inventions, because the next generation of technology will speak to us, understand us, and perceive our behavior. It will enter every home and office and intercede between us and much of the information and experience we receive. The design of such intimate technology is an aesthetic issue as much as an engineering one. We must recognize this if we are to understand and choose what we become as a result of what we have made.

Alan Kay

"User Interface: A Personal View" (1989)

<< 13 >>

Alan Kay. Courtesy of Alan Kay.

"Therefore, let me argue that the actual dawn of user interface design first happened when computer designers finally noticed, not just that end users had functioning minds, but that a better understanding of how those minds worked would completely shift the paradigm of interaction."

<< During the late 1960s, while a graduate student at the University of Utah, Alan Kay studied with Ivan Sutherland, the pioneering scientist whose work launched the field of interactive computer graphics. In 1968 he attended Douglas Engelbart's historic presentation of the oNLine System, which introduced the mouse to computing, suggesting for the first time the possibility of "navigating" information space. In the early 1970s, after forming the Learning Research Group at the newly founded Xerox PARC (Palo Alto Research Center), Kay synthesized these influences into what is considered the most crucial advancement of human-computer interactivity, the graphical user interface (GUI). Kay developed the idea of iconic, graphical representations of computing functions—the folders, menus, and overlapping windows found on the desktop—based on his research into the intuitive processes of learning and creativity.

The GUI was an entirely new approach to interactive computing. The fundamentals of the computer interface had received scant attention since the birth of the first digital computer, ENIAC, twenty-five years before. At the time, computers were programmed and operated almost exclusively by scientists. Kay realized early on that designing an intuitive, easy-to-use interface required a subtle understanding of the dynamics of human perception. Norbert Wiener's concept of the digital computer had been modeled after the brain, with its sophisticated network of message passing and synoptic relays. Kay went deeper in this direction. He drew on the theories of psychologists Jean Piaget, Seymour Papert, and Jerome Bruner, who had studied the intuitive capacities for learning present in the child's mind, and the role that images and symbols play in the building of complex concepts. Kay came to understand, as he put it, that "doing with images makes symbols." This was the premise behind the GUI, which enabled computers users to formulate ideas in real time by manipulating icons on the computer screen.

Kay's approach made computers accessible to nonspecialists. More important, it transformed the computer into a vehicle for popular creative expression. In this essay, Kay describes his realization that the computer was not just a computational device but a medium in its own right. Computers, Kay recognized, might one day replace books. This led him to build a prototype for a personal computer, the Dynabook, which would be capable of dynamically representing images and concepts. This notebook-sized computer, able to provide unprecedented access to information, images, sound, and animation, was a revolutionary departure not only from the stasis of the book but also from the passivity of television. While the Dynabook was never built, it led to the Xerox Alto, the first true multimedia machine. >>

When I was asked to write this chapter, my first reaction was "A book on user interface design—does that mean it's now a real subject?" Well, as of 1989, it's still yes and no. User interface has certainly been a hot topic for discussion since the advent of the Macintosh. Everyone seems to want user interface but they are not sure whether they should order it by the yard or by the ton. Many are just now discovering that user interface is not a sandwich spread—applying the Macintosh style to poorly designed applications and machines is like trying to put Béarnaise sauce on a hotdog!

Of course the practice of user interface design has been around at least since humans invented tools. The unknown designer who first put a haft on a hand axe was trying not just to increase leverage but also to make it an extension of the *arm,* not just the fist. The evolutionary designer whom Richard Dawkins calls the Blind Watchmaker has been at it much longer; all of life's startling interfitness is the result. A more recent byproduct of the industrial revolution called *ergonomics* in Europe and *human factors* in the U.S. has studied how the human body uses senses and limbs to work with tools. From the earliest use of interactive computing in the fifties—mostly for air traffic control and defense—there have been attempts at user interface design and application of ergonomic principles. Many familiar components of modern user interface design appeared in the fifties and early sixties, including pointing devices, windows, menus, icons, gesture recognition, hypermedia, the first personal computer, and more. There was even a beautifully designed user interface for an end-user system in JOSS—but its significance was appreciated only by its designer and users.

Therefore, let me argue that the actual dawn of user interface design first happened when computer designers finally noticed, not just that end users had functioning minds, but that a better understanding of how those minds worked would completely shift the paradigm of interaction.

This enormous change in point of view happened to many computerists in the late sixties, especially in the ARPA research community. Everyone had their own catalyst. For me it was the FLEX machine, an early desktop personal computer of the late sixties designed by Ed Cheadle and myself.

Based on much previous work by others, it had a tablet as a pointing device, a high-resolution display for text and animated graphics, and multiple windows, and it directly executed a high-level object-oriented end-user simulation language. And of course it had a "user interface," but one that repelled end users instead of drawing them closer to the hearth. I recently revisited the FLEX machine design and was surprised to find how modern its components were—even a use of icon-like structures to access previous work.

But the combination of ingredients didn't gel. It was like trying to bake a pie from random ingredients in a kitchen: baloney instead of apples, ground-up Cheerios instead of flour, etc.

Then, starting in the summer of 1968, I got hit on the head randomly but repeatedly by some really nifty work. The first was just a little piece of glass at the University of Illinois. But the glass had tiny glowing dots that showed text characters. It was the first flat-screen display. I and several other grad students wondered when the surface could become large and inexpensive enough to be a useful display. We also wondered when the FLEX machine silicon could become small enough to fit on the back of the display. The answer to both seemed to be the late seventies or early eighties. Then we could all have an inexpensive powerful notebook computer—I called it a "personal computer" then, but I was thinking *intimacy*.

I read McLuhan's *Understanding Media* [1964] and understood that the most important thing about any communications medium is that message receipt is really message recovery; anyone who wishes to receive a message embedded in a medium must first have internalized the medium so it can be "subtracted" out to leave the message behind. When he said "the medium is the message" he meant that you have to *become* the medium if you use it.

That's pretty scary. It means that even though humans are the animals that shape tools, it is in the nature of tools and man that learning to use tools reshapes us. So the "message" of the printed book is, first, its availability to individuals, hence, its potential detachment from extant social processes; second, the uniformity, even coldness, of noniconic type, which detaches readers from the vividness of the now and the slavery of commonsense thought to propel them into a far more abstract realm in which ideas that don't have easy visualizations can be treated.

McLuhan's claim that the printing press was the dominant force that transformed the hermeneutic Middle Ages into our scientific society should not be taken too lightly—especially because the main point is that the press didn't do it just by making books more available, it did it by changing the thought patterns of those who learned to read.

Though much of what McLuhan wrote was obscure and arguable, the sum total to me was a shock that reverberates even now. The computer is a medium! I had always thought of it as a tool, perhaps a vehicle—a much weaker conception. What McLuhan was saying is that if the personal computer is a truly new medium then the very use of it would actually change the thought patterns of an entire civilization. He had certainly been right about the effects of the electronic stained-glass window that was television—a remedievalizing tribal influence at best. The intensely interactive and involving nature of the personal computer seemed an antiparticle that could annihilate the passive boredom invoked by television. But it also promised to surpass the book to bring about a new kind of renaissance by going beyond static representations to dynamic simulation. What kind of a thinker would you become if you grew up with an active simulator connected, not just to one point of view, but to all the points of view of the ages represented so they could be dynamically tried out and compared? I named the notebook-sized computer idea the Dynabook to capture McLuhan's metaphor in the silicon to come.

Shortly after reading McLuhan, I visited Wally Feurzeig, Seymour Papert, and Cynthia Solomon at one of the earliest LOGO tests within a school. I was amazed to see children writing programs (often recursive) that generated poetry, created arithmetic environments, and translated English into Pig Latin. And they were just starting to work with the new wastepaper-basket-sized turtle that roamed over sheets of butcher paper making drawings with its pen.

I was possessed by the analogy between print literacy and LOGO. While designing the FLEX machine I had believed that end users needed to be able to program before the computer could become truly theirs—but here was a real demonstration, and with children! The ability to "read" a medium means you can *access* materials and tools created by others. The ability to "write" in a medium means you can *generate* materials and tools for others. You must have both to be literate. In print writing, the tools you generate are rhetorical; they demonstrate and convince. In computer writing, the tools you generate are processes; they simulate and decide.

If the computer is only a vehicle, perhaps you can wait until high school to give "driver's ed" on it—but if it's a medium, then it must be extended all the way into the world of the child. How to do it? Of course it has to be done on the intimate notebook-sized Dynabook! But how would anyone "read" the Dynabook, let alone "write" on it?

LOGO showed that a special language designed with the end user's characteristics in mind could be more successful than a random hack. How had Papert learned about the nature of children's thought? From Jean Piaget, the doyen of European cognitive psychologists. One of his most important contributions is the idea that children go through several distinctive intellectual stages as they develop from birth to maturity. Much can be accomplished if the nature of the stages is heeded and much grief to the child can be caused if the stages are ignored. Piaget noticed a kinesthetic stage, a visual stage, and a symbolic stage. An example is that children in the visual stage, when shown a squat glass of water poured into a tall thin one, will say there is more water in the tall thin one even though the pouring was done in front of their eyes.

One of the ways Papert used Piaget's ideas was to realize that young children are not well equiped to do "standard" symbolic mathematics until the age of 11 or 12, but that even very young children can do other kinds of math, even advanced math such as topology and differential geometry, when it is presented in a form that is well matched to their current thinking processes. The LOGO turtle with its local coordinate system (like the child, it is always at the center of its universe) became a highly successful "microworld" for exploring ideas in differential geometry.

This approach made a big impression on me and got me to read many more psychology books. Most (including Piaget's) were not very useful, but then I discovered Jerome Bruner's *Towards a Theory of Instruction* [1966]. He had repeated and verified many of Piaget's results, and in the process came up with a different and much more powerful way to interpret what Piaget had seen. For example, in the water-pouring experiment, after the child asserted there was more water in the tall thin glass, Bruner covered it up with a card and asked again. This time the child said, "There must be the same because where would the water go?" When Bruner took away the card to again reveal the tall thin glass, the child immediately changed back to saying there was more water.

When the cardboard was again interposed the child changed yet again. It was as though one set of processes was doing the reasoning when the child could see the situation, and another set was employed when the child could not see. Bruner's interpretation of experiments like these is one of the most important foundations for human-related design. Our mentalium seems to be made up of multiple separate mentalities with very different characteristics. They reason differently, have different skills, and often are in conflict. Bruner identified a separate mentality with each of Piaget's stages: he called them *enactive, iconic, symbolic.* While not ignoring the existence of other mentalities, he concentrated on these three to come up with what are still some of the strongest ideas for creating learning-rich environments.

The work of Papert convinced me that whatever user interface design might be, it was solidly intertwined with learning. Bruner convinced me that learning takes place best environmentally and roughly in stage order—it is best to learn something kinesthetically, then iconically, and finally the intuitive knowledge will be in place that will allow the more powerful but less vivid symbolic processes to work at their strongest. This led me over the years to the pioneers of environmental learning: Montessori Method, Suzuki Violin, and Tim Gallwey's *Inner Game of Tennis,* to name just a few.

My point here is that as soon as I was ready to look deeply at the human element, and especially after being convinced that the heart of the matter lay with Bruner's multiple mentality model, I found the knowledge landscape positively festooned with already accomplished useful work. It was like the man in Moliere's *Bourgeois Gentilhomme* who discovered that all his life he had been speaking prose! I suddenly remembered McLuhan: "I don't know who discovered water but it wasn't a fish." Because it is in part the duty of consciousness to represent ourselves to ourselves as simply as possible, we should sorely distrust our commonsense self view. It is likely that this mirrors-within-mirrors problem in which we run into a misleading commonsense notion about ourselves at every turn is what forced psychology to be one of the most recent sciences—if indeed it yet is.

Now, if we agree with the evidence that the human cognitive facilities are made up of a *doing* mentality, an *image* mentality, and a *symbolic* mentality, then any user interface we construct should at least cater to the mechanisms that seem to be there. But how? One approach is to realize that no single mentality offers a complete answer to the entire range of thinking and problem solving. User interface design should integrate them at least as well as Bruner did in his spiral curriculum ideas.

One of the implications of the Piaget-Bruner decomposition is that the mentalities originated at very different evolutionary times and there is little probability that they can intercommunicate and synergize in more than the most rudimentary fashion. In fact, the mentalities are more likely to interfere with each other as they compete for control. The study by Hadamard on math and science creativity [1945] and others on music and the arts indicate strongly that creativity in these areas is not at all linked to the symbolic mentality (as most theories of teaching suppose), but that the important work in creative areas is done in the initial two mentalities—most in the iconic (or figurative) and quite a bit in the enactive. The groundbreaking work by Tim Gallwey on the "inner game" [1974] showed what could be done if interference were removed (mentalities not relevant to the learning were distracted) and attention was facilitated (the mentalities that could actually do the learning were focused more strongly on the environment).

Finally, in the sixties a number of studies showed just how modeful was a mentality that had "seized control"—particularly the analytical-problem-solving one (which identifies most strongly with the Bruner symbolic mentality). For example, after working on five analytic tasks in a row, if a problem was given that was trivial to solve figuratively, the solver could be blocked for hours trying to solve it symbolically. This makes quite a bit of sense when you consider that the main jobs of the three mentalities are:

enactive	know where you are, manipulate
iconic	recognize, compare, configure, concrete
symbolic	tie together long chains of reasoning, abstract

The visual system's main job is to be interested in everything in a scene, to dart over it as one does with a bulletin board, to change context. The symbolic system's main job is to stay with a context and to make indirect connections. Imagine what it would be like if it were reversed. If the visual system looked at the object it first saw in the morning for five hours straight! Or if the symbolic system couldn't hold a thought for more than a few seconds at a time!

It is easy to see that one of the main reasons that the figurative system is so creative is that it tends not to get blocked because of the constant flitting and darting. The chance of finding an interesting pattern is very high. It is not surprising, either, that many people who are "figurative" have extreme difficulty getting anything finished—there is always something new and interesting that pops up to distract. Conversely, the "symbolic" person is good at getting things done, because of the long focus on single contexts, but has a hard time being creative, or even being a good problem solver, because of the extreme tendency to get blocked. In other words, because none of the mentalities is supremely useful to the exclusion of the others, the best strategy would be to try to gently force synergy between them in the user interface design.

Out of all this came the main slogan I coined to express this goal:

Doing with Images makes Symbols

The slogan also implies—as did Bruner—that one should start with—be grounded in—the concrete "Doing with Images," and be carried into the more abstract "makes Symbols."

All the ingredients were already around. We were ready to notice what the theoretical frameworks from other fields of Bruner, Gallwey, and others were trying to tell us. What is surprising to me is just how long it took to put it all together. After Xerox PARC provided the opportunity to turn these ideas into reality, it still took our group about five years and experiments with hundreds of users to come up with the first practical design that was in accord with Bruner's model and really worked.

DOING	mouse	*enactive*	know where you are, manipulate
with IMAGES	icons, windows	*iconic*	recognize, compare, configure, concrete
makes SYMBOLS	Smalltalk	*symbolic*	tie together long chains of reasoning, abstract

Part of the reason perhaps was that the theory was much better at confirming that an idea was good than at actually generating the ideas. In fact, in certain areas like "iconic programming," it actually held back progress, for example, the simple use of icons as signs, because the siren's song of trying to do symbolic thinking iconically was just too strong.

Some of the smaller areas were obvious and found their place in the framework immediately. Probably the most intuitive was the idea of multiple overlapping windows. NLS had multiple panes, FLEX had multiple windows, and the bit-map display that we thought was too small, but that was made from individual pixels, led quickly to the idea that windows could appear to overlap. The contrastive ideas of Bruner suggested that there should always be a way to compare. The flitting-about nature of the iconic mentality suggested that having as many resources showing on the screen as possible would be a good way to encourage creativity and problem solving and prevent blockage. An *intuitive* way to use the windows was to activate the window that the mouse was in and bring it to the "top." This interaction was *modeless* in a special sense of the word. The active window constituted a mode to be sure—one window might hold a painting kit, another might hold text—but one could get to the next window to do something in *without any special termination.* This is what *modeless* came to mean for me—the user could always get to the next thing desired without any backing out. The contrast of the nice modeless interactions of windows with the clumsy command syntax of most previous systems directly suggested that everything should be made modeless. Thus began a campaign to "get rid of modes."

The object-oriented nature of Smalltalk was very suggestive. For example, *object-oriented* means that the object knows what it can do. In the abstract symbolic arena, it means we should first write the object's name (or whatever will fetch it) and then follow with a message it can understand that asks it to do something. In the concrete user-interface arena, it suggests that we should select the object first. It can then furnish us with a menu of what it is willing to do. In both cases we have the *object* first and the *desire* second. This unifies the concrete with the abstract in a highly satisfying way.

The most difficult area to get to be modeless was a very tiny one, that of elementary text editing. How to get rid of "insert" and "replace" modes that had plagued a decade of editors? Several people arrived at the solution simultaneously. My route came as the result of several beginning programmer adults who were having trouble building a paragraph editor in Smalltalk, a problem I thought should be easy. Over a weekend I built a sample paragraph editor whose main simplification was that it eliminated the distinction between insert, replace, and delete by allowing selections to extend *between* the characters. Thus, there could be a zero-width selection, and thus every operation could be a replace. "Insert" meant replace the zero-width selection. "Delete" meant replace the selection with a zero-width string of characters. I got the tiny one-page program running in Smalltalk and came in crowing over the victory. Larry Tesler thought it was great and showed me the idea, already working in his new Gypsy editor (which he implemented on the basis of a suggestion from Peter Deutsch). So much for creativity and invention when ideas are in the air. As Goethe noted, the most important thing is to enjoy the thrill of discovery rather than to make vain attempts to claim priority! . . .

The only stumbling place for this onrushing braver new world is that all of its marvels will be very difficult to communicate with, because, as always, the user interface design that could make it all simple lags far, far behind. If *communication* is the watchword, then what do we communicate with and how do we do it? We communicate with:

- Our selves
- Our tools
- Our colleagues and others
- Our agents

Until now, personal computing has concentrated mostly on the first two. Let us now extend everything we do to be part of a *grand collaboration*—with one's self, one's tools, other humans, and, increasingly, with *agents:* computer processes that act as guide, as coach, and as amanuensis. The user interface design will be the critical factor in the success of this new way to work and play on the computer. One of the implications is that the "network" will not be seen at all, but rather "felt" as a shift in capacity and range from that experienced via one's own hard disk. . . .

Well, there are so many more new issues that must be explored as well. I say thank goodness for that. How do we navigate in once-again uncharted waters? I have always believed that of all the ways to approach the future, the vehicle that gets you to the most interesting places is Romance. The notion of tool has always been a romantic idea to humankind—from swords to musical instruments to personal computers, it has been easy to say: "The best way to predict the future is to invent it!" The romantic dream of "how nice it would be if . . ." often has the power to bring the vision to life. Though the notion of management of complex processes has less cachet than that of the hero singlehandedly wielding a sword, the real romance of management is nothing less than the creation of civilization itself. What a strange and interesting frontier to investigate. As always, the strongest weapon we have to explore this new world is the one between our ears—providing it's loaded!

<< part III >>

Hypermedia

Vannevar Bush

"As We May Think" (1945)

<< **14** >>

Vannevar Bush and the differential analyzer. Courtesy of MIT Computer Museum.

"Presumably man's spirit should be elevated if he can better review his shady past and analyse more completely and objectively his present problems. He has built a civilization so complex that he needs to mechanize his records more fully if he is to push his experiment to its logical conclusion and not merely become bogged down part way there by overtaxing his limited memory. His excursions may be more enjoyable if he can reacquire the privilege of forgetting the manifold things he does not need to have immediately at hand, with some assurance that he can find them again if they prove important."

<< Vannevar Bush rose to prominence during World War II as chief scientific adviser to Franklin Roosevelt and director of the government's Office of Scientific Research and Development, where he supervised the research that led to the creation of the atomic bomb and other military technologies. By orchestrating this ambitious collaboration between the military, scientific, and academic communities, Bush is considered the founder of what came to be known as the military-industrial complex. His contribution to the evolution of the computer ranges far and wide: from the invention in 1930 of the Differential Analyzer, one of the first automatic electronic computers, to his concept of the "memex," the prototypical hypermedia machine, as he describes it in this article.

After World War II, many in the scientific community turned their attention to applying the technologies that won the war to further the cause of peace. In 1945 the *Atlantic Monthly* invited Bush to contribute an article on this theme, and the result was the landmark essay "As We May Think." He used this high-profile forum to propose a solution to what he considered the paramount challenge of the day: how information would be gathered, stored, and accessed in an increasingly information-saturated world. This article had a profound influence on the scientists and theorists responsible for the evolution of the personal computer and the Internet.

Although he addresses this subject from the vantage point of 1940s technology—relying on film processing, microfilm storage, and mechanical retrieval—Bush introduces many of the concepts central to hypermedia. The machine that he proposes, the memex, is a new approach to the storing and sharing of information—a "memory extender" (hence "memex") that could organize diverse materials according to an individual's own personal associations. Conceived as a vast encyclopedia of text, images, and sounds that is able to mimic the mind's capability to link between ideas freely, the memex would effectively remember the leaps of thought someone had while researching a particular topic, and then make that trail of associations available to others. Bush never used the word *hyperlink,* but in this essay he invented the notion. >>

This has not been a scientist's war; it has been a war in which all have had a part. The scientists, burying their old professional competition in the demand of a common cause, have shared greatly and learned much. It has been exhilarating to work in effective partnership. Now, for many, this appears to be approaching an end. What are the scientists to do next?

For the biologists, and particularly for the medical scientists, there can be little indecision, for their war has hardly required them to leave the old paths. Many indeed have been able to carry on their war research in their familiar peacetime laboratories. Their objectives remain much the same.

It is the physicists who have been thrown most violently off stride, who have left academic pursuits for the making of strange destructive gadgets, who have had to devise new methods for their unanticipated assignments. They have done their part on the devices that made it possible to turn back the enemy, have worked in combined effort with the physicists of our allies. They have felt within themselves the stir of achievement. They have been part of a great team. Now, as peace approaches, one asks where they will find objectives worthy of their best.

1

Of what lasting benefit has been man's use of science and of the new instruments which his research brought into existence? First, they have increased his control of his material environment. They have improved his food, his clothing, his shelter; they have increased his security and released him partly from the bondage of bare existence. They have given him increased knowledge of his own biological processes so that he has had a progressive freedom from disease and an increased span of life. They are illuminating the interactions of his physiological and psychological functions, giving the promise of an improved mental health.

Science has provided the swiftest communication between individuals; it has provided a record of ideas and has enabled man to manipulate and to make extracts from that record so that knowledge evolves and endures throughout the life of a race rather than that of an individual.

There is a growing mountain of research. But there is increased evidence that we are being bogged down today as specialization extends. The investigator is staggered by the findings and conclusions of thousands of other workers—conclusions which he cannot find time to grasp, much less to remember, as they appear. Yet specialization becomes increasingly necessary for progress, and the effort to bridge between disciplines is correspondingly superficial.

Professionally our methods of transmitting and reviewing the results of research are generations old and by now are totally inadequate for their purpose. If the aggregate time spent in writing scholarly works and in reading them could be evaluated, the ratio between these amounts of time might well be startling. Those who conscientiously attempt to keep abreast of current thought, even in restricted fields, by close and continuous reading might well shy away from an examination calculated to show how much of the previous month's efforts could be produced on call. Mendel's concept of the laws of genetics was lost to the world for a generation because his publication did not reach the few who were capable of grasping and extending it; and this sort of catastrophe is undoubtedly being repeated all about us, as truly significant attainments become lost in the mass of the inconsequential.

The difficulty seems to be, not so much that we publish unduly in view of the extent and variety of present day interests, but rather that publication has been extended far beyond our present ability to make real use of the record. The summation of human experience is being expanded at a prodigious rate, and the means we use for threading through the consequent maze to the momentarily important item is the same as was used in the days of square-rigged ships.

But there are signs of a change as new and powerful instrumentalities come into use. Photocells capable of seeing things in a physical sense, advanced photography which can record what is seen or even what is not, thermionic tubes capable of controlling potent forces under the guidance of less power than a mosquito uses to vibrate his wings, cathode ray tubes rendering visible an occurrence so brief that by comparison a microsecond is a long time, relay combinations which will carry out involved sequences of movements more reliably than any human operator and thousands of times as fast—there are plenty of mechanical aids with which to effect a transformation in scientific records.

Two centuries ago Leibnitz invented a calculating machine which embodied most of the essential features of recent keyboard devices, but it could not then come into use. The economics of the situation were against it: the labor involved in constructing it, before the days of mass production, exceeded the labor to be saved by its use, since all it could accomplish could be duplicated by sufficient use of pencil and paper. Moreover, it would have been subject to frequent breakdown, so that it could not have been depended upon; for at that time and long after, complexity and unreliability were synonymous.

Babbage, even with remarkably generous support for his time, could not produce his great arithmetical machine. His idea was sound enough, but construction and maintenance costs were then too heavy. Had a Pharaoh been given detailed and explicit designs of an automobile, and had he understood them completely, it would have taxed the resources of his kingdom to have fashioned the thousands of parts for a single car, and that car would have broken down on the first trip to Giza.

Machines with interchangeable parts can now be constructed with great economy of effort. In spite of much complexity, they perform reliably. Witness the humble typewriter, or the movie camera, or the automobile. Electrical contacts have ceased to stick when thoroughly understood. Note the automatic telephone exchange, which has hundreds of thousands of such contacts, and yet is reliable. A spider web of metal, sealed in a thin glass container, a wire heated to brilliant glow, in short, the thermionic tube of radio sets, is made by the hundred million, tossed about in packages, plugged into sockets—and it works! Its gossamer parts, the precise location and alignment involved in its construction, would have occupied a master craftsman of the guild for months; now it is built for thirty cents. The world has arrived at an age of cheap complex devices of great reliability; and something is bound to come of it.

2

A record, if it is to be useful to science, must be continuously extended, it must be stored, and above all it must be consulted. Today we make the record conventionally by writing and photography, followed by printing; but we also record on film, on wax disks, and on magnetic wires. Even if utterly new recording procedures do not appear, these present ones are certainly in the process of modification and extension.

Certainly progress in photography is not going to stop. Faster material and lenses, more automatic cameras, finer-grained sensitive compounds to allow an extension of the minicamera idea, are all imminent. Let us project this trend ahead to a logical, if not inevitable, outcome. The camera hound of the future wears on his forehead a lump a little larger than a walnut. It takes pictures 3 millimeters square, later to be projected or enlarged, which after all involves only a factor of 10 beyond present practice. The lens is of universal focus, down to any distance accommodated by the unaided eye, simply because it is of short focal length. There is a built-in photocell on the walnut such as we now have on at least one camera, which automatically adjusts exposure for a wide range of illumination. There is film in the walnut for a hundred exposures, and the spring for operating its shutter and shifting its film is wound once for all when the film clip is inserted. It produces its result in full color. It may well be stereoscopic, and record with two spaced glass eyes, for striking improvements in stereoscopic technique are just around the corner.

The cord which trips its shutter may reach down a man's sleeve within easy reach of his fingers. A quick squeeze, and the picture is taken. On a pair of ordinary glasses is a square of fine lines near the top of one lens, where it is out of the way of ordinary vision. When an object appears in that square, it is lined up for its picture. As the scientist of the future moves about the laboratory or the field, every time he looks at something worthy of the record, he trips the shutter and in it goes, without even an audible click. Is this all fantastic? The only fantastic thing about it is the idea of making as many pictures as would result from its use.

Will there be dry photography? It is already here in two forms. When Brady made his Civil War pictures, the plate had to be wet at the time of exposure. Now it has to be wet during development instead. In the future perhaps it need not be wetted at all. There have long been films impregnated with diazo dyes which form a picture without development, so that it is already there as soon as the camera has been operated. An exposure to ammonia gas destroys the unexposed dye, and the picture can then be taken out into the light and examined. The process is now slow, but someone may speed it up, and it has no grain difficulties such as now keep photographic researchers busy. Often it would be advantageous to be able to snap the camera and to look at the picture immediately.

Another process now in use is also slow, and more or less clumsy. For fifty years impregnated papers have been used which turn dark at every point where an electrical contact touches them, by reason of the chemical change thus produced in an iodine compound included in the paper. They have been used to make records, for a pointer moving across them can leave a trail behind. If the electrical potential on the pointer is varied as it moves, the line becomes light or dark in accordance with the potential.

This scheme is now used in facsimile transmission. The pointer draws a set of closely spaced lines across the paper one after another. As it moves, its potential is varied in accordance with a varying current received over wires from a distant station, where these variations are produced by a photocell which is similarly scanning a picture. At every instant the darkness of the line being drawn is made equal to the darkness of the point on the picture being observed by the photocell. Thus, when the whole picture has been covered, a replica appears at the receiving end.

A scene itself can be just as well looked over line by line by the photocell in this way as can a photograph of the scene. This whole apparatus constitutes a camera, with the added feature, which can be dispensed with if desired, of making its picture at a distance. It is slow, and the picture is poor in detail. Still, it does give another process of dry photography, in which the picture is finished as soon as it is taken.

It would be a brave man who would predict that such a process will always remain clumsy, slow, and faulty in detail. Television equipment today transmits sixteen reasonably good pictures a second, and it involves only two essential differences from the process described above. For one, the record is made by a moving beam of electrons rather than a moving pointer, for the reason that an electron beam can sweep across the picture very rapidly indeed. The other difference involves merely the use of a screen which glows momentarily when the electrons hit, rather than a chemically treated paper or film which is permanently altered. This speed is necessary in television, for motion pictures rather than stills are the object.

Use chemically treated film in place of the glowing screen, allow the apparatus to transmit one picture only rather than a succession, and a rapid camera for dry photography results. The treated film needs to be far faster in action than present examples, but it probably could be. More serious is the objection that this scheme would involve putting the film inside a vacuum chamber, for electron beams behave normally only in such a rarefied environment. This difficulty could be avoided by allowing the electron beam to play on one side of a partition, and by pressing the film against the other side, if this partition were such as to allow the electrons to go through perpendicular to its surface, and to prevent them from spreading out sideways. Such partitions, in crude form, could certainly be constructed, and they will hardly hold up the general development.

Like dry photography, microphotography still has a long way to go. The basic scheme of reducing the size of the record, and examining it by projection rather than directly, has possibilities too great to be ignored. The combination of optical projection and photographic reduction is already producing some results in microfilm for scholarly purposes, and the potentialities are highly suggestive. Today, with microfilm, reductions by a linear factor of 20 can be employed and still produce full clarity when the material is re-enlarged for examination. The limits are set by the graininess of the film, the excellence of the optical system, and the efficiency of the light sources employed. All of these are rapidly improving.

Assume a linear ratio of 100 for future use. Consider film of the same thickness as paper, although thinner film will certainly be usable. Even under these conditions there would be a total factor of 10,000 between the bulk of the ordinary record on books and its microfilm replica. The *Encyclopaedia Britannica* could be reduced to the volume of a matchbox. A library of a million volumes could be compressed into one end of a desk. If the human race has produced since the invention of movable type a total record, in the form of magazines, newspapers, books, tracts, advertising blurbs, correspondence, having a volume corresponding to a billion books, the whole affair, assembled and compressed, could be lugged off in a moving van. Mere compression, of course, is not enough; one needs not only to make and store a record but also be able to consult it, and this aspect of the matter comes later. Even the modern great library is not generally consulted; it is nibbled at by a few.

Compression is important, however, when it comes to costs. The material for the microfilm *Britannica* would cost a nickel, and it could be mailed anywhere for a cent. What would it cost to print a million copies? To print a sheet of newspaper, in a large edition, costs a small fraction of a cent. The entire material of the *Britannica* in reduced microfilm form would go on a sheet eight and one-half by eleven inches. Once it is available, with the photographic reproduction methods of the future, duplicates in large quantities could probably be turned out for a cent apiece beyond the cost of materials. The preparation of the original copy? That introduces the next aspect of the subject.

3

To make the record, we now push a pencil or tap a typewriter. Then comes the process of digestion and correction, followed by an intricate process of typesetting, printing, and distribution. To consider the first stage of the procedure, will the author of the future cease writing by hand or typewriter and talk directly to the record? He does so indirectly, by talking to a stenographer or a wax cylinder; but the elements are all present if he wishes to have his talk directly produce a typed record. All he needs to do is to take advantage of existing mechanisms and to alter his language.

At a recent World Fair a machine called a Voder was shown. A girl stroked its keys and it emitted recognizable speech. No human vocal chords entered into the procedure at any point; the keys simply combined some electrically produced vibrations and passed these on to a loud-speaker. In the Bell Laboratories there is the converse of this machine, called a Vocoder. The loudspeaker is replaced by a microphone, which picks up sound. Speak to it, and the corresponding keys move. This may be one element of the postulated system.

The other element is found in the stenotype, that somewhat disconcerting device encountered usually at public meetings. A girl strokes its keys languidly and looks about the room and sometimes at the speaker with a disquieting gaze. From it emerges a typed strip which records in a phonetically simplified language a record of what the speaker is supposed to have said. Later this strip is retyped into ordinary language, for in its nascent form it is intelligible only to the initiated. Combine these two elements, let the Vocoder run the stenotype, and the result is a machine which types when talked to.

Our present languages are not especially adapted to this sort of mechanization, it is true. It is strange that the inventors of universal languages have not seized upon the idea of producing one which better fitted the technique for transmitting and recording speech. Mechanization may yet force the issue, especially in the scientific field; whereupon scientific jargon would become still less intelligible to the layman.

One can now picture a future investigator in his laboratory. His hands are free, and he is not anchored. As he moves about and observes, he photographs and comments. Time is automatically recorded to tie the two records together. If he goes into the field, he may be connected by radio to his recorder. As he ponders over his notes in the evening, he again talks his comments into the record. His typed record, as well as his photographs, may both be in miniature, so that he projects them for examination.

Much needs to occur, however, between the collection of data and observations, the extraction of parallel material from the existing record, and the final insertion of new material into the general body of the common record. For mature thought there is no mechanical substitute. But creative thought and essentially repetitive thought are very different things. For the latter there are, and may be, powerful mechanical aids.

Adding a column of figures is a repetitive thought process, and it was long ago properly relegated to the machine. True, the machine is sometimes controlled by a keyboard, and thought of a sort enters in reading the figures and poking the corresponding keys, but even this is avoidable. Machines have been made which will read typed figures by photocells and then depress the corresponding keys; these are combinations of photocells for scanning the type, electric circuits for sorting the consequent variations, and relay circuits for interpreting the result into the action of solenoids to pull the keys down.

All this complication is needed because of the clumsy way in which we have learned to write figures. If we recorded them positionally, simply by the configuration of a set of dots on a card, the automatic reading mechanism would become comparatively simple. In fact if the dots are holes, we have the punched-card machine long ago produced by Hollorith for the purposes of the census, and now used throughout business. Some types of complex businesses could hardly operate without these machines.

Adding is only one operation. To perform arithmetical computation involves also subtraction, multiplication, and division, and in addition some method for temporary storage of results, removal from storage for further manipulation, and recording of final results by printing. Machines for these purposes are now of two types: keyboard machines for accounting and the like, manually controlled for the insertion of data, and usually automatically controlled as far as the sequence of operations is concerned; and punched-card machines in which separate operations are usually delegated to a series of machines, and the cards then transferred bodily from one to another. Both forms are very useful; but as far as complex computations are concerned, both are still in embryo.

Rapid electrical counting appeared soon after the physicists found it desirable to count cosmic rays. For their own purposes the physicists promptly constructed thermionic-tube equipment capable of counting electrical impulses at the rate of 100,000 a second. The advanced arithmetical machines of the future will be electrical in nature, and they will perform at 100 times present speeds, or more.

Moreover, they will be far more versatile than present commercial machines, so that they may readily be adapted for a wide variety of operations. They will be controlled by a control card or film, they will select their own data and manipulate it in accordance with the instructions thus inserted, they will perform complex arithmetical computations at exceedingly high speeds, and they will record results in such form as to be readily available for distribution or for later further manipulation. Such machines will have enormous appetites. One of them will take instructions and data from a whole roomful of girls armed with simple keyboard punches, and will deliver sheets of computed results every few minutes. There will always be plenty of things to compute in the detailed affairs of millions of people doing complicated things.

4

The repetitive processes of thought are not confined, however, to matters of arithmetic and statistics. In fact, every time one combines and records facts in accordance with established logical processes, the creative aspect of thinking is concerned only with the selection of the data and the process to be employed and the manipulation thereafter is repetitive in nature and hence a fit matter to be relegated to the machine. Not so much has been done along these lines, beyond the bounds of arithmetic, as might be done, primarily because of the economics of the situation. The needs of business and the extensive market obviously waiting assured the advent of mass-produced arithmetical machines just as soon as production methods were sufficiently advanced.

With machines for advanced analysis no such situation existed; for there was and is no extensive market; the users of advanced methods of manipulating data are a very small part of the population. There are, however, machines for solving differential equations—and functional and integral equations, for that matter. There are many special machines, such as the harmonic synthesizer which predicts the tides. There will be many more, appearing certainly first in the hands of the scientist and in small numbers.

If scientific reasoning were limited to the logical processes of arithmetic, we should not get far in our understanding of the physical world. One might as well attempt to grasp the game of poker entirely by the use of the mathematics of probability. The abacus, with its beads strung on parallel wires, led the Arabs to positional numeration and the concept of zero many centuries before the rest of the world; and it was a useful tool—so useful that it still exists.

It is a far cry from the abacus to the modern keyboard accounting machine. It will be an equal step to the arithmetical machine of the future. But even this new machine will not take the scientist where he needs to go. Relief must be secured from laborious detailed manipulation of higher mathematics as well, if the users of it are to free their brains for something more than repetitive detailed transformations in accordance with established rules. A mathematician is not a man who can readily manipulate figures; often he cannot. He is not even a man who can readily perform the transformations of equations by the use of calculus. He is primarily an individual who is skilled in the use of symbolic logic on a high plane, and especially he is a man of intuitive judgment in the choice of the manipulative processes he employs.

All else he should be able to turn over to his mechanism, just as confidently as he turns over the propelling of his car to the intricate mechanism under the hood. Only then will mathematics be practically effective in bringing the growing knowledge of atomistics to the useful solution of the advanced problems of chemistry, metallurgy, and biology. For this reason there still come more machines to handle advanced mathematics for the scientist. Some of them will be sufficiently bizarre to suit the most fastidious connoisseur of the present artifacts of civilization.

5

The scientist, however, is not the only person who manipulates data and examines the world about him by the use of logical processes, although he sometimes preserves this appearance by adopting into the fold anyone who becomes logical, much in the manner in which a British labor leader is elevated to knighthood. Whenever logical processes of thought are employed—that is, whenever thought for a time runs along an accepted groove—there is an opportunity for the machine. Formal logic used to be a keen instrument in the hands of the teacher in his trying of students' souls. It is readily possible to construct a machine which will manipulate premises in accordance with formal logic, simply by the clever use of relay circuits. Put a set of premises into such a device and turn the crank, and it will readily pass out conclusion after conclusion, all in accordance with logical law, and with no more slips than would be expected of a keyboard adding machine.

Logic can become enormously difficult, and it would undoubtedly be well to produce more assurance in its use. The machines for higher analysis have usually been equation solvers. Ideas are beginning to appear for equation transformers, which will rearrange the relationship expressed by an equation in accordance with strict and rather advanced logic. Progress is inhibited by the exceedingly crude way in which mathematicians express their relationships. They employ a symbolism which grew like Topsy and has little consistency; a strange fact in that most logical field.

A new symbolism, probably positional, must apparently precede the reduction of mathematical transformations to machine processes. Then, on beyond the strict logic of the mathematician, lies the application of logic in everyday affairs. We may some day click off arguments on a machine with the same assurance that we now enter sales on a cash register. But the machine of logic will not look like a cash register, even of the streamlined model.

So much for the manipulation of ideas and their insertion into the record. Thus far we seem to be worse off than before—for we can enormously extend the record; yet even in its present bulk we can hardly consult it. This is a much larger matter than merely the extraction of data for the purposes of scientific research; it involves the entire process by which man profits by his inheritance of acquired knowledge. The prime action of use is selection, and here we are halting indeed. There may be millions of fine thoughts, and the account of the experience on which they are based, all encased within stone walls of acceptable architectural form; but if the scholar can get at only one a week by diligent search, his syntheses are not likely to keep up with the current scene.

Selection, in this broad sense, is a stone adze in the hands of a cabinetmaker. Yet, in a narrow sense and in other areas, something has already been done mechanically on selection. The personnel officer of a factory drops a stack of a few thousand employee cards into a selecting machine, sets a code in accordance with an established convention, and produces in a short time a list of all employees who live in Trenton and know Spanish. Even such devices are much too slow when it comes, for example, to matching a set of fingerprints with one of five million on file. Selection devices of this sort will soon be speeded up from their present rate of reviewing data at a few hundred a minute. By the use of photocells and microfilm they will survey items at the rate of a thousand a second, and will print out duplicates of those selected.

This process, however, is simple selection: it proceeds by examining in turn every one of a large set of items, and by picking out those which have certain specified characteristics. There is another form of selection best illustrated by the automatic telephone exchange. You dial a number and the machine selects and connects just one of a million possible stations. It does not run over them all. It pays attention only to a class given by a first digit, then only to a subclass of this given by the second digit, and so on; and thus proceeds rapidly and almost unerringly to the selected station. It requires a few seconds to make the selection, although the process could be speeded up if increased speed were economically warranted. If necessary, it could be made extremely fast by substituting thermionic-tube switching for mechanical switching, so that the full selection could be made in one one-hundredth of a second. No one would wish to spend the money necessary to make this change in the telephone system, but the general idea is applicable elsewhere.

Take the prosaic problem of the great department store. Every time a charge sale is made, there are a number of things to be done. The inventory needs to be revised, the salesman needs to be given credit for the sale, the general accounts need an entry, and, most important, the customer needs to be charged. A central records device has been developed in which much of this work is done conveniently. The salesman places on a stand the customer's identification card, his own card, and the card taken from the article sold—all punched cards. When he pulls a lever, contacts are made through the holes, machinery at a central point makes the necessary computations and entries, and the proper receipt is printed for the salesman to pass to the customer.

But there may be ten thousand charge customers doing business with the store, and before the full operation can be completed someone has to select the right card and insert it at the central office. Now rapid selection can slide just the proper card into position in an instant or two, and return it afterward. Another difficulty occurs, however. Someone must read a total on the card, so that the machine can add its computed item to it. Conceivably the cards might be of the dry photography type I have described. Existing totals could then be read by photocell, and the new total entered by an electron beam.

The cards may be in miniature, so that they occupy little space. They must move quickly. They need not be transferred far, but merely into position so that the photocell and recorder can operate on them. Positional dots can enter the data. At the end of the month a machine can readily be made to read these and to print an ordinary bill. With tube selection, in which no mechanical parts are involved in the switches, little time need be occupied in bringing the correct card into use—a second should suffice for the entire operation. The whole record on the card may be made by magnetic dots on a steel sheet if desired, instead of dots to be observed optically, following the scheme by which Poulsen long ago put speech on a magnetic wire. This method has the advantage of simplicity and ease of erasure. By using photography, however, one can arrange to project the record in enlarged form and at a distance by using the process common in television equipment.

One can consider rapid selection of this form, and distant projection for other purposes. To be able to key one sheet of a million before an operator in a second or two, with the possibility of then adding notes thereto, is suggestive in many ways. It might even be of use in libraries, but that is another story. At any rate, there are now some interesting combinations possible. One might, for example, speak to a microphone, in the manner described in connection with the speech-controlled typewriter, and thus make his selections. It would certainly beat the usual file clerk.

6

The real heart of the matter of selection, however, goes deeper than a lag in the adoption of mechanisms by libraries, or a lack of development of devices for their use. Our ineptitude in getting at the record is largely caused by the artificiality of systems of indexing. When data of any sort are placed in storage, they are filed alphabetically or numerically, and information is found (when it is) by tracing it down from subclass to subclass. It can be in only one place, unless duplicates are used; one has to have rules as to which path will locate it, and the rules are cumbersome. Having found one item, moreover, one has to emerge from the system and re-enter on a new path.

The human mind does not work that way. It operates by association. With one item in its grasp, it snaps instantly to the next that is suggested by the association of thoughts, in accordance with some intricate web of trails carried by the cells of the brain. It has other characteristics, of course; trails that are not frequently followed are prone to fade, items are not fully permanent, memory is transitory. Yet the speed of action, the intricacy of trails, the detail of mental pictures, is awe-inspiring beyond all else in nature.

Man cannot hope fully to duplicate this mental process artificially, but he certainly ought to be able to learn from it. In minor ways he may even improve, for his records have relative permanency. The first idea, however, to be drawn from the analogy concerns selection. Selection by association, rather than indexing, may yet be mechanized. One cannot hope thus to equal the speed and flexibility with which the mind follows an associative trail, but it should be possible to beat the mind decisively in regard to the permanence and clarity of the items resurrected from storage.

Consider a future device for individual use, which is a sort of mechanized private file and library. It needs a name, and, to coin one at random, "memex" will do. A memex is a device in which an individual stores all his books, records, and communications, and which is mechanized so that it may be consulted with exceeding speed and flexibility. It is an enlarged intimate supplement to his memory.

It consists of a desk, and while it can presumably be operated from a distance, it is primarily the piece of furniture at which he works. On the top are slanting translucent screens, on which material can be projected for convenient reading. There is a keyboard, and sets of buttons and levers. Otherwise it looks like an ordinary desk.

In one end is the stored material. The matter of bulk is well taken care of by improved microfilm. Only a small part of the interior of the memex is devoted to storage, the rest to mechanism. Yet if the user inserted 5,000 pages of material a day it would take him hundreds of years to fill the repository, so he can be profligate and enter material freely.

Most of the memex contents are purchased on microfilm ready for insertion. Books of all sorts, pictures, current periodicals, newspapers, are thus obtained and dropped into place. Business correspondence takes the same path. And there is provision for direct entry. On the top of the memex is a transparent platen. On this are placed longhand notes, photographs, memoranda, all sorts of things. When one is in place, the depression of a lever causes it to be photographed onto the next blank space in a section of the memex film, dry photography being employed.

There is, of course, provision for consultation of the record by the usual scheme of indexing. If the user wishes to consult a certain book, he taps its code on the keyboard, and the title page of the book promptly appears before him, projected onto one of his viewing positions. Frequently used codes are mnemonic, so that he seldom consults his code book; but when he does, a single tap of a key projects it for his use. Moreover, he has supplemental levers. On deflecting one of these levers to the right he runs through the book before him, each page in turn being projected at a speed which just allows a recognizing glance at each. If he deflects it further to the right, he steps through the book 10 pages at a time; still further at 100 pages at a time. Deflection to the left gives him the same control backwards.

A special button transfers him immediately to the first page of the index. Any given book of his library can thus be called up and consulted with far greater facility than if it were taken from a shelf. As he has several projection positions, he can leave one item in position while he calls up another. He can add marginal notes and comments, taking advantage of one possible type of dry photography, and it could even be arranged so that he can do this by a stylus scheme, such as is now employed in the telautograph seen in railroad waiting rooms, just as though he had the physical page before him.

7

All this is conventional, except for the projection forward of present-day mechanisms and gadgetry. It affords an immediate step, however, to associative indexing, the basic idea of which is a provision whereby any item may be caused at will to select immediately and automatically another. This is the essential feature of the memex. The process of tying two items together is the important thing.

When the user is building a trail, he names it, inserts the name in his code book, and taps it out on his keyboard. Before him are the two items to be joined, projected onto adjacent viewing positions. At the bottom of each there are a number of blank code spaces, and a pointer is set to indicate one of these on each item. The user taps a single key, and the items are permanently joined. In each code space appears the code word. Out of view, but also in the code space, is inserted a set of dots for photocell viewing; and on each item these dots by their positions designate the index number of the other item.

Thereafter, at any time, when one of these items is in view, the other can be instantly recalled merely by tapping a button below the corresponding code space. Moreover, when numerous items have been thus joined together to form a trail, they can be reviewed in turn, rapidly or slowly, by deflecting a lever like that used for turning the pages of a book. It is exactly as though the physical items had been gathered together from widely separated sources and bound together to form a new book. It is more than this, for any item can be joined into numerous trails.

The owner of the memex, let us say, is interested in the origin and properties of the bow and arrow. Specifically he is studying why the short Turkish bow was apparently superior to the English long bow in the skirmishes of the Crusades. He has dozens of possibly pertinent books and articles in his memex. First he runs through an encyclopedia, finds an interesting but sketchy article, leaves it projected. Next, in a history, he finds another pertinent item, and ties the two together. Thus he goes, building a trail of many items. Occasionally he inserts a comment of his own, either linking it into the main trail or joining it by a side trail to a particular item. When it becomes evident that the elastic properties of available materials had a great deal to do with the bow, he branches off on a side trail which takes him through textbooks on elasticity and tables of physical constants. He inserts a page of longhand analysis of his own. Thus he builds a trail of his interest through the maze of materials available to him.

And his trails do not fade. Several years later, his talk with a friend turns to the queer ways in which a people resist innovations, even of vital interest. He has an example, in the fact that the outraged Europeans still failed to adopt the Turkish bow. In fact he has a trail on it. A touch brings up the code book. Tapping a few keys projects the head of the trail. A lever runs through it at will, stopping at interesting items, going off on side excursions. It is an interesting trail, pertinent to the discussion. So he sets a reproducer in action, photographs the whole trail out, and passes it to his friend for insertion in his own memex, there to be linked into the more general trail.

8

Wholly new forms of encyclopedias will appear, ready made with a mesh of associative trails running through them, ready to be dropped into the memex and there amplified. The lawyer has at his touch the associated opinions and decisions of his whole experience, and of the experience of friends and authorities. The patent attorney has on call the millions of issued patents, with familiar trails to every point of his client's interest. The physician, puzzled by a patient's reactions, strikes the trail established in studying an earlier similar case, and runs rapidly through analogous case histories, with side references to the classics for the pertinent anatomy and histology. The chemist, struggling with the synthesis of an organic compound, has all the chemical literature before him in his laboratory, with trails following the analogies of compounds, and side trails to their physical and chemical behavior.

The historian, with a vast chronological account of a people, parallels it with a skip trail which stops only on the salient items, and can follow at any time contemporary trails which lead him all over civilization at a particular epoch. There is a new profession of trail blazers, those who find delight in the task of establishing useful trails through the enormous mass of the common record. The inheritance from the master becomes not only his additions to the world's record, but for his disciples the entire scaffolding by which they were erected.

Thus science may implement the ways in which man produces, stores, and consults the record of the race. It might be striking to outline the instrumentalities of the future more spectacularly, rather than to stick closely to methods and elements now known and undergoing rapid development, as has been done here. Technical difficulties of all sorts have been ignored, certainly, but also ignored are means as yet unknown which may come any day to accelerate technical progress as violently as did the advent of the thermionic tube. In order that the picture may not be too commonplace, by reason of sticking to present-day patterns, it may be well to mention one such possibility, not to prophecy but merely to suggest, for prophecy based on extension of the known has substance, while prophecy founded on the unknown is only a doubly involved guess. All our steps in creating or absorbing material of the record proceed through one of the senses—the tactile when we touch keys, the oral when we speak or listen, the visual when we read. Is it not possible that some day the path may be established more directly?

We know that when the eye sees, all the consequent information is transmitted to the brain by means of electrical vibrations in the channel of the optic nerve. This is an exact analogy with the electrical vibrations which occur in the cable of a television set: they convey the picture from the photocells which see it to the radio transmitter from which it is broadcast. We know further that if we can approach that cable with the proper instruments, we do not need to touch it; we can pick up those vibrations by electrical induction and thus discover and reproduce the scene which is being transmitted, just as a telephone wire may be tapped for its message.

The impulses which flow in the arm nerves of a typist convey to her fingers the translated information which reaches her eye or ear, in order that the fingers may be caused to strike the proper keys. Might not these currents be intercepted, either in the original form in which information is conveyed to the brain, or in the marvelously metamorphosed form in which they then proceed to the hand?

By bone conduction we already introduce sounds: into the nerve channels of the deaf in order that they may hear. Is it not possible that we may learn to introduce them without the present cumbersomeness of first transforming electrical vibrations to mechanical ones, which the human mechanism promptly transforms back to the electrical form? With a couple of electrodes on the skull the encephalograph now produces pen-and-ink traces which bear some relation to the electrical phenomena going on in the brain itself. True, the record is unintelligible, except as it points out certain gross misfunctioning of the cerebral mechanism; but who would now place bounds on where such a thing may lead?

In the outside world, all forms of intelligence, whether of sound or sight, have been reduced to the form of varying currents in an electric circuit in order that they may be transmitted. Inside the human frame exactly the same sort of process occurs. Must we always transform to mechanical movements in order to proceed from one electrical phenomenon to another? It is a suggestive thought, but it hardly warrants prediction without losing touch with reality and immediateness.

Presumably man's spirit should be elevated if he can better review his shady past and analyze more completely and objectively his present problems. He has built a civilization so complex that he needs to mechanize his records more fully if he is to push his experiment to its logical conclusion and not merely become bogged down part way there by overtaxing his limited memory. His excursions may be more enjoyable if he can reacquire the privilege of forgetting the manifold things he does not need to have immediately at hand, with some assurance that he can find them again if they prove important.

The applications of science have built man a well-supplied house, and are teaching him to live healthily therein. They have enabled him to throw masses of people against one another with cruel weapons. They may yet allow him truly to encompass the great record and to grow in the wisdom of race experience. He may perish in conflict before he learns to wield that record for his true good. Yet, in the application of science to the needs and desires of man, it would seem to be a singularly unfortunate stage at which to terminate the process, or to lose hope as to the outcome.

Ted Nelson

<<15>> *Computer Lib/ Dream Machines* (1974)

Ted Nelson, Computer Lib/Dream Machines.
© Ted Nelson. Courtesy of Ted Nelson.

"Everything is deeply intertwingled."

<< Ted Nelson describes himself as a person who invents paradigms and then makes up words to express them. Nelson had little formal training in computer science. However, as a graduate student in philosophy in the late 1950s and early 1960s, Nelson had two critical intellectual encounters that led him to become one of the most influential figures in computing. One was with Vannevar Bush's article "As We May Think," which convinced Nelson that emerging information technologies could extend the power of human memory. The second was with Samuel Taylor Coleridge's poem "Xanadu," "a magic place of literary memory," in Nelson's words, that provided him with the image of a vast storehouse of memories, and which served as the inspiration for his life's work. From these influences, Nelson began his quest to build creative tools that would transform the way we read and write, and in 1963 he coined the words *hypertext* and *hypermedia* to describe the new paradigms that these tools would make possible.

In 1974, he self-published his landmark tome on personal computing, *Computer Lib/Dream Machines,* which instantly drew an underground following of computer hackers, media theorists, and experimental artists. The oversized volume had a playful, do-it-yourself quality, with typesetting done by typewriter, distinctive hand-drawn illustrations, and a layout that suggested the nonlinear nature of hypermedia. Written with the author's trademark droll humor, the book is a stream-of-consciousness, randomly generated sequence of bombastic truths and speculations, a patchworklike journal that covers everything from hardware systems and new art forms to the decline of western civilization. Nelson was particularly concerned with the complex nature of the creative impulse, and he saw the computer as the tool that would make explicit the interdependence of ideas, drawing out connections between literature, art, music, and science, since, as he put it, everything is "deeply intertwingled."

Nelson's critical breakthrough was to call for a system of nonsequential writing that would allow the reader to aggregate meaning in snippets ("lexias," as Roland Barthes defined them), in the order of his or her choosing, rather than according to a preestablished structure fixed by the author. Nelson's vision for a new form of interactive literature was a hybrid concept that married Vannevar Bush's notion of a memory bank organized around associative links with Douglas Engelbart's research in human-computer interface and the augmentation of human intellect. It was from this conceptual framework that he drew his ambitious plans for a global repository of cultural information, Xanadu. Intended to become the ideal new publishing system for the digital age, this visionary design was later eclipsed by Tim Berners-Lee's invention of the World Wide Web in 1989. >>

The Tissue of Thought

Uneducated people typically think of education as the learning of a lot of facts and skills. While facts and skills certainly have their merits, "higher education" is also largely concerned with tying ideas together, and especially alternative structures of such tying-together: with showing you the vast uncertainties of things.

A wonderful Japanese film of the fifties was called *Rasho-Mon.* It depicted a specific event—a rape—as told by five different people. As the audience watches the five separate stories, they must try to judge what *really* happened.

The Rasho-Mon Principle: everything is like that. The complete truth about something is never known.

Nobody tells the complete truth, though some try. Nobody knows the complete truth. Nowhere may we find *printed* the complete truth. There are only different views, assertions, supposed facts that support one view or another but are disputed by disbelievers in the particular views; and so on. There are "agreed-on facts," but their *meaning* is often in doubt.

The great compromise of the western world is that we go by the rule: *assume* that we never know the final truth about anything. There are continuing Issues, Mysteries, Continuing Dialogues. What about flying saucers, "why Rome fell," was there a Passover Plot, and Did Roosevelt know Pearl Harbor would be attacked?

Outsiders find the intellectual world pompous, vague in its undecided issues, stuffy in its quotes and citations. But in a way these are the sounds of battle. The clash of theories is what many find exhilarating about the intellectual world. The Scholarly Arena is simply a Circus Maximus in which these battles are scheduled.

Many people think "science" is free from all this. These are people who do not know much about science. More and more is scientifically known, true; but it is repeatedly discovered that some scientific "knowledge" is *un*true, and this problem is built into the system. The important thing about science is not that everything will be known, or that everything unanimously believed by scientists is necessarily true, but that science contains a *system* for seeking untruth and purging it.

This is the great tradition of western civilization. The Western World is, in an important sense, a continuing dialogue among people who have thought different things. "Scholarship" is the tradition of trying to improve, collate and resolve uncertainties. The fundamental ground rules are that no issue is ever closed, no interconnection is impossible. It all comes down to *what is written,* because the thoughts and minds themselves, of course, do not last. (The apparatus of citation and footnote are simply a combination of hat-tipping, go-look-if-you-don't-believe-me, and you-might-want-to-read-this-yourself.)

"Knowledge," then—and indeed most of our civilization and what remains of those previous—is a vasty cross-tangle of ideas and evidential materials, not a pyramid of truth. So that preserving its structure, and improving its accessibility, is important to us all.

Which is one reason we need hypertexts and thinkertoys.

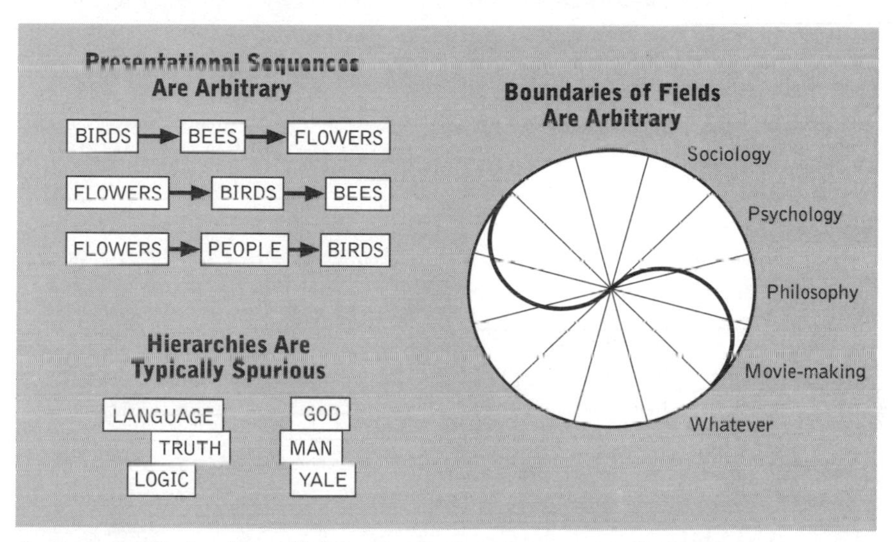

Compartmentalized and Stratified Teaching Produces Compartmentalized and Stratified Minds.

The Heritage

The past is like the receding view out the back of an automobile: the most recent is more conspicuous, and everything seems eventually to be lost.

We know we could save things, but what? Those with the *job* of saving things—the libraries and museums—save so many of the wrong things, the fashionable and expensive and high-toned things esteemed by a given time, and most of the rest slips past. Each generation seems to ridicule the things held in esteem by times before, but of course this can never be a guide to what *should* be saved. And there is so much to save: music, writing, sinking Venice, vanishing species.

But why should things be saved? Everything is deeply intertwingled. We save for knowledge and nostalgia, but what we thought was knowledge often turns to nostalgia, and nostalgia often brings us deeper insights that cut across our lives and very selves.

Computers offer an interesting daydream: that we may be able to store things *digitally* instead of *physically*. In other words, turn the libraries to digital storage; digitize paintings and photographs; even digitize the genetic codes of animals, so that species can be restored at future dates.

Digital storage possesses several special advantages. Digitally stored materials may be copied by automatic means; corrective measures are possible, to prevent errors from creeping in—i.e., "no deterioration" in principle; and they could be kept in various places, lessening mankind's dependence on its eggs being all in one basket (like the Library at Alexandria, whose burning during the occupation of Julius Caesar was one of the greatest losses in human history).

But this would of course require far more compact and reliable forms of digital storage than exist right now.

Nevertheless, we better start thinking about it. Those who fear a coming holocaust had best think about pulling some part of mankind through, with some part of what he used to have.

Hypermedia

In recent years a very basic change has occurred in presentational systems of all kinds. We may summarize it under the name *branching,* although there are many variants. Essentially, today's systems for presenting pictures, texts and whatnot can bring you different things automatically depending on what you do. Selection of this type is generally called *branching.* (I have suggested the generic term *hypermedia* for presentational media which perform in this (and other) multidimensional ways.)

A number of branching media exist or are possible.

Branching *movies* or hyperfilms (see nearby).

Branching *texts* or hypertexts (see nearby).

Branching audio, music, etc.
Branching slide-shows.
Wish we could get into some of that stuff here.

Hypertext

By "hypertext" I mean non-sequential writing.

Ordinary writing is sequential for two reasons. First, it grew out of speech and speech-making, which have to be sequential; and second, because books are not convenient to read except in a sequence.

But *the structures of ideas* are not sequential. They tie together every which-way. And when we write, we are always trying to tie things together in non-sequential ways. The *footnote* is a break from sequence; but it cannot really be extended (though some, like Will Cuppy, have toyed with the technique).

I have run into perhaps a dozen people who understood this instantly when I talked to them about it. Most people, however, act more bemused, thinking I'm trying to tell them something technical or pointlessly philosophical. It's not pointless at all: the point is, writers do better if they don't have to *write* in sequence (but may create multiple structures, branches and alternatives), and readers do better if they don't have to *read* in sequence, but may establish impressions, jump around, and try different pathways until they find the ones they want to study most closely.

(The astute reader, and anybody who's gotten to this point must be, will have noticed that this book is in "magazine" layout, organized visually by ideas and meanings, for that precise reason. I will be interested to hear whether that has worked.)

And the pity of it is that (like the man in the French play who was surprised to learn that he had been "speaking prose all his life and never known it"), we've been speaking *hypertext* all our lives and never known it.

Now, many writers have *tried* to break away from sequence. I think of Nabokov's *Pale Fire,* of *Tristram Shandy* and an odd novel of Julio Cortázar called *Hopscotch,* made up of sections ending with numbers telling you where you can branch to. There are many more; and large books generally use many tricks to get around the problem of indexing and reviewing what has and hasn't been said or done already.

However, in my view, a new day is dawning. Computer storage and screen display mean that we no longer *have* to have things in sequence; totally arbitrary structures are possible, and I think that after we've tried them enough people will see how desirable they are.

Types of Hypertext

Let's assume that you have a high-power display—and storage displays *won't do,* because you have to see things *move* in order to understand where they come from and what they mean. (Especially text.) So it has to be a refreshed CRT.

Basic or chunk style hypertext offers choices, either as footnote-markers (like asterisks) or labels at the end of a chunk. Whatever you point at then comes to the screen.

Collateral hypertext means compound annotations or parallel text.

Stretchtext changes *continuously.* This requires very unusual techniques, but exemplifies how "continuous" hypertext might work.

Ideally, chunk and continuous and collateral hypertext could all be combined (and in turn collaterally linked).

A "fresh" or "specific" hypertext—I don't have a better term at the moment—would consist of material especially written for some purpose. An *anthological* hypertext, however, would consist of materials brought together from all over, like an anthological book.

A *grand hypertext,* then, folks, would be a hypertext consisting of "everything" written about a subject, or vaguely relevant to it, tied together by editors (and NOT by "programmers," dammit), in which you may read *in all the directions you wish to pursue.* There can be alternative pathways for people who think different ways. People who *have* to have one thing explained to them at a time—many have insisted to me that this is normal, although I contend that it is a pathological condition—may have that; others, learning like true human beings, may gather and sift impressions until the ideas become clear.

And then, of course, you see the real dream.

The real dream is for "everything" to be in the hypertext.

Everything you read, you read from the screen (and can always get back to right away); everything you write, you write at the screen (and can cross-link to whatever you read).

Paper moulders. Microfilm is inconvenient. In the best libraries it takes at least minutes to get a particular thing. But as to linking them together—footnoting Aeschylus with Marcus Aurelius, linking genetic data to 15th-century accounts of Indian tribes—well, you can only do it on paper by writing something *new* that ties them together. Isn't that ridiculous? When you could do it all electronically in seconds?

Now that we *have* all these wonderful devices, it should be the goal of society to put them in the service of truth and learning. And this is the way I propose. *Not* through obscure forms of "information retrieval"; *not* through newly oppressive forms of "computer-assisted instruction;" and *not* through a purported science of "artificial intelligence" that will create new personalisms to irk us. All these obstructive oddities, I think, have developed as separate ideals because of the grand preposterosity of Professionalism that has created a world-wide cult of mutual incomprehensibility and disconnected special goals. Now we need to get everybody together again. We want to go back to the roots of our civilization—the ability, which we once had, for everybody who could read to be able to read everything. We *must* once again become a community of common access to a shared heritage.

This was of course what Vannevar Bush said in 1945, in an article everybody cites but nobody reads.

The hypertext solution in many ways obviates some of these other approaches, and in addition retains and puts back together the great traditions of literature and scholarship, traditions based on the fact that dividing things up arbitrarily just generally doesn't work.

The Burning Bush

In fact hypertexts were foreseen very clearly in 1945 by Vannevar Bush, Roosevelt's science advisor. When the war was in the bag, he published a little article on various groovy things that had become possible by that time.

"As We May Think" (*Atlantic Monthly*, July 1945) is most notable for its clear description of various hypertext techniques—that is, linkages between documents which may be brought rapidly to the screen according to their linkages. (So what if he thought they'd be on microfilm.)

How characteristic of Professionalism. Bush's article has been taken as the starting point for the field of Information Retrieval, but its actual contents have been ignored by acclamation. Information Retrieval folk have mostly done very different things, yet thought they were in the tradition.

Now people are "rediscovering" the article. If there's another edition of this book I hope I can run it in entirety.

Doug Engelbart and "The Augmentation of Intellect"

Douglas Engelbart is a saintly man at Stanford Research Institute whose dream has been to make people smarter and bring them together. His system, on which millions of dollars have been spent, is a wonder and a glory.

He began as an engineer of CRTs; but his driving thought was, quite correctly, that these remarkable objects could be used to expand man's mind and improve each shining hour.

Doug Engelbart's vision has never been restricted to narrow technical issues. From the beginning his concern was not merely to plank people down at display consoles, but in the most profound sense to expand man's mind. "The Augmentation of Human Intellect," he calls it, by which he means making minds work better by giving them better tools to work with.

An obvious example is writing: before people could write things down, men could only learn what they experienced or were told by others in person; writing changed all that. Within the computer-screen fraternity, the next step is obvious; screens can double and redouble our intellectual capacities. But this is not obvious to everybody. Engelbart, patiently instructing those outside, came up with a beautiful example. To show what he meant by the Augmentation of Intellect, Engelbart *tied a pencil to a brick.* Then he actually made someone write with it. The result, which was of course dreadful, Engelbart solemnly put into a published report. Not yet being able to demonstrate the augmentation of intellect, since he had as yet no system to show off, he had masterfully demonstrated the *dis*augmentation of intellect: what happens if you make man's tools for working out his thoughts *worse* instead of better. As this poor guy was with his brickified pencil, explained Engelbart, so are we all among our bothersome, inflexible systems of paper.

Starting small, Engelbart programmed up a small version of what most fans call "The Engelbart System" some ten years ago. One version has it that when it came to looking for grants, management thought he acted too kooky, and so assigned a Front Man to make the presentation. But, as the story goes, the man from ARPA pointed at Engelbart and said, "We want to back *him.*"

A small but dedicated group at SRI has built up a system from scratch. First they used little CDC 1700 minicomputers; then, various grants later, they were able to set up their own PDP-10, in which the system now resides, and from which it reaches out across the country.

Doug calls his system NLS, or "oN-Line System." Basically it is a highly responsive, deeply-structured text system, feeding out to display terminals. From a terminal you may read anything you or others have written, and write with as-yet-unmatched flexibility.

The display terminals are all over. The project has gone national, though at great expense: through the ARPA net of computers, you can in principle become a user of NLS for something like $50,000 a year.

MOUSE?

The Engelbart Folks have built a pointing device, for telling the system where you're pointing on the screen, that is considerably faster and handier than a lightpen. (Unfortunately, I don't believe it's commercially available.) It's called The Mouse.

The Engelbart Mouse is a little box with hidden wheels underneath and a cable to the terminal. As you roll it, the wheel's turns are signalled to the computer and the computer moves the cursor on the screen. It's *fast and accurate,* and in fact beats a lightpen hands down in working speed.

Through the command language, NLS allows users to create programs that respond in all sorts of ways; thus the fact that certain texthandling styles are standard results more from tradition than necessity.

The same apparently is true of the data structure. I used to be somewhat disturbed at the way Engelbart's text systems seem to be rigorously hierarchical. This in fact is the case, in the sense that having multiple discrete levels is built deep into the system. But it turns out to be harmless. The stored text is divided by the storage techniques into multiple levels, corresponding to a Harvard outline. Think of it as something like this:

1. HIERARCHICAL FORMAT
   A. STORAGE
   B. DISPLAY
   C. LANGUAGE

But let's expand this example a little:

1. HIERARCHICAL FORMAT
   A. STORAGE
      A1. Everything in NLS is stored with hierarchical codes.
      A2. Their effect depends on the display.

B. DISPLAY

B1. The hierarchical codes of NLS have no consequences in *particular.*

B2. The hierarchical codes for NLS can splay the material out into a variety of display arrangements.

B2A. They can be displayed in outline form.

B2B. They can be displayed in normal text form.

B2C. These dratted numbers can even be made to disappear.

C. LANGUAGE

C1. The command language determines what the display shows of the hierarchical structure.

C2. *What* is shown can be determined by a program in the command language. (For instance, "how many levels down" it is being shown).

C2A. This is four levels down. (The earlier example wasn't.)

C3. The display format all depends on what display program you use, in the NLS command language.

That's enough of that. I can't help remarking that I still don't like that sort of structuring, but it is deep in NLS, and if you don't like it either (poor deprived lucky user of NLS) you can program it to disappear, so it's hardly in your way.

By the Beard of the Prophet!

Engelbart in German means Angelbeard; Doug Engelbart is indeed on the side of the angels. In building his mighty system he points a new way for humanity. The sooner the better. Any history of the twentieth century will certainly hold him high. Few great men are also such nice guys.

A Very Basic Hypertext System

Hypertext is non-sequential writing. It's no good to us, though, unless we can go instantly in a choice of directions from a given point.

This of course can only mean on computer display screens.

Engelbart's system, now, was mainly designed for people who wanted to immerse themselves in it and learn its conventions. Indeed, it might be said to have been designed for a community of people in close contact, a sort of system of blackboards and collaborative talking papers.

A more elemental system, with a different slant, was put together at Brown U. on IBM equipment. We'll refer to it here as "Carmody's System," after the young programmer whose name came first on the writeup.

Carmody's system runs on an IBM 360 with 2250 display. While the 2250 is a fine piece of equipment, the quirks of the 360's operating system often delay the user by making him wait, e.g., for someone else's cards to get punched before it responds to his more immediate uses; this is like making ice-skaters wait for oxcarts.

Anyway, the system essentially imposes no structure on the material; it may consist of text segments of any length and ties and links between them. An asterisk appearing anywhere in one piece of text signals a possible jump, but the reader doesn't necessarily know where to; sapping the asterisk with the lightpen takes you there, however.

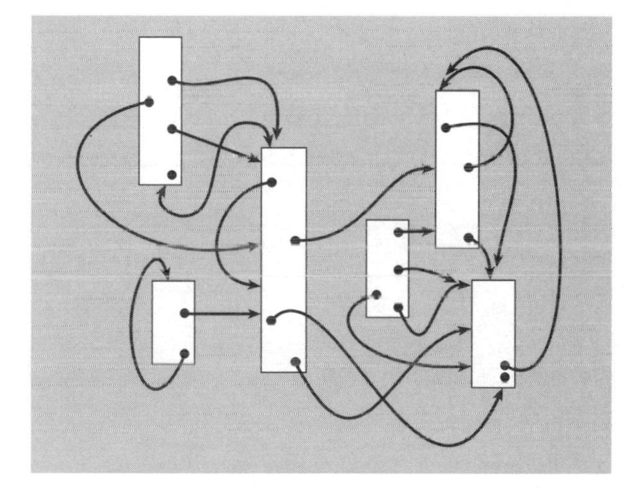

This is stark and simple. It could also get you good and lost. However, a simple technique took care of that: every time the user jumped, the address of his previous location was saved on a *stack.* The user also had a RETURN button: when he wanted to go back to where he had last jumped from, the system would pop the last address off the top of the stack, and take him there. (This feature was adapted from my 1967 Stretchtext paper, and turned out to work out quite well in practice.)

The system also had handy features for light-pen text editing, and various nice printout techniques. All told, it was a clean and powerful design. While it lacked higher-level visualization facilities, like Engelbart's display of Levels or collateral display, it was in some ways suited for naive users; that is, it was eventually fairly safe to use, and could in large part be taught to rank beginners in a couple of hours—provided they didn't have to know about JCL cards.

It is left for the reader to figure out interesting uses for it. How would you do collateral structures? How could you signal to a reader which of several pieces of text a jump was to?

(At least one real hypertext was actually written on this system. It tied together a lot of patents for multilayer electrodes. Readers agreed that they could learn more from it about multilayer electrodes than they had imagined wanting to know.)

Alan Kay and Adele Goldberg

"Personal Dynamic Media" (1977)

<< 16 >>

Dynabook. Courtesy of Alan Kay.

"Imagine having your own self-contained knowledge manipulator in a portable package the size and shape of an ordinary notebook. Suppose it had enough power to outrace your senses of sight and hearing, enough capacity to store for later retrieval thousands of page-equivalents of reference materials, poems, letters, recipes, records, drawings, animations, musical scores, waveforms, dynamic simulations, and anything else you would like to remember and change."

<< Alan Kay's goal for the Learning Research Group he founded at Xerox PARC in the early 1970s was no less than to invent the future of communication, and the knowledge tools to make it possible. This article is condensed from a PARC technical report that documents the six years of research behind the first workable hypermedia system. Together with his assistant Adele Goldberg, Kay outlined the genesis of Smalltalk, an object-oriented software language that led to the invention of the graphical user interface. Their research took root in the conviction that hypermedia, or "dynamic media," as they called it, represented a profound departure from static media such as painting, television, photography, print publishing, and film. They saw in hypermedia the radical interactivity that would characterize communications in the future.

The prototype that was the focus of Kay and Goldberg's research, the Dynabook, was conceived as a "dynamic medium for creative thought," capable of synthesizing all media—pictures, animation, sound, and text—through the intimacy and responsiveness of the personal computer. The Dynabook was, in part, an interpretation of Vannevar Bush's information storage device, the memex. It incorporated the navigational conventions of Douglas Engelbart's oNLine System (the mouse, windows, and menus in particular). It also borrowed from Ted Nelson's concept of hypermedia. While Kay and his team never built the prototype of the Dynabook, by 1974 its key ideas were incorporated into the Xerox Alto computer running the Smalltalk operating system. As a result, the Alto is regarded as the first personal computer, and the original desktop multimedia system.

The Dynabook was meant for everyone, enabling people of all ages to discover and express their creativity. It included software such as a text editor, an animation system, a paint program, and music scoring systems. It was, as the authors describe it, a "metamedium," containing within it all other media. The Dynabook also brought together divergent media through a single interface, making desktop hypermedia a reality. To prove the accessibility of the device, children were recruited to test the prototype machine, and to write original software applications such as music and painting programs. Kay's work came to the attention of young entrepreneurs in the computer industry, including Steve Jobs of Apple Computer and Bill Gates of Microsoft, who profited enormously from the powerful operating system that was designed to give free reign to the intuitive mechanisms of a child's mind. >>

Introduction

The Learning Research Group at Xerox Palo Alto Research Center is concerned with all aspects of the communication and manipulation of knowledge. We design, build, and use dynamic media which can be used by human beings of all ages. Several years ago, we crystallized our dreams into a design idea for a personal dynamic medium the size of a notebook (the *Dynabook*) which could be owned by everyone and could have the power to handle virtually all of its owner's information-related needs. Towards this goal we have designed and built a communications system: the Smalltalk language, implemented on small computers we refer to as "interim Dynabooks." We are exploring the use of this system as a programming and problem solving tool; as an interactive memory for the storage and manipulation of data; as a text editor; and as a medium for expression through drawing, painting, animating pictures, and composing and generating music.

We offer this paper as a perspective on our goals and activities during the past years. In it, we explain the Dynabook idea, and describe a variety of systems we have already written in the Smalltalk language in order to give broad images of the kinds of information-related tools that might represent the kernel of a personal computing medium.

Background

Humans and media. "Devices" which variously store, retrieve, or manipulate information in the form of messages embedded in a medium have been in existence for thousands of years. People use them to communicate ideas and feelings both to others and back to themselves. Although thinking goes on in one's head, external media serve to materialize thoughts and, through feedback, to augment the actual paths the thinking follows. Methods discovered in one medium provide metaphors which contribute new ways to think about notions in other media.

For most of recorded history, the interactions of humans with their media have been primarily nonconversational and passive in the sense that marks on paper, paint on walls, even "motion" pictures and television, do not change in response to the viewer's wishes. A mathematical formulation—which may symbolize the essence of an entire universe—once put down on paper, remains static and requires the reader to expand its possibilities.

Every message is, in one sense or another, a *simulation* of some idea. It may be representational or abstract. The essence of a medium is very much dependent on the way messages are embedded, changed, and viewed. Although digital computers were originally designed to do arithmetic computation, the ability to simulate the details of any descriptive model means that the computer, viewed as a medium itself, can be *all other media* if the embedding and viewing methods are sufficiently well provided. Moreover, this new "metamedium" is *active*—it can respond to queries and experiments—so that the messages may involve the learner in a two-way conversation. This property has never been available before except through the medium of an individual teacher. We think the implications are vast and compelling.

A dynamic medium for creative thought: the Dynabook. Imagine having your own self-contained knowledge manipulator in a portable package the size and shape of an ordinary notebook. Suppose it had enough power to outrace your senses of sight and hearing, enough capacity to store for later retrieval thousands of page-equivalents of reference materials, poems, letters, recipes, records, drawings, animations, musical scores, waveforms, dynamic simulations, and anything else you would like to remember and change.

We envision a device as small and portable as possible which could both take in and give out information in quantities approaching that of human sensory systems. Visual output should be, at the least, of higher quality than what can be obtained from newsprint. Audio output should adhere to similar high-fidelity standards.

There should be no discernible pause between cause and effect. One of the metaphors we used when designing such a system was that of a musical instrument, such as a flute, which is owned by its user and responds instantly and consistently to its owner's wishes. Imagine the absurdity of a one-second delay between blowing a note and hearing it!

These "civilized" desires for flexibility, resolution, and response lead to the conclusion that a user of a dynamic personal medium needs several hundred times as much power as the average adult now typically enjoys from timeshared computing. This means that we should either build a new resource several hundred times the capacity of current machines and share it (very difficult and expensive), or we should investigate the possibility of giving each person his own powerful machine. We chose the second approach.

Design background. The first attempt at designing this metamedium (the FLEX machine) occurred in 1967–69. Much of the hardware and software was successful from the standpoint of computer science state-of-the-art research, but lacked sufficient expressive power to be useful to an ordinary user. At that time we became interested in focusing on children as our "user community." We were greatly encouraged by the Bolt Beranek and Newman/MIT Logo work that uses a robot turtle that draws on paper, a CRT version of the turtle, and a single music generator to get kids to program.

Considering children as the users radiates a compelling excitement when viewed from a number of different perspectives. First, the children really can write programs that do serious things. Their programs use symbols to stand for objects, contain loops and recursions, require a fair amount of visualization of alternative strategies before a tactic is chosen, and involve interactive discovery and removal of "bugs" in their ideas.

Second, the kids love it! The interactive nature of the dialogue, the fact that *they* are in control, the feeling that they are doing *real* things rather than playing with toys or working out "assigned" problems, the pictorial and auditory nature of their results, all contribute to a tremendous sense of accomplishment to their experience. Their attention spans are measured in hours rather than minutes.

Another interesting nugget was that children really needed as much or more computing power than adults were willing to settle for when using a timesharing system. The best that timesharing has to offer is slow control of crude wire-frame green-tinted graphics and square-wave musical tones. The kids, on the other hand, are used to finger-paints, water colors, color television, real musical instruments, and records. If the "medium is the message," then the message of low-bandwidth timesharing is "blah."

An Interim Dynabook

We have designed an interim version of the Dynabook on which several interesting systems have been written in a new medium for communication, the Smalltalk programming language. We have explored the usefulness of the systems with more than 200 users, most notably setting up a learning resource center in a local junior high school.

The interim Dynabook is a completely self-contained system. To the user, it appears as a small box in which a disk memory can be inserted; each disk contains about 1500 page-equivalents of manipulable storage. The box is connected to a very crisp high-resolution black and white CRT or a lower-resolution high-quality color display. Other input devices include a typewriter keyboard, a "chord" keyboard, a pointing device called a "mouse" which inputs position as it is moved about on the table, and a variety of organ-like keyboards for playing music. New input devices such as these may be easily attached, usually without building a hardware interface for them. Visual output is through the display, auditory output is obtained from a built-in digital-to-analog converter connected to a standard hi-fi amplifier and speakers.

We will attempt to show some of the kinds of things that can be done with a Dynabook; a number of systems developed by various users will be briefly illustrated.

Remembering, seeing and hearing. The Dynabook can be used as an interactive memory or file cabinet. The owner's context can be entered through a keyboard and active editor, retained and modified indefinitely, and displayed on demand in a font of publishing quality.

Drawing and painting can also be done using a pointing device and an iconic editor which allows easy modification of pictures. A picture is thus a manipulable object and can be animated dynamically by the Dynabook's owner.

A book can be read through the Dynabook. It need not be treated as a simulated paper book since this is a new medium with new properties. A dynamic search may be made for a particular context. The non-sequential nature of the file medium and the use of dynamic manipulation allows a story to have many accessible points of view; Durrell's *Alexandria Quartet,* for instance, could be one book in which the reader may pursue many paths through the narrative.

Different fonts for different effects. One of the goals of the Dynabook's design is *not* to be *worse* than paper in any important way. Computer displays of the past have been superior in matters of dynamic writing and erasure, but have failed in contrast, resolution, or ease of viewing. There is more to the problem than just the display of text in a high-quality font. Different fonts create different moods and cast an aura that influences the subjective style of both writing and reading. The Dynabook is supplied with a number of fonts which are contained on the file storage.

The Dynabook as a personal medium is flexible to the point of allowing an owner to choose his own ways to view information. Any character font can be described as a matrix of black and white dots. The owner can draw in a character font of his own choosing. He can then immediately view font changes within the context of text displayed in a window. With the Dynabook's fine grain of display, the rough edges disappear at normal viewing distance to produce high-quality characters.

Editing. Every description or object in the Dynabook can be displayed and edited. Text, both sequential and structured, can easily be manipulated by combining pointing and a simple "menu" for commands, thus allowing deletion, transposition, and structuring. Multiple windows allow a document (composed of text, pictures, musical notation) to be created and viewed simultaneously at several levels of refinement. Editing operations on other viewable objects (such as pictures and fonts) are handled in analogous ways.

Filing. The multiple-window display capability of Smalltalk has inspired the notion of a dynamic *document.* A document is a collection of objects that have a sensory display and have something to do with each other; it is a way to store and retrieve related information. Each subpart of the document, or *frame,* has its own editor which is automatically invoked when pointed at by the "mouse." These frames may be related sequentially, as with ordinary paper usage, or *inverted* with respect to properties, as in cross-indexed file systems. *Sets* which can automatically map their contents to secondary storage with the ability to form unions, negations, and intersections are part of this system, as is a "modeless" text editor with automatic right justification.

The current version of the system is able to automatically cross-file several thousand multifield records (with formats chosen by the user), which include ordinary textual documents indexed by content, the Smalltalk system, personal files, diagrams, and so on.

Drawing/painting. The many small dots required to display high-quality characters (about 500,000 for an 8-1/2″ × 11″ sized display) also allow sketching-quality drawing, "halftone painting," and animation. The subjective effect of gray scale is caused by the eye fusing an area containing a mixture of small black and white dots. A brush can be grabbed with the "mouse," dipped into a paint pot, and then the halftone can be swabbed on as a function of the size, shape, and velocity of the brush.

Curves are drawn by a *pen* on the display screen. (Straight lines are curves with zero curvature.) In the Dynabook, *pens* are members of a class that can selectively draw with black or white (or colored) ink and change the thickness of the trace. Each *pen* lives in its own *window,* careful not to traverse its window boundaries but to adjust as its window changes size and position.

Animation and music. Animation, music, and programming can be thought of as different *sensory views* of dynamic processes. The structural similarities among them are apparent in Smalltalk, which provides a common framework for expressing those ideas.

All of the systems are equally controllable by hand or by program. Thus, drawing and painting can be done using a pointing device or in conjunction with programs which draw curves, fill in areas with tone, show perspectives of three-dimensional models, and so on. Any graphic expression can be animated, either by reflecting a simulation or by example (giving an "animator" program a sample trace or a route to follow).

Music is controlled in a completely analogous manner. The Dynabook can act as a "super synthesizer" getting direction either from a keyboard or from a "score." The keystrokes can be captured, edited, and played back. Timbres, the "fonts" of musical expression, contain the quality and mood which different instruments bring to an orchestration. They may be captured, edited, and used dynamically.

Simulation

In a very real sense, simulation is the central notion of the Dynabook. Each of the previous examples has shown a simulation of visual or auditory media. Here are a number of examples of interesting simulations done by a variety of users.

An animation system programmed by animators. Several professional animators wanted to be able to draw and paint pictures which could then be animated in real time by simply showing the system roughly what was wanted. Desired changes would be made by iconically editing the animation sequences.

Much of the design of SHAZAM, their animation tool, is an automation of the media with which animators are familiar: *movies* consisting of sequences of *frames* which are a composition of transparent *cels* containing foreground and background drawings. Besides retaining these basic concepts of conventional animation, SHAZAM incorporates some creative supplementary capabilities.

Animators know that the main action of animation is due not to an individual frame, but to the change from one frame to the next. It is therefore much easier to plan an animation if it can be seen moving as it is being created. SHAZAM allows any cel of any frame in an animation to be edited while the animation is in progress. A library of already-created cels is maintained. The animation can be single-stepped; individual cels can be repositioned, reframed, and redrawn; new frames can be inserted; and a frame sequence can be created at any time by attaching the cel to the pointing device, then *showing* the system what kind of movement is desired. The cels can be stacked for background parallax; *holes* and *windows* are made with *transparent* paint. Animation objects can be painted by programs as well as by hand. The control of the animation can also be easily done from a Smalltalk simulation. For example, an animation of objects bouncing in a room is most easily accomplished by a few lines of Smalltalk that express the class of bouncing objects in physical terms.

A drawing and painting system programmed by a child. One young girl, who had never programmed before, decided that a pointing device *ought* to let her draw on the screen. She then built a sketching tool without ever seeing ours. She constantly embellished it with new features including a menu for brushes selected by pointing. She later wrote a program for building tangram designs.

This girl has taught her own Smalltalk class; her students were seventh-graders from her junior high school. One of them designed an even more elaborate system in which pictures are constructed out of geometric shapes created by pointing to a menu of commands for creating regular polygons. The polygons can then be relocated, scaled, and copied; their color and line width can change.

An audio animation system programmed by musicians. Animation can be considered to be the coordinated parallel control through time of images conceived by an animator. Likewise, a system for representing and controlling musical images can be imagined which has very strong analogies to the visual world. Music is the design and control of images (pitch and duration changes) which can be *painted* different *colors* (timbre choices); it has synchronization and coordination, and a very close relationship between audio and spatial visualization.

The Smalltalk model created by several musicians, called TWANG, has the notion of a *chorus* which contains the main control directions for an overall piece. A chorus is a kind of *rug* with a warp of parallel sequences of "pitch, duration, and articulation" commands, and a woof of synchronizations and global directives. The control and the *player* are separate: in SHAZAM, a given movie sequence can animate many drawings; in TWANG, a given chorus can tell many different kinds of instrumentalists what should be played. These *voices* can be synthetic timbres or timbres captured from real instruments. Musical effects such as vibrato, portamento, and diminuation are also available.

A chorus can be *drawn* using the pointing device, or it can be *captured* by playing it on a keyboard. It can be played back in real time and dynamically edited in a manner very similar to the animation system.

We use two methods for real-time production of high-quality timbres; both allow arbitrary transients and many independent parallel voices, and are completely produced by programs. One of these allows independent dynamic control of the spectrum, the frequency, the amplitude, and the particular collection of partials which will be heard.

For children, this facility has a number of benefits: the strong similarities between the audio and visual worlds are emphasized because a single vernacular *which actually works* in both worlds is used for description; and second, the arts and skills of composing can be learned at the same time since tunes may be drawn in by hand and played by the system. A line of music may be copied, stretched, and shifted in time and pitch; individual notes may be edited. Imitative counterpoint is thus easily created by the fledgling composer.

A musical score capture system programmed by a musician. OPUS is a musical score capture system that produces a display of a conventional musical score from data obtained by playing a musical keyboard. OPUS is designed to allow incremental input of an arbitrarily complicated score (full orchestra with chorus, for example), editing pages of the score, and hard copy of the final result with separate parts for individual instruments.

Conclusion

What would happen in a world in which everyone had a Dynabook? If such a machine were designed in a way that *any* owner could mold and channel its power to his own needs, then a new kind of medium would have been created: a metamedium, whose content would be a wide range of already-existing and not-yet-invented media.

An architect might wish to simulate three-dimensional space in order to peruse and edit his current designs, which could be conveniently stored and cross-referenced.

A doctor could have on file all of his patients, his business records, a drug reaction system, and so on, all of which could travel with him wherever he went.

A composer could hear his composition while it was in progress, particularly if it were more complex than he was able to play. He could also bypass the incredibly tedious chore of redoing the score and producing the parts by hand.

Learning to play music could be aided by being able to capture and hear one's own attempts and compare them against expert renditions. The ability to express music in visual terms which could be filed and played means that the acts of composition and self-evaluation could be learned without having to wait for technical skill in playing.

Home records, accounts, budgets, recipes, reminders, and so forth, could be easily captured and manipulated.

Those in business could have an active briefcase which travelled with them, containing a working simulation of their company, the last several weeks of correspondence in a structured cross-indexed form—a way to instantly calculate profiles for their futures and help make decisions.

For educators, the Dynabook could be a new world limited only by their imagination and ingenuity. They could use it to show complex historical interrelationships in ways not possible with static linear books. Mathematics could become a living language in which children could cause exciting things to happen. Laboratory experiments and simulations too expensive or difficult to prepare could easily be demonstrated. The production of stylish prose and poetry could be greatly aided by being able to easily edit and file one's own compositions.

These are just a few ways in which we envision using a Dynabook. But if the projected audience is to be "everyone," is it possible to make the Dynabook generally useful, or will it collapse under the weight of trying to be too many different tools for too many people? The total range of possible users is so great that any attempt to specifically anticipate their needs in the design of the Dynabook would end in a disastrous feature-laden hodgepodge which would not be really suitable for anyone.

Some mass items, such as cars and television sets, attempt to anticipate and provide for a variety of applications in a fairly inflexible way; those who wish to do something different will have to put in considerable effort. Other items, such as paper and clay, offer many dimensions of possibility and high resolution; these can be used in an unanticipated way by many, though *tools* need to be made or obtained to stir some of the medium's possibilities while constraining others.

We would like the Dynabook to have the flexibility and generality of this second kind of item, combined with tools which have the power of the first kind. Thus a great deal of effort has been put into providing both endless possibilities and easy tool-making through the Smalltalk programming language.

Our design strategy, then, divides the problem. The burden of system design and specification is transferred to the user. This approach will only work if we do a very careful and comprehensive job of providing a general medium of communication which will allow ordinary users to casually and easily describe their desires for a specific tool. We must also provide enough already-written general tools so that a user need not start from scratch for most things she or he may wish to do.

We have stated several specific goals. In summary, they are:

- to provide coherent, powerful examples of the use of the Dynabook in and across subject areas;
- to study how the Dynabook can be used to help expand a person's visual and auditory skills;
- to provide exceptional freedom of access so kids can spend a lot of time probing for details, searching for a personal key to understanding processes they use daily; and
- to study the unanticipated use of the Dynabook and Smalltalk by children in all age groups.

Marc Canter

"The New Workstation: CD ROM Authoring Systems" (1986)

<< 17 >>

Marc Canter. © Chip Simons.

"Authoring software should aim to shorten the feedback loop between the computer and the user—between the idea itself and its actualization. The overall outcome is a more direct connection to creativity for the user."

<< Marc Canter emerged in the 1980s as an amalgamation of opera singer, rock musician, software programmer, and entrepreneur. With academic credentials from Oberlin College in intermedia art and electronic music, and after a brief stint producing music videos in New York, Canter applied his unusual array of talents to engineering the first commercial multimedia authoring system. He launched his software company, Macromind, in 1984, when the Apple Macintosh computer made the GUI and its potential for hypermedia applications widely available. Later rechristened Macromedia, its first products, SoundWorks and VideoWorks, introduced multimedia production to the desktop computer. In 1988, after the first color Macintosh appeared, Canter released the now ubiquitous Director authoring software. What followed was an explosion of new creative possibilities, along with a good deal of Madison Avenue–style hype. By the close of the decade, desktop multimedia grew into a global phenomenon, with Canter at the center of the excitement, transforming the studios of artists, architects, and designers, reinventing the classroom, and altering the business plans of executives from Silicon Valley to Singapore.

Canter's pioneering approach to multimedia authoring tools became the standard for all forms of new media development. At the core of his approach was a notational system that looks quite similar to a musical score, an intuitive format that could be used easily by the artist. Canter saw the digital artist of the future as a "composer" of all forms of media, orchestrating fragments of graphics, animation, and text, in juxtaposition with sound and musical passages, into a single artwork. His authoring system made use of "action codes" derived from Alan Kay's Smalltalk language, giving artists without extensive programming experience the ability to incorporate hyperlinks into their creations by manipulating icons on the computer screen. The metaphors Canter chose to describe this system have become widely adopted, including his meaning for the terms *stage, cast,* and *score.* This predisposition toward theater and music belies Canter's roots in live performance, and reinforces his vision that desktop multimedia would evolve into the digital *Gesamtkunstwerk.* >>

As the reality of CD ROM draws nearer, so does the need for development tools that will help programmers and artists to create CD applications fast and efficiently.

Currently, most CD ROM applications are text-based, displaying mainly words and numbers on the screen. But as the technology moves forward, graphics and music will be stored on CD ROMs, resulting in a brand-new generation of applications.

The tools we use in developing such applications will play an important role in making these projects feasible. As we begin to conceive of complex new uses for CD ROM, the ease of use and versatility of our tools will make or break our concepts from day one.

Among the tools necessary for these tasks, authoring systems will probably be the most important. Integral to these systems will be a powerful, easy-to-use notational system that will unify entire multimedia systems on one score, just as an orchestra is unified by the symphonic score used by its conductor. These scores will be capable of representing any sort of data, including the "action codes" necessary for interactive programming (or authoring).

No longer will it be necessary to "program" a sequence of data or to devise another "search" algorithm. Authoring systems provide the necessary functions to nonprogrammers to control the flow of the program, the level of interactivity and the production of the data itself.

For the purposes of this discussion, we will concentrate on an authoring system for text, art, animation, and music. It will require the simulation of a fairly complex output device (a premastering system) and a very fast (as real-time as possible) development system. This does not mean that this system will not be able to work with text-based CD applications. It just means that the technology of the future offers us a lot more than just static text, and we must be ready to create and control new media as they become available.

In the following discussion, "user" refers to the user of the authoring system and "end user" refers to the actual consumer in the home or office using the final product.

What Is an Authoring System?

An authoring system is a set of hardware and software tools for designing interactive programs. The hardware makes it possible to convert information into a machine-readable format; the software makes it possible for nonprogrammers (which will often be artists and musicians) to create complex programs by defining decision points, branches, and subroutines.

Authoring systems originated in the mainframe and minicomputer world, but many have begun to appear for microcomputers in the past few years. The more sophisticated authoring systems of yesterday and today produce complicated logical flowcharts that programmers use to implement their interactive programs. Many of these programs have "intelligent front ends" that ask the users questions to help them develop their programs, like "What sort of question would you like to ask?" or "How many dialog boxes would you like and what will they look like?" Using a development system based on this sort of authoring "language" often takes months, and sometimes years, to produce interactive videodisks or training programs.

The real problem with these types of authoring systems is that they are very hard to use and they do not produce a final product. They produce only an outline: a structure for a programmer to follow. And they are often entirely text-based, making them ideal for text-based applications but not for graphically oriented or video-based applications.

Any language or program that can control events through time is an authoring system. Computer-controlled lighting systems, slide shows, and video editing are all authoring systems in that (1) they can control sequences and events through time, (2) these sequences can be edited or changed, and (3) these sequences can be saved and retrieved from disc.

Authoring systems have crept into the entertainment world as a form of automation, making the director's life easier by guaranteeing that certain actions will happen, no matter what. This sort of automation may have cost a number of operators' jobs, but on the whole it has greatly enhanced the quality of special effects in the past few years.

Most of these large expensive systems (including the ones controlling TV studios) are text-based, with the operator/programmer using a CRT to edit sequences, in conjunction with dedicated buttons, switches, and levers. But the increased interest in graphics and graphic interfaces is spawning a new generation of authoring systems that are graphically based.

The MacroMind Sound Vision™ authoring system is graphically based and uses a notational "scoring" system for controlling the flow of the program through time. It has no language at all but uses the notational system to represent sequential, branching, or even simultaneous events that occur in the program. The events may be a combination of text, graphics, music, or animation. This sort of system is a multimedia authoring system, since it can deal with all sorts of data, not just text.

On SoundVision's score, time moves horizontally from left to right, with multiple channels of information stacked up vertically. These channels represent the text, graphics, or music stored in the score.

The scoring system is used by programmers or artists to create sequences of information that can easily be edited or transformed. This replaces the traditional means of programming: typing text into a document and compiling it. All the data represented on the score of the authoring system can be edited at any time. In other words, there is no source code or compiler to convert the code into executable instructions. Any text or piece of artwork can be edited or changed even after the application is finished.

The notational system is the unifying element in the authoring system. Text, graphics, animation, and music can all be synchronized and edited on the score. The "action codes," or programming codes, necessary for creating the interactive programs can be embedded into the score and edited just like any piece of animation or music. Data generated from any word processor, paint program, or MIDI (musical instrument digital interface) sequencer can be used in the SoundVision system.

Simulating the programmed interaction is another necessary part of the authoring process and is achieved via a software/hardware system called a "premastering" system. These premastering units are very expensive, but they make it possible to test the applications before committing them to disc or ROM. Typically, an entire project would have one, maybe two, premastering units to use as simulators.

Creating the Applications

Education and training applications are certain to become more common on optical disc once the production costs can be brought down by using authoring systems and other development tools. Their existence in schools and industry will undoubtedly change the way educators develop curricula, perhaps giving rise to new professions within the educational arena.

- By being able to customize educational courseware, regional needs can be catered to and accelerated or disabled learners can concentrate on specialized material.
- Educators can focus on a particular subject by designing a custom lesson around that subject.
- Current news can be explained by integrating recent video footage from TV with encyclopedia references to that area of the world or to that world leader.

- Questions relating to the subject matter can be placed on the screen, and each answer associated with a particular screen "button." Behind each button is an action, such as "Jump to an animated sequence for 10 seconds and then return" or "Jump to another document to hear the company theme song and exit" or "Jump to another screen of questions, answer one of them, and branch depending on what answer is given."

These sorts of applications are typical of what an authoring system can create. It is very important to be able to model or try out some of these interactive ideas before committing them to disc, and an authoring system is designed to do just that: model an application.

Because the people who come up with these interactive ideas are usually not programmers, it is equally important that they are able to model these environments with as little effort and ability as possible. Since teachers, trainers, marketing people, technical writers, and managers will probably be the biggest users of authoring systems, the systems must adhere to easy-to-learn user interface standards, such as those found in the Macintosh or MS-Windows environments.

Equipment Used in an Authoring System

The workstation for a CD ROM authoring system would vary according to the actual work being done there. Each workstation will specialize in some function and thereby require different types of equipment, though very likely some equipment will be common to every workstation.

The basic workstation will include:

- An IBM AT (or better) with 4M of memory and a 30M hard disk
- A premastering system to simulate an output device (or access to it)
- A mouse or tablet along with a full-feature keyboard
- A network of other workstations, and 2400-baud modem line

A text entry workstation would include optical scanners, voice recognition hardware, and monochrome monitors. Besides a host of text processors, the software available in a workstation like this would integrate data from various sources: standard word processors, distant databases via phone lines, ASCII, SYLK, dBase, and other standardized data protocols. This would all be necessary to facilitate the task of entering in hundreds of thousands of words per day.

A graphics workstation would include video digitizers, tablets, high-resolution scanners, and so on. There should also be a photo and video studio with camera and lighting equipment at the disposal of the artists. High-quality color monitors must be standard equipment in a graphics workstation.

Paint programs of every make, color, and size would be standard software in the graphics workstation. Some of the other necessary software "equipment" would include video digitizing and touch-up, graphics processing (like skew, rotate, and distort), optical scanning, and 3D graphics generators. Other paint programs and graphics databases will be compatible with this system.

In this sort of environment, with the right management, artists could crank out a meg of art a day (1,024,000 bytes).

A music or sound effects workstation will require high-quality audio gear (speakers, tape recorders, amplifier, preamplifier) for both recording and playback. Patch bays, MIDI keyboards, processing equipment, sound digitizers, mixers, and microphones should also be standard equipment. Color monitors for displaying composing programs are also necessary.

Special facilities should be prepared with sound dampening, especially in the workstations designated as "audio studios." Each workstation will be equipped with sequencer and composition software with a variety of notational scoring systems available, e.g., Conventional Music Notation and the MusicWorks grid system. Sound effects generation through algorithmic control will probably also be desired, as well as sampled sound editors.

A multimedia scoring system (like SoundVision) will be used to synchronize the music or sound effects to the animation. Access to other music databases, synthesis patches, or algorithms will also be possible with this workstation. A 10-minute composition could be produced in a day at one of these workstations.

An animation workstation requires much the same equipment as the graphics workstation, with some differences. Multiple monitors can facilitate the simultaneous viewing, previewing, and editing of data. Videotape recorders, video SEGs (or mixers), and audio equipment are also important to animators who wish to transfer their work onto other media (VHS, Beta, 3/4-inch tape).

The animation workstation of an authoring system is where all the elements of the system are brought together and incorporated into the score. This is where the text and artwork are "animated," the music is synchronized, and the action codes are embedded to control the flow of the interactive programs.

The SoundVision scoring system is the main software element of an animation workstation like this. Other types of development tools could be used to integrate incompatible data or to configure the premastering system, but creating and editing documents on the score would be the main task of an animator.

So you can see there is more than just one type of workstation involved in a CD ROM project. Managers, designers, and executives could theoretically work in an animation workstation. Artists could roam between graphics and animation stations, and producers and directors would have to know them all equally well.

Tailoring Your Workstation

Each workstation should be customized to the particular project and application being produced. One of the best ways to utilize workers' abilities would be to use remote workstations. You could place them anywhere (especially at home) so that work could continue seven days a week and on holidays. And be fun!

How to Use an Authoring System

Stage One

The first stage in using an authoring system is to block out the approximate timings and interaction desired in the application. A typical application would have a menu at the beginning, at least two or three other menus somewhere else, and some sort of ending section.

Text and artwork (which should have been created earlier), along with action codes, are placed into a document's score at the approximate locations. Dialog boxes, menus, and buttons, which start off as artwork, soon turn into interactive controls when an action code is associated with them. The connection between artwork and action code is all done on the score, at the exact frame desired. (The exact logistics of how to assign an action code in the score is explained below.)

Once a rough draft has been worked out, initial viewings can be used to detect mistakes, wrong codes, or ill-conceived notions. This process of trying things out as soon as possible allows the author to correct any grievous errors before they become uncontrollable.

The material can also be edited at any time, so different sections of the project can be polished at different times, and conceivably by different people.

Stage Two

The second stage in using an authoring system (once the timings have been frozen) is to add more action codes. These codes may link several documents together, branch to other sections and return, or even start other applications. Music and animation can also be added at this stage of the development process.

Memory limitations and access times are some of the typical problems encountered when developing applications with authoring systems. Once a writer releases the tremendous potential of the system, he usually goes overboard and asks the system to do too much. For instance, jump from a section on DNA synthesis, to a musical selection by Beethoven, to a survey of Picasso paintings, all in 0:15 second, and then return to the chemistry section and continue the lesson.

Authoring systems can help organize and control a huge amount of data, but it is very easy to overextend yourself. Early on in your design you must determine how much data is enough. Just how many sequences of animation demonstrating the principles of fluid mechanics or wave dynamics will be sufficient to get the point across? Whatever you decide, it is very important to not be too ambitious. Tackle only a small part of a huge task at a time. There will always be room for more later.

It's important to realize that authoring systems will evolve and grow with technology. In the future, new types of authoring systems will be designed to take advantage of new technology. Besides simple menu choices and passive viewing, new standards of interacting will develop, such as text entry or real-time input.

Stage Three

The final stage of using an authoring system is the debugging of the application. Do all the questions make sense? Does the program flow smoothly? Are there any interactions that don't go anywhere? Is the artwork less than perfect? Going back to the score and changing action codes is as easy as editing text or touching up graphics.

This stage is probably the most important one, as it's always that last 5% of an application that can really make or break it.

Because of the nature of authoring systems, applications can evolve and change as they are worked on. Sections that make perfect sense on paper often turn out to be less than expected. Because of this, an authoring system user should always keep an open mind, ever ready to shift gears and do things differently. A good authoring system should facilitate this need by making it as easy as possible to edit and change the application at any time. . . .

Important Issues for CD ROM Authors

Authoring software should aim to shorten the feedback loop between the computer and the user—between the idea itself and its actualization. The overall outcome is a more direct connection to creativity for the user and a higher level of productivity.

This type of interaction is exactly what computers were made for: real-time tools. An example of this real-time interaction can be seen in MusicWorks, where the notes are heard as they are placed onto the score, or in VideoWorks, where you can edit an image as you watch it animate.

Widespread use of such authoring workstations will enable large groups of artists and programmers to produce very large databases of text, animation, and music. By keeping the notational system as generalized as possible, workstations like this have the potential of controlling Broadway productions or entire television studios. . . .

Tim Berners-Lee

"Information Management: A Proposal" (1989)

<< 18 >>

Tim Berners-Lee. Photo by Fabian Bachrach. Courtesy of Tim Berners-Lee.

"An important part . . . is the integration of a hypertext system with existing data, so as to provide a universal system, and to achieve critical usefulness. . . ."

<< In the early 1980s, the British engineer Tim Berners-Lee began to develop a computer system for the electronic publishing of project management notes at CERN, the particle physics laboratory in Geneva, Switzerland. This system, named Enquire, was to have enabled the storing, retrieval, and hyperlinking of documents. It was never completed, but Berners-Lee expanded on its underlying concepts to explore how a hypertext system might work in conjunction with the Internet. The result is the unassuming proposal reprinted here, which is nothing less than a prospectus for the creation of the World Wide Web.

Berners-Lee's colleagues at CERN largely ignored his proposal, however. Working under his own initiative, in the fall of 1990 Berners-Lee completed the first Web browser and server software. In 1991, he began to distribute his software, now named the World Wide Web, to scientists over the Internet. After two years of gradual adoption among scientific colleagues, who, like Berners-Lee, used the Next computer, the Web began to take off when Marc Andreesen and a group of graduate students at the University of Illinois, Campagne-Urbana, adopted the browser for the Macintosh and PC. In 1993, their version of the browser, Mosaic, was widely distributed on the Internet, transforming the World Wide Web into a mass medium.

Berners-Lee's Web is a software system that unites research, documents, programs, laboratories, and scientists in a fluid, open, hypermedia environment. He proposes a decentralized network to which new servers can be added at any time without the approval of a centralized authority. Berners-Lee's Web is inherently dynamic, capable of expanding at an explosive rate; this was a significant departure from the hierarchical data systems that had previously been the standard. Another cornerstone of the system is the user's ability to annotate documents himself—to write his own "private links," as Berners-Lee phrases it here, creating an infinitely expanding variety of connections. While this paper addresses only how the Web might benefit CERN, Berners-Lee was well aware of his system's potential to link documents across the globe, and to transform our information culture. It is worth noting that while Berners-Lee focuses here on hypertext, he also addresses the potential to incorporate sound and image. From the start he saw the Web's eventual embrace of multimedia, which could well prove to be its enduring legacy. >>

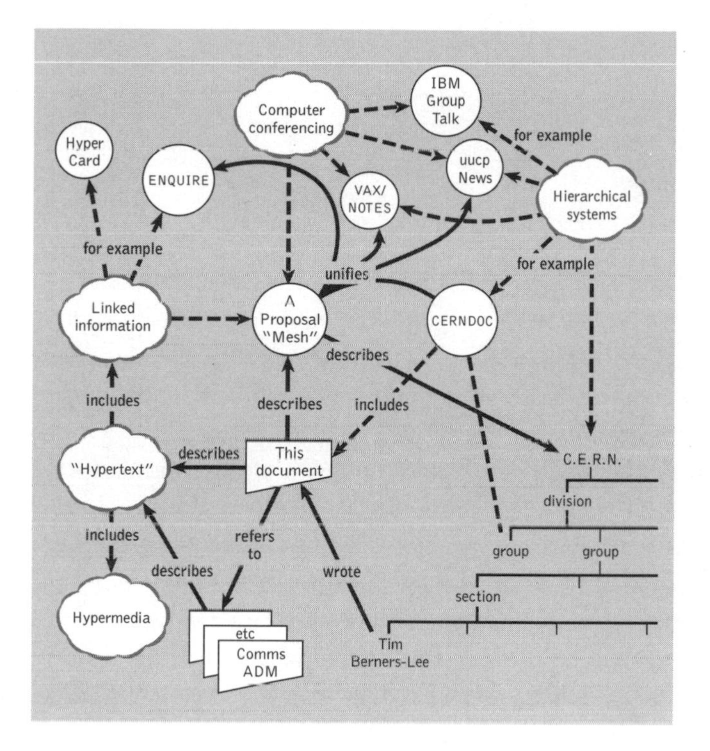

This proposal concerns the management of general information about accelerators and experiments at CERN. It discusses the problems of loss of information about complex evolving systems and derives a solution based on a distributed hypertext system.

Overview

Many of the discussions of the future at CERN and the LHC era end with the question—"Yes, but how will we ever keep track of such a large project?" This proposal provides an answer to such questions. Firstly, it discusses the problem of information access at CERN. Then, it introduces the idea of linked information systems, and compares them with less flexible ways of finding information.

It then summarises my short experience with non-linear text systems known as "hypertext", describes what CERN needs from such a system, and what industry may provide. Finally, it suggests steps we should take to involve ourselves with hypertext now, so that individually and collectively we may understand what we are creating.

Losing Information at CERN

CERN is a wonderful organisation. It involves several thousand people, many of them very creative, all working toward common goals. Although they are nominally organised into a hierarchical management structure, this does not constrain the way people will communicate, and share information, equipment and software across groups.

The actual observed working structure of the organisation is a multiply connected "web" whose interconnections evolve with time. In this environment, a new person arriving, or someone taking on a new task, is normally given a few hints as to who would be useful people to talk to. Information about what facilities exist and how to find out about them travels in the corridor gossip and occasional newsletters, and the details about what is required to be done spread in a similar way. All things considered, the result is remarkably successful, despite occasional misunderstandings and duplicated effort.

A problem, however, is the high turnover of people. When two years is a typical length of stay, information is constantly being lost. The introduction of the new people demands a fair amount of their time and that of others before they have any idea of what goes on. The technical details of past projects are sometimes lost forever, or only recovered after a detective investigation in an emergency. Often, the information has been recorded, it just cannot be found.

If a CERN experiment were a static once-only development, all the information could be written in a big book. As it is, CERN is constantly changing as new ideas are produced, as new technology becomes available, and in order to get around unforeseen technical problems. When a change is necessary, it normally affects only a small part of the organisation. A local reason arises for changing a part of the experiment or detector. At this point, one has to dig around to find out what other parts and people will be affected. Keeping a book up to date becomes impractical, and the structure of the book needs to be constantly revised.

The sort of information we are discussing answers, for example, questions like

- Where is this module used?
- Who wrote this code? Where does he work?

- What documents exist about that concept?
- Which laboratories are included in that project?
- Which systems depend on this device?
- What documents refer to this one?

The problems of information loss may be particularly acute at CERN, but in this case (as in certain others), CERN is a model in miniature of the rest of world in a few years' time. CERN now meets some problems which the rest of the world will have to face soon. In 10 years, there may be many commercial solutions to the problems above, while today we need something to allow us to continue.

Linked Information Systems

In providing a system for manipulating this sort of information, the hope would be to allow a pool of information to develop which could grow and evolve with the organisation and the projects it describes. For this to be possible,

the method of storage must not place its own restraints on the information.

This is why a "web" of notes with links (like references) between them is far more useful than a fixed hierarchical system. When describing a complex system, many people resort to diagrams with circles and arrows. Circles and arrows leave one free to describe the interrelationships between things in a way that tables, for example, do not. The system we need is like a diagram of circles and arrows, where circles and arrows can stand for anything.

We can call the circles nodes, and the arrows links. Suppose each node is like a small note, summary article, or comment. I'm not overly concerned here with whether it has text or graphics or both. Ideally, it represents or describes one particular person or object. Examples of nodes can be

- People
- Software modules
- Groups of people
- Projects
- Concepts
- Documents
- Types of hardware
- Specific hardware objects

The arrows which link circle A to circle B can mean, for example, that A depends on B

- is part of B
- made B
- refers to B
- uses B
- is an example of B

These circles and arrows, nodes and links, have different significance in various sorts of conventional diagrams:

DIAGRAM	NODES ARE	ARROWS MEAN
Family tree	People	"Is parent of"
Dataflow diagram	Software modules	"Passes data to"
Dependency	Module	"Depends on"
PERT chart	Tasks	"Must be done before"
Organisational chart	People	"Reports to"

The system must allow any sort of information to be entered. Another person must be able to find the information, sometimes without knowing what he is looking for.

In practice, it is useful for the system to be aware of the generic types of the links between items (dependences, for example), and the types of nodes (people, things, documents . . .) without imposing any limitations.

The Problem with Trees

Many systems are organised hierarchically. The CERNDOC documentation system is an example, as is the Unix file system, and the VMS/HELP system. A tree has the practical advantage of giving every node a unique name. However, it does not allow the system to model the real world. For example, in a hierarchical HELP system such as VMS/HELP, one often gets to a leaf on a tree such as

HELP COMPILER SOURCE__FORMAT PRAGMAS DEFAULTS

only to find a reference to another leaf: "Please see

HELP COMPILER COMMAND OPTIONS DEFAULTS PRAGMAS"

and it is necessary to leave the system and re-enter it. What was needed was a link from one node to another, because in this case *the information was not naturally organised into a tree.*

Another example of a tree-structured system is the uucp News system (try "rn" under Unix). This is a hierarchical system of discussions ("newsgroups"), each containing articles contributed by many people. It is a very useful method of pooling expertise, but suffers from the inflexibility of a tree. Typically, a discussion under one newsgroup will develop into a different topic, at which point it ought to be in a different part of the tree. (See Figure 1).

The Problem with Keywords

Keywords are a common method of accessing data for which one does not have the exact coordinates. The usual problem with keywords, however, is that two people never choose the same keywords. The keywords then become useful only to people who already know the application well.

Practical keyword systems (such as that of VAX/NOTES for example) require keywords to be registered. This is already a step in the right direction.

A linked system takes this to the next logical step. Keywords can be nodes which stand for a concept. A keyword node is then no different from any other node. One can link documents, etc., to keywords. One can then find keywords by finding any node to which they are related. In this way, documents on similar topics are indirectly linked, through their key concepts.

A keyword search then becomes a search starting from a small number of named nodes, and finding nodes which are close to all of them.

It was for these reasons that I first made a small linked information system, not realising that a term had already been coined for the idea: hypertext.

A Solution: Hypertext

Personal Experience with Hypertext

In 1980, I wrote a program for keeping track of software with which I was involved in the PS control system. Called *Enquire,* it allowed one to store snippets of information, and to link related pieces together in any way. To find information,

```
From mcvax!uunet!pyrdc!pyrnj!rutgers!bellcore!geppetto!duncan Thu
Mar . . .
Article 93 of alt.hypertext:
Path:
cernvax!mcvax!uunet!pyrdc!pyrnj!rutgers!bellcore!geppetto!duncan
>From: duncan@geppetto.ctt.bellcore.com (Scott Duncan)
Newsgroups: alt.hypertext
Subject: Re: Threat to free information networks
Message-ID: <14646@bellcore.bellcore.com>
Date: 10 Mar 89 21:00:44 GMT
References: <1784.2416BB47@isishq.FIDONET.ORG> <3437@uhccux.
uhcc . . .
Sender: news@bellcore.bellcore.com
Reply-To: duncan@ctt.bellcore.com (Scott Duncan)
Organization: Computer Technology Transfer, Bellcore
Lines: 18

Doug Thompson has written what I felt was a thoughtful article on
censorship—my acceptance or rejection of its points is not
particularly germane to this posting, however.

In reply Greg Lee has somewhat tersely objected.

My question (and reason for this posting) is to ask where we might
logically take this subject for more discussion. Somehow
alt.hypertext does not seem to be the proper place.

Would people feel it appropriate to move to alt.individualism or
even one of the soc groups. I am not so much concerned with the
specific issue of censorship of rec.humor.funny, but the views
presented in Greg's article.

Speaking only for myself, of course, I am . . .
Scott P. Duncan (duncan@ctt.bellcore.com OR . . .
!bellcore!ctt!duncan)
(Bellcore, 444 Hoes Lane RRC 1H-210, Piscataway, NJ . . .)
(201-699-3910 (w) 201-463-3683 (h))
```

FIGURE 1. *An article in the UUCP News scheme. The Subject field allows notes on the same topic to be linked together within a "newsgroup". The name of the newsgroup (alt.hypertext) is a hierarchical name. This particular note expresses a problem with the strict tree structure of the scheme: this discussion is related to several areas. Note that the "References," "From" and "Subject" fields can all be used to generate links.*

one progressed via the links from one sheet to another, rather like in the old computer game "adventure." I used this for my personal record of people and modules. It was similar to the application *Hypercard* produced more recently by Apple for the Macintosh. A difference was that *Enquire,* although lacking the

```
Documentation of the RPC project (concept)

Most of the documentation is available on VMS, with the two
principle manuals being stored in the CERNDOC system.

1) includes: The VAX/NOTES conference VXCERN::RPC
2) includes: Test and Example suite
3) includes: RPC BUG LISTS
4) includes: RPC System: Implementation Guide
Information for maintenance, porting, etc.
5) includes: Suggested Development Strategy for RPC Applications
6) includes: "Notes on RPC", Draft 1, 20 feb 86
7) includes: "Notes on Proposed RPC Development" 18 Feb 86
8) includes: RPC User Manual
How to build and run a distributed system.
9) includes: Draft Specifications and Implementation Notes
10) includes: The RPC HELP facility
11) describes: THE REMOTE PROCEDURE CALL PROJECT in DD/OC

Help Display Select Back Quit Mark Goto_mark Link Add Edit
```

FIGURE 2. *A screen in an Enquire scheme.*

fancy graphics, ran on a multiuser system, and allowed many people to access the same data.

This example is basically a list, so the list of links is more important than the text on the node itself. Note that each link has a type ("includes," for example) and may also have comment associated with it. (The bottom line is a menu bar.)

Soon after my re-arrival at CERN in the DD division, I found that the environment was similar to that in PS, and I missed *Enquire.* I therefore produced a version for the VMS, and have used it to keep track of projects, people, groups, experiments, software modules and hardware devices with which I have worked. I have found it personally very useful. I have made no effort to make it suitable for general consumption, but have found that a few people have successfully used it to browse through the projects and find out all sorts of things of their own accord.

Hot Spots

Meanwhile, several programs have been made exploring these ideas, both commercially and academically. Most of them use "hot spots" in documents, like icons, or highlighted phrases, as sensitive areas. Touching a hot spot with a mouse brings up the relevant information, or expands the text on the screen to include it. Imagine, then, the references in this document, all being associated with the network address of the thing to which they referred, so that while reading this document you could skip to them with a click of the mouse.

"Hypertext" is a term coined in the 1960s by Ted Nelson,[1] which has become popular for these systems, although it is used to embrace two different ideas. One idea (which is relevant to this problem) is the concept:

"Hypertext": Human-readable information linked together in an unconstrained way.

The other idea, which is independent and largely a question of technology and time, is of multimedia documents which include graphics, speech and video. I will not discuss this latter aspect further here, although I will use the word "hypermedia" to indicate that one is not bound to text.

It has been difficult to assess the effect of a large hypermedia system on an organisation, often because these systems never had seriously large-scale use. For this reason, we require large amounts of existing information should be accessible using any new information management system.

CERN Requirements

To be a practical system in the CERN environment, there are a number of clear practical requirements.

Remote access across networks	CERN is distributed, and access from remote machines is essential.
Heterogeneity	Access is required to the same data from different types of systems (VM/CMS, Macintosh, VAX/VMS, Unix).
Non-centralisation	Information systems start small and grow. They also start isolated and then merge. A new system must allow existing systems to be linked together without requiring any central control or coordination.
Access to existing data	If we provide access to existing databases as though they were in hypertext form, the system will get off the ground quicker. This is discussed further below.
Private links	One must be able to add one's own private

	links to and from public information. One must also be able to annotate links, as well as nodes, privately.
Bells and whistles	Storage of ASCII text, and display on 24×80 screens, is in the short term sufficient, and essential. Addition of graphics would be an optional extra with very much less penetration for the moment.

Data Analysis

An intriguing possibility, given a large hypertext database with typed links, is that it allows some degree of automatic analysis. It is possible to search, for example, for anomalies such as undocumented software or divisions which contain no people. It is possible to generate lists of people or devices for other purposes, such as mailing lists of people to be informed of changes.

It is also possible to look at the topology of an organisation or a project, and draw conclusions about how it should be managed, and how it could evolve. This is particularly useful when the database becomes very large, and groups of projects, for example, so interwoven as to make it difficult to see the wood for the trees.

In a complex place like CERN, it's not always obvious how to divide people into groups. Imagine making a large three-dimensional model, with people represented by little spheres, and strings between people who have something in common at work.

Now imagine picking up the structure and shaking it, until you make some sense of the tangle: perhaps, you see tightly knit groups in some places, and in some places weak areas of communication spanned by only a few people. Perhaps a linked information system will allow us to see the real structure of the organisation in which we work.

Live Links

The data to which a link (or a hot spot) refers may be very static, or it may be temporary. In many cases at CERN information about the state of systems is changing all the time. Hypertext allows documents to be linked into "live" data so that every time the link is followed, the information is retrieved. If one sacrifices portability, it is possible to make following a link fire up a special application, so that diagnostic programs, for example, could be linked directly into the maintenance guide.

Nonrequirements

Discussions on hypertext have sometimes tackled the problem of copyright enforcement and data security. These are of secondary importance at CERN, where information exchange is still more important than secrecy. Authorisation and accounting systems for hypertext could conceivably be designed which are very sophisticated, but they are not proposed here.

In cases where reference must be made to data which is in fact protected, existing file protection systems should be sufficient.

Specific Applications

The following are three examples of specific places in which the proposed system would be immediately useful. There are many others.

Development Project Documentation

The Remote procedure Call project has a skeleton description using *Enquire*. Although limited, it is very useful for recording who did what, where they are, what documents exist, etc. Also, one can keep track of users, and can easily append any extra little bits of information which come to hand and have nowhere else to be put. Cross-links to other projects, and to databases which contain information on people and documents, would be very useful, and save duplication of information.

Document Retrieval

The CERNDOC system provides the mechanics of storing and printing documents. A linked system would allow one to browse through concepts, documents, systems and authors, also allowing references between documents to be stored. (Once a document had been found, the existing machinery could be invoked to print it or display it.)

The "Personal Skills Inventory"

Personal skills and experience are just the sort of thing which need hypertext flexibility. People can be linked to projects they have worked on, which in turn can be linked to particular machines, programming languages, etc.

THE STATE OF THE ART IN HYPERMEDIA

An increasing amount of work is being done in hypermedia research at universities and commercial research labs, and some commercial systems have resulted. There have been two conferences, Hypertext '87 and '88, and in Washington D.C., the National Institute of Standards and Technology (NST) hosted a workshop on standardisation in hypertext, a followup of which will occur during 1990.

The *Communications of the ACM* special issue on hypertext contains many references to hypertext papers.[2] A bibliography on hypertext is given in *Proceedings of the Hypertext Standardisation Workshop, January 16–18, 1990,*[3] and a uucp newsgroup alt.hypertext exists.[4] I do not, therefore, give a list here.

Browsing Techniques

Much of the academic research is into the human interface side of browsing through a complex information space. Problems addressed are those of making navigation easy, and avoiding a feeling of being "lost in hyperspace." Whilst the results of the research are interesting, many users at CERN will be accessing the system using primitive terminals, and so advanced window styles are not so important for us now.

Interconnection or Publication?

Most systems available today use a single database. This is accessed by many users by using a distributed file system. There are few products which take Ted Nelson's idea of a wide "docuverse" literally by allowing links between nodes in different databases. In order to do this, some standardisation would be necessary. However, at the standardisation workshop, the emphasis was on standardisation of the format for exchangeable media, nor for networking. This is prompted by the strong push toward publishing of hypermedia information, for example on optical disk. There seems to be a general consensus about the abstract data model which a hypertext system should use.

Many systems have been put together with little or no regard for portability, unfortunately. Some others, although published, are proprietary software which is not for external release. However, there are several interesting projects and more are appearing all the time. Digital's "Compound Document Architecture" (CDA), for example, is a data model which may be extendible into a hypermedia model, and there are rumours that this is a way Digital would like to go.

Incentives and CALS

The U.S. Department of Defence has given a big incentive to hypermedia research by, in effect, specifying hypermedia documentation for future procurement. This means that all manuals for parts for defence equipment must be provided in hypermedia form. The acronym CALS stands for Computer-aided Acquisition and Logistic Support.

There is also much support from the publishing industry, and from librarians whose job it is to organise information.

What Will the System Look Like?

Let us see what components a hypertext system at CERN must have.

The only way in which sufficient flexibility can be incorporated is to separate the information storage software from the information display software, with a well-defined interface between them. Given the requirement for network access, it is natural to let this clean interface coincide with the physical division between the user and the remote database machine.

This division also is important in order to allow the heterogeneity which is required at CERN (and would be a boon for the world in general).

Therefore, **an important phase in the design of the system is to define this interface.** After that, the development of various forms of display program and of database server can proceed in parallel. This will have been done well if many different information sources, past, present and future, can be mapped onto the definition, and if many different human interface programs can be written over the years to take advantage of new technology and standards.

Accessing Existing Data

The system must achieve a critical usefulness early on. Existing hypertext systems have had to justify themselves solely on new data. If, however, there was an existing base of data of personnel, for example, to which new data could be linked, the value of each new piece of data would be greater.

What is required is a gateway program which will map an existing structure onto the hypertext model, and allow limited (perhaps read-only) access to it. This takes the form of a hypertext server written to provide existing information in a form matching the standard interface. One would not imagine the server actually generating a hypertext database from an existing one: rather, it would generate a hypertext view of an existing database.

Some examples of systems which could be connected in this way are

uucp News	This is a Unix electronic conferencing system. A server for uucp news could makes links between notes on the same subject, as well as showing the structure of the conferences.
VAX/Notes	This is Digital's electronic conferencing system. It has a fairly wide following in FermiLab, but much less in CERN. The topology of a conference is quite restricting.[5]
CERNDOC	This is a document registration and distribution system running on CERN's VM machine. As well as documents, categories and projects, keywords and authors lend themselves to representation as hypertext nodes.[6]
File systems	This would allow any file to be linked to from other hypertext documents.
The Telephone Book	Even this could be viewed as hypertext, with links between people and sections, sections and groups, people and floors of buildings, etc.
The unix manual	This is a large body of computer-readable text, currently organised in a flat way, but which also

	contains link information in a standard format ("See also . . .").
Databases	A generic tool could perhaps be made to allow any database which uses a commercial DBMS to be displayed as a hypertext view.

In some cases, writing these servers would mean unscrambling or obtaining details of the existing protocols and/or file formats. It may not be practical to provide the full functionality of the original system through hypertext. In general, it will be more important to allow read access to the general public: it may be that there is a limited number of people who are providing the information, and that they are content to use the existing facilities.

It is sometimes possible to enhance an existing storage system by coding hypertext information in, if one knows that a server will be generating a hypertext representation. In "news" articles, for example, one could use (in the text) a standard format for a reference to another article. This would be picked out by the hypertext gateway and used to generate a link to that note. This sort of enhancement will allow greater integration between old and new systems.

There will always be a large number of information management systems—we get a lot of added usefulness from being able to cross-link them. However, we will lose out if we try to constrain them, as we will exclude systems and hamper the evolution of hypertext in general.

Conclusion

We should work toward a universal linked information system, in which generality and portability are more important than fancy graphics techniques and complex extra facilities.

The aim would be to allow a place to be found for any information or reference which one felt was important, and a way of finding it afterwards. The result should be sufficiently attractive to use that the information contained would grow past a critical threshold, so that the usefulness of the scheme would in turn encourage its increased use.

The passing of this threshold accelerated by allowing large existing databases to be linked together and with new ones.

A Practical Project

Here I suggest the practical steps to go to in order to find a real solution at CERN. After a preliminary discussion of the requirements listed above, a survey of what is available from industry is obviously required. At this stage, we will be looking for systems which are future-proof:

- Portable, or supported on many platforms.
- Extendible to new data formats.

We may find that with a little adaptation, pars of the system we need can be combined from various sources: for example, a browser from one source with a database from another.

I imagine that two people for 6 to 12 months would be sufficient for this phase of the project.

A second phase would almost certainly involve some programming in order to set up a real system at CERN on many machines. An important part of this, discussed below, is the integration of a hypertext system with existing data, so as to provide a universal system, and to achieve critical usefulness at an early stage.

(. . . and yes, this would provide an excellent project with which to try our new object oriented programming techniques!)

George Landow and Paul Delany

<< 19 >> "Hypertext, Hypermedia and Literary Studies: The State of the Art" (1991)

Mark Amerika, Grammatron. *©Mark Amerika. Courtesy of Mark Amerika.*

"Hypertext . . . changes our sense of authorship, authorial property, and creativity (or originality) by moving away from the constrictions of page-bound technology. In so doing, it promises to have an effect on cultural and intellectual disciplines as important as those produced by earlier shifts in the technology of cultural memory that followed the invention of writing and printing."

<< George Landow has done much to pave the way for the critical acceptance of hypermedia as a medium for creative and academic writing. A noted literary theorist from Brown University—a hotbed for hypertext since Ted Nelson was a fellow there during the 1960s—Landow introduced the potential of hypermedia to a generation of writers and scholars eager to explore the medium's possibilities. His protégés include noted hyperfiction authors Michael Joyce and Mark Amerika, who began pioneering forms of interactive writing while studying hypertext at Brown. Landow also helped develop the Intermedia software system in the late 1980s, one of the first authoring tools for creating interactive texts. Intermedia was popular among hypermedia enthusiasts at the time and served as a model for Tim Berners-Lee's initial development of the World Wide Web.

Hypertext erodes the rigidity of print by encouraging the reader to navigate its contents freely, clicking through it in multiple directions, creating a unique, participatory reading experience. Landow, with Paul Delaney from Simon Fraser University, discusses hypermedia's impact on reading and writing, and how this ability to reassemble the text is an almost "embarrassingly literal" example of the deconstruction of meaning explored in late-twentieth-century critical theory.

Landow and Delaney examine literary conventions, such as footnotes and indexes, to illustrate how electronic means of linking texts have roots in traditional writing practice. They describe how hypertext extends these devices in such a way that they become central to the reading, transforming the text into a complex network of paragraphs, sentences, and fragments of "textual units" or "lexias," as they are called. The reader's ability to explore this network freely, and disrupt the linearity of the text, enables a dramatic shift from what they regard as hierarchical, centralized, and author-dominated literary forms. Landow and Delanay describe this new form as "intertextualities," a terrain where the boundaries between literary works dissolve as they join into a single, vast "docuverse." While they note that this literature of the future may introduce chaos into the experience of reading, they see a greater potential for an enhanced "technology of cultural memory," inspired by Vannevar Bush and Douglas Engelbart, that calls for a revitalized cultural dialogue and collaborative exchange between author and reader. >>

Hypertext, Hypermedia and the History of the Text

The written text is the stable record of thought, and to achieve this stability the text had to be based on a physical medium: clay, papyrus or paper; tablet, scroll or book.[1] But the text is more than just the shadow or trace of a thought already shaped; in a literate culture, the textual structures that have evolved over the centuries *determine* thought almost as powerfully as the primal structure that shapes all expression, language. So long as the text was married to a physical media, readers and writers took for granted three crucial attributes: that the text was *linear, bounded,* and *fixed.* Generations of scholars and authors internalized these qualities as the rules of thought, and they had pervasive social consequences. We can define *Hypertext* as the use of the computer to transcend the linear, bounded and fixed qualities of the traditional written text.[2] Unlike the static form of the book, a hypertext can be composed, and read, non-sequentially; it is a variable structure, composed of blocks of text (or what Roland Barthes terms *lexia*) and the electronic links that join them.[3] Although conventional reading habits apply within each block, once one starts to follow links from one block to another new rules and new experience apply. Instead of facing a stable object—the book—enclosing an entire text and held between two hands, the hypertext reader sees only the image of a single block of text on the computer screen. Behind that image lies a variable textual structure that can be represented on the screen in different ways, according to the reader's choice of links to follow. Metaphors that can help us to visualise the structure "behind" the screen include a network, a tree diagram, a nest of Chinese boxes, or a web.

The immediate ancestor of modern hypertext was described in a pioneering article by Vannevar Bush in the 1945 *Atlantic Monthly.*[4] Bush called for mechanically linked information-retrieval machines to help scholars and decision makers in the midst of what was already an explosion of information. In the 1960s Douglas C. Englebart and Theodor H. Nelson began to design and implement computer systems that could implement some of these notions of linked texts, and today hypertext as a term refers almost exclusively to computerized hypertext programs, and to the textual structures that can be composed with their aid. Hypertext programs began to be widely available on personal computers in the late 1980s; current examples are *Guide* and *Linkway* for IBM-compatible PCs, *Intermedia* for Macintoshes running A/UX, and *Writing Space* and *HyperCard* for most Macintoshes.

Because hypertext breaks down our habitual way of understanding and experiencing texts, it radically challenges students, teachers, and theorists of literature. But it can also provide a revelation, by making visible and explicit mental processes that have always been part of the total experience of reading. For the text as the reader *imagined* it—as opposed to the physical text objectified in the book—never had to be linear, bounded or fixed. A reader could jump to the last page to see how the story ended; could think of relevant passages in other works; could re-order texts by cutting and pasting. Still, the stubborn materiality of the text constrained such operations: they required some physical task such as flipping pages, pulling another book from a shelf, or dismembering the original text beyond repair.[5] Over the centuries, readers developed a repertoire of aids to textual management; these aids operated both within a single volume, and in the relations between volumes. They constituted a *proto-hypertext,* in which we can find important models for hypertext design today—though the special powers of the computer allow us to look beyond the textual aids that evolved during the long history of writing and printing.

Within the individual volume of the traditional book we may find such *internal* hypertextual functions as tables of contents, page-numbers, chapters, verses, rubrications, footnotes, and indexes. Some of these may be assigned by the original author, others by specialists in textual organization such as indexers or printers, or by later generations of scholars. *External* hypertextual functions have traditionally been post-authorial, supplied by librarians and bibliographers. Indeed, once one extends the idea of "a text" to include a collection of volumes the object of study ceases to be bounded, linear or fixed, and some kind of implicitly hypertextual organization will always be necessary.

How does hypertext actually work? Two examples, already familiar to most scholars, will help to show its underlying principles. Take first the elementary, but not trivial, question faced by large research libraries: we have a hundred and fifty kilometers or more of shelves—in what order should the books be placed on them? The primitive solution is an accession system, like the British Library, where each new volume is simply added on at the "end" of the shelf, and volumes are classified by the *physical* address of their original placement. Modern systems try to establish a *logical* order that places related books together, independent of any particular arrangement of shelves. This creates a kind of hypertextual linking, but done on a fixed and one-dimensional basis (so that, for example, one must choose between uniting all the books on the same subject, or all by the same author).

On-line library catalogues, now coming into general use, can be thought of as coarse-grained and rudimentary hypertext systems that support a "virtual" rearrangement and retrieval of individual volumes at the terminal. However, they are usually limited to a few standardized search categories, such as author, title and subject; and they cannot discriminate between any textual units smaller than a complete book. More sophisticated systems, such as the on-line MLA Bibliography, can perform subtler searches, using more detailed descriptors and dealing with articles as well as books. But neither deserve to be called true hypertext systems, because they operate on textual classifications rather than on the actual underlying complete texts. True hypertext must be able to define textual units, and link them in various ways, within an overall textbase or, to use another term now gaining currency, "docuverse."

Our second example of the principles underlying hypertext could be any standard scholarly article in the humanities. In reading an article on, say, Joyce's *Ulysses,* one reads through the main text, encounters a symbol that indicates the presence of a footnote, and leaves the main text to read that note, which can contain a citation of passages in *Ulysses* that supposedly support the argument in question as well as information about sources, influences, historical background, or related articles. In each case, the reader can follow the link to another text and thus move entirely outside the scholarly article itself. Having completed reading the note—and perhaps some of the texts to which it refers—one returns to the main text and continues reading until one encounters another note, and again leaves the main text.

This kind of reading constitutes a mental model of hypertext. Suppose now that one could simply touch the page where the symbol of a note, reference, or annotation appeared, and that act instantly brought into view the material contained in a note or even the entire other text—here all of *Ulysses*—to which that note refers. Scholarly articles situate themselves within a network of textual relations, most of which the print medium keeps out of sight and relatively difficult to follow—because the referenced (or linked) materials lie spatially distant from the reference mark. Electronic hypertext, in contrast, makes individual references easy to follow and the entire field of interconnections explicit and easy to navigate. Instant access to the whole network of textual references radically changes both the experience of reading and, ultimately, the nature of that which is read. If our putative Joyce article was linked, through hypertext, to all the other materials it cited, it would exist as part of a much larger system in which the totality might count more than the individual document; the article would now appear woven more tightly into its context than would a print-technology counterpart. The ease with which readers traverse such a system has further consequences: for as they move through this web or network of texts, they continually shift the center—and hence focus or organizing principle—of their investigation and experience. Hypertext provides an infinitely re-centerable system whose provisional point of focus depends upon the choices made by a truly active reader.

Hypertext thus presages a potential revolution in literary studies. However, an almost unlimited power to manipulate texts brings with it conceptual problems of extreme difficulty, which can be summed up by the questions: "What is a unit of text?" and "What are the relevant links between units?" Among traditional textual units the best-recognized and most functional ones are the word, the sentence, and the book. To think of them as commensurable units on a linear scale of magnitude is natural, but misleading. A word is a conceptual unit, a sentence a syntactical one, a book a unit whose identity is largely determined by its traditional status as a physical object. Nonetheless, they are units that can be handled by many kinds of textual aids developed over the whole period of literacy. But between the sentence and the book (in terms of magnitude) we find such units as footnotes, paragraphs, chapters and essays; these are less amenable to definition, because they are largely informal means of organizing thought.[6] For the same reason, however, they are likely to be important elements for building hypertext structures: they are the kind of mental "chunks" that we use to break a complex issue into components and make it intelligible. In addition, we are beginning to imagine new textual units, not yet codified or even named, that will be specific to the hypertext environment. They will come into being and pass away in the dynamic virtual text of the computer, products of such broad cognitive principles as identity, association and structure.[7]

These deep theoretical implications of hypertext converge with some major points of contemporary literary and semiological theory, particularly with Derrida's emphasis on decentering, with Barthes's conception of the readerly versus the writerly text, with post-modernism's rejection of sequential narratives and unitary perspectives, and with the issue of "intertextuality." In fact, hypertext creates an almost embarrassingly literal embodiment of such concepts. . . .

Finally, hypertext can be expected to have important institutional as well as intellectual effects, for it is at the same time a form of electronic text, a radically new information technology, a mode of publication, and a resource for collaborative work. "Both an author's tool and a reader's medium, a hypertext document system allows authors or groups of authors to link information together, create paths through a corpus of related material, annotate existing texts, and create notes that point readers to either bibliographic data or the body of the referenced text. . . . Readers can browse through linked, cross-referenced, annotated texts in an orderly but nonsequential manner."[8] Such electronic linking shifts the boundaries between individual works as well as those between author and reader and between teacher and student. It also has radical effects upon our experience of author, text, and work, revealing that many of our most cherished, most commonplace, ideas and attitudes towards literary production are the result of the particular technology of information and cultural memory that has provided the setting for them. This technology—that of the printing-press, the book, and the library—engenders certain notions of authorial property, authorial uniqueness, and a physically isolated text that hypertext makes untenable. Hypertext historicizes many of our most commonplace assumptions, forcing them to descend from the ethereality of abstraction and appear as corollary to a particular technology and historical era. We can be sure that a new era of computerized textuality has begun; but what it will be like we are just beginning to imagine.

From Hypertext to Hypermedia

Expository prose, with its linear and propositional structures, has been too much identified with the privileged form of reason itself. Hypertext provides a better model for the mind's ability to re-order the elements of experience by changing the links of association or determination between them. But hypertext, like the traditional text from which it derives, is still a radical reduction—to a schematic visual code—of what was originally a complex physical and intellectual experience, engaging all the five senses. *Hypermedia* takes us even closer to the complex interrelatedness of everyday consciousness; it extends hypertext by re-integrating our visual and auditory faculties into textual experience, linking graphic images, sound and video to verbal signs. Hypermedia seeks to approximate the way our waking minds always make a synthesis of information received from all five senses. Integrating or (re-integrating) touch, taste and smell seems the inevitable consummation of the hypermedia concept.

Consciousness itself is a continuous linking and re-structuring of images selected from past, present and future; from the real and the imaginary; from the internal and external realms of experience. Current hypermedia programs have taken only a few, faltering steps towards electronic representation of human memory, fantasy and cognition. Nonetheless, hypermedia is, in conception at least, a much better model of the mind's typical activities than exists in the severely restricted code of linear prose. We can argue, therefore, for a natural progression from the printed word to hypertext and hypermedia—analogous to the progression from painting to still photography, to silent movies, and now to movies with color and sound. This is not to claim that the newer media will altogether supersede the older ones. The black and white photograph remains viable, but is no longer the absolute standard of representation that it was in the nineteenth century. Similarly, the printed book will remain a central element of culture even as the new ways of interacting with texts make their own claims on our attention. . . .

Reconfiguring the Text

Dispersing the Traditional Text

Although in some not-so-distant future all individual texts may electronically link to one another, thus creating metatextual structures of a kind only partly imaginable at present, less far-reaching forms of hypertextuality have already appeared.[9] We already have works composed in hypertext that join blocks of text by electronic links to each other and to such graphic supplements as illustrations, maps, diagrams, visual directories and overviews. Second, there are the metatexts formed by interlinking individual sections of individual works. A third case is the adaptation for hypertextual presentation of material conceived in book technology. Such adaptations can work with textual units already given by the author, such as the individual sections of *In Memoriam*.[10] Conversely, one may, in the manner of Barthes's treatment of "Sarrasine," impose one's own *lexias* upon a work not explicitly divided into sections.

A fourth kind of hypertext puts a classical linear text, with its order and fixity, at the center of the structure. The composer then links various supplementary texts to this center, including critical commentary, textual variants, and chronologically anterior and later texts. In this case, the original text, which retains its old form, becomes an unchanging axis from which radiate linked texts that surround it, modifying the reader's experience of this original text-in-a-new-context.[11]

When compared to text as it exists in print technology, all these forms of hypertext evince varying combinations of atomization and dispersal. Unlike the spatial fixity of printed text, no one state of an electronic text is ever final; it can always be changed. Hypertext builds in a second fundamental mode of variation, since electronic links or reading pathways among individual blocks permit different paths through a text. The numerating rhetoric of "first, second, third" so well suited to linear text may appear within individual blocks of text but cannot control the unfolding of understanding in a medium that encourages readers to choose various paths, rather than following a fixed and linear one.

From a literary perspective based on book technology, the effects of electronic linking may appear harmful and dangerous. The notion of an individual, discrete work becomes increasingly undermined and untenable within this form of information technology, as it already has within much contemporary critical theory. Hypertext linking, reader control, and continual re-structuring not only militate against modes of argumentation to which we have become accustomed, but they have other, more general effects. The reader is now faced by a kind of textual randomness. The writer, conversely, loses certain basic controls over his text: the text appears to break down, to fragment and atomize into constituent elements (the *lexia* or block of text), and these reading units take on a life of their own as they become more self-contained because they are less dependent on what comes before or after in a linear succession.

At the same time that the individual hypertext block has looser, or less determining bonds to other blocks from the *same work* (to use a terminology that now threatens to become obsolete), it also can bond freely with text created by other authors. In fact, it bonds with whatever text links to it, thereby dissolving notions of the intellectual separation of one text from others as some chemicals destroy the cell membrane of an organism. Destroying the cell membrane will kill the cell; but destroying our conventional notions of textual separation has no fatal consequences. However, it will reconfigure the text and our expectations of it. As an individual block loses its physical and intellectual separation from others when linked electronically to them, it also finds itself dispersed into them. The necessary contextualization and intertextuality produced by situating individual reading units within a network of easily navigable pathways weaves texts, including those by different authors and those in nonverbal media, tightly together. One effect is to weaken and even destroy altogether any sense of textual uniqueness, for what is essential in any text appears intermingled with other texts. Such notions are hardly novel to contemporary literary theory, but here again hypertext creates an almost embarrassingly literal reification or actualization of a principle or quality that had seemed particularly abstract and difficult in its earlier statement. Since much of the appeal, even charm, of these theoretical insights lies in their difficulty and even preciousness, this more literal presentation promises to disturb theoreticians, in part, of course, because it disturbs status and power relations within their—our—field of expertise. . . .

Hypertext and the Author

Collaborative Work Hypertext demands new modes of reading, writing, teaching, and learning. In so doing it creates new understanding of collaborative learning and collaborative work. To most people, "collaboration" suggests two or more scientists, songwriters, or the like working side by side on the same endeavor, continually conferring as they pursue a project in the same place at the same time. Landow has worked on an essay with a fellow scholar in this manner. One of us, he relates, would type a sentence, at which point the other would approve, qualify, or rewrite it, and then we would proceed to the next sentence. But probably a far more common form of collaboration (and the one used in this introduction, whose authors live two thousand miles apart) is "versioning," in which one worker produces a draft that another person then later edits by modifying and adding. Both of these models require considerable ability to work productively with other people, and evidence suggests that many people lack this quality. According to those who have carried out experiments in collaborative work, a third form proves more common than the first two: the assembly-line or segmentation model of working together, in which individual workers divide up the overall task and work entirely independently. This last mode is the form that most people engaged in collaborative work choose.[12]

Networked hypertext systems like Intermedia offer a fourth model of collaborative work that combines aspects of the three just described. By emphasizing the cooperative interaction of blocks of text, networked hypertext makes all additions to a system simultaneously a matter of versioning and of the assembly-line mode. Once ensconced within a network of electronic links, a document no longer exists by itself. It always has an active relation to other documents in a way that a book or printed document never can. From this crucial shift in the way texts exist in relation to others derive two principles that determine this fourth form of collaboration: First, any document placed on a networked system that supports electronically linked materials potentially exists in collaboration with any and all other documents on that system; second, any document electronically linked to any other document collaborates with it.

According to the *American Heritage Dictionary,* to *collaborate* can mean either "to work together, especially in a joint intellectual effort" or "to cooperate treasonably, as with an enemy occupying one's country." The combination of labor, political power, and aggressiveness that appears in this dictionary definition well indicates some of the problems that arise when one discusses collaborative work. On the one hand, the notion of collaboration embraces notions of working together with others, of forming a community of action. This meaning recognizes that we all exist within social groups, and must make our contributions to them. On the other hand, collaboration also invokes a deep suspicion of working with others, something both aesthetically as well as emotionally engrained since the advent of romanticism, which exalts the idea of individual effort to such a degree that it often fails to recognize or even suppresses the fact that artists and writers work collaboratively with texts created by others.

Most of our intellectual endeavors involve collaboration, but we do not always recognize it. The rules of our culture, particularly those that define intellectual property and authorship, do not encourage such recognition; further, information technology from Gutenberg to the present—the technology of the book—systematically hinders full recognition of collaborative authorship. Hypertext, however, foregrounds this element of collaboration that other technologies of cultural memory suppress. It changes our sense of authorship, authorial property, and creativity (or originality) by moving away from the constrictions of page-bound technology. In so doing, it promises to have an effect on cultural and intellectual disciplines as important as those produced by earlier shifts in the technology of cultural memory that followed the invention of writing and printing. . . .[13]

<< part IV >>

Immersion

Morton Heilig

"The Cinema of the Future" (1955)

<< 20 >>

Morton Heilig, Sensorama. *Courtesy of Scott Fisher.*

"Thus, individually and collectively, by thoroughly applying the methodology of art, the cinema of the future will become the first art form to reveal the new scientific world to man in the full sensual vividness and dynamic vitality of his consciousness."

<< Morton Heilig, through a combination of ingenuity, determination, and sheer stubbornness, was the first person to attempt to create what we now call virtual reality. In the 1950s it occurred to him that all the sensory splendor of life could be simulated with "reality machines." Heilig was a Hollywood cinematographer, and it was as an extension of cinema that he thought such a machine might be achieved. With his inclination, albeit amateur, toward the ontological aspirations of science, Heilig proposed that an artist's expressive powers would be enhanced by a scientific understanding of the senses and perception. His premise was simple but striking for its time: if an artist controlled the multisensory stimulation of the audience, he could provide them with the illusion and sensation of first-person experience, of actually "being there."

Inspired by short-lived curiosities such as Cinerama and 3-D movies, it occurred to Heilig that a logical extension of cinema would be to immerse the audience in a fabricated world that engaged all the senses. He believed that by expanding cinema to involve not only sight and sound but also taste, touch, and smell, the traditional fourth wall of film and theater would dissolve, transporting the audience into an inhabitable, virtual world; a kind of "experience theater."

Unable to find support in Hollywood for his extraordinary ideas, Heilig moved to Mexico City in 1954, finding himself in a fertile mix of artists, filmmakers, writers, and musicians. There he elaborated on the multidisciplinary concepts found in this remarkable essay, "The Cinema of the Future." Though not widely read, it served as the basis for two important inventions that Heilig patented in the 1960s. The first was the Telesphere Mask. The second, a quirky, nickelodeon-style arcade machine Heilig aptly dubbed *Sensorama,* catapulted viewers into multisensory excursions through the streets of Brooklyn, and offered other adventures in surrogate travel. While neither device became a popular success, they influenced a generation of engineers fascinated by Heilig's vision of inhabitable movies. >>

Pandemonium reigns supreme in the film industry. Every studio is hastily converting to its own "revolutionary" system—Cinerama, Colorama, Panoramic Screen, Cinemascope, Three-D, and Stereophonic Sound. A dozen marquees in Time Square are luring customers into the realm of a "sensational new experience."

Everywhere we see the "initiated" holding pencils before the winked eyes of the "uninitiated" explaining the mysteries of 3-D. The critics are lining up pro and con concluding their articles profoundly with "after all, it's the story that counts." Along with other filmgoers desiring orientation, I have been reading these articles and have sadly discovered that they reflect this confusion rather than illuminate it. It is apparent that the inability to cope with the problem stems from a refusal to adopt a wider frame of reference, and from a meager understanding of the place art has in life generally.

All living things engage, on a higher or lower level, in a continuous cycle of orientation and action. For example, an animal on a mountain ledge hears a rumbling sound and sees an avalanche of rocks descending on it. It cries with terror and makes a mighty leap to another ledge. Here in small is the essence of a process that in animals and man is so automatic—so rapid—as to seem one indivisible act. By careful introspection, however, men have been able to stop its rapid flow, bring it into the light of consciousness, and divide it into three basic phases. The first, observation (the noise and image of the boulders in our example), is the reception of isolated impressions or facts. The second, integration, is the combining of these isolated facts with the inner needs of the life force into an emotional unity that prompts and controls action (the animal's sensation of danger and terror). The third, action (the leap to safety), is a change in the creature's physical relation to the world.

With the forming of society, different men concentrated on one of these three phases, and by learning to cast the results of their labor into concrete forms (that could be passed from man to man, and generation to generation) they created science, art, and industry. These three have the same methods and aims in the social body as the mind, heart, and muscle do in individual man. Their goals are clear. For science it is to bestow the maximum knowledge on humanity. For art it is to digest this knowledge into the deeper realms of feeling, generating emotions of beauty and love that will guide the crude energies of mankind to constructive actions. And for industry it is to act on the material world so as to procure more living energy for mankind. The success with which each field can approach its goal depends on its understanding of method. Science has come the closest because it has uncovered the individual's scientific thought processes and codified it into a clear and systematic method of experimentation. Consciously applying this method, it makes more discoveries in one year than previously were made in millenniums. Writing, international mail, and international conferences have long been efficient ways of distributing its findings to humanity.

Industry, within the last one hundred years, has also made great strides toward its goal because production geniuses like Ford have rationalized it to the last degree. They have instigated assembly line, mass production techniques that pour out more food, machines, and fuels in one year than were produced in centuries. The problem of distributing its bulky goods has been solved theoretically and only awaits practical application.

It is the middle field, art, which today is furthest from its goal. The world is woefully barren of peaceful, tolerant, humanitarian feelings and the art that should create them. And this is because, as yet, art has evolved no clear-cut methodology to make it as efficient as science and industry in creating its product. Art is now struggling feverishly to achieve this, and only in the light of this struggle and the laws it seeks to establish will we be able to understand the innovations that prompted this article.

The laws of art, like those of science and industry, lie hidden in the subconscious of man. When a primitive man desired to convey to another man the complete emotional texture of an experience that occurred to him he tried to reproduce, as closely as possible, the elements that generated his own emotions. His art was very simple, being limited to the means provided by his own body. He used his voice to growl like the bear that attacked him, pumped his arms and legs to show how he climbed a tree, and then he blew on his listener's face to make him feel the hot breath of the bear. If he were a good storyteller, he would arrange these effects in more or less the same order they originally happened to him. Of course, his listener would feel everything more intensely if he, and not his friend, were attacked by the bear. But aside from being impossible, this is not advisable, for by listening to his friend's story, he can have all the excitement, learn all the lessons, without paying the price for them.

With time language became more complete. A specific word-sound became associated with the impressions, objects, and feelings in man's experience. Words were useful in conveying the general structure of an event to the mind, but could rarely quicken the listener's pulse the way fresh and direct contact with the original sense elements could—that is unless spoken by a very skilled narrator. And even then not a thousand of his choicest words could convey the sensation of yellow better than one glance at yellow, or high C better than listening for one second to high C. And so side by side with verbal language they evolved more direct forms of communication—painting, sculpture, song and dance. They found they could bore deeper into experience by concentrating all their powers of observation on one of nature's aspects and mastering the limited materials necessary to its expression.

Materials became more complex and techniques more refined as each art form sought to exploit the full range and delicacy of its own domain. The few lines scratched on a rock developed into the full glory of painting. The singing voice evolved into symphonic music and the few words into the rich fabric of poetry. For all the apparent variety of the art forms created, there is one thread uniting all of them. And that is man, with his particular organs of perception and action. For all their ingenuity, a race of blind men could never have evolved painting. Similarly, no matter how much they appreciated movement through their eyes a race of limbless men could never have developed dancing. Thus art is like a bridge connecting what man can do to what he can perceive.

What we commonly refer to as the "pure arts" are those whose materials are so simple, so pliable, that one artist can master them sufficiently to express his inner feelings to perfection. The painter fashions color, the musician notes, the poet words. Each additional impression of their artistic form is like an electric charge driving the spectator higher and higher to peaks of pure and intense feeling that he rarely experiences in his daily life. The simple materials of the pure arts are apprehensible through only one sense, but this sense is not a necessary condition of purity. Precision and subtlety of form achieved through *control* is the decisive factor.

Desiring to convey the full richness of experience in more lifelike form, men have combined the pure arts into forms known as the "combined" or "secondary arts," such as opera, ballet, and theatre. Their effects were fuller, more spectacular, but rarely deeper. The essential factor of control was missing. Not only did the artist have to master visual, musical, choreographic, and verbal materials, not only did he have to limit the scope of his imagination to the practical limitations of a theatre and depend on the collaborations of dozens of singers, painters, dancers, musicians, and actors, but even after he had masterminded every detail and rehearsed the cast into perfect form he had absolutely no way of fixing his creation so that it could remain exactly the same whenever and wherever played. This was an impasse the artist could never surmount and never did, until the arrival of a strange newcomer on the scene—the machine. The machine with its genius for tireless repetition and infinite exactitude was an extension of the limbs and will of man. It could be trusted to perform all his purely mechanical operations, freeing his energies for more creative tasks.

In the form of the printing press, lithograph, radio, phonograph, and now television, the machine has rapidly solved the second part of art's age-old problem—distribution. Painting, poetry, music, drama, and ballet can now reach millions of people about the globe as they never could before. But the machine has done more. It has entered, as it has done in industry and science, into the very sphere of artistic creation itself, providing the artist with a much wider palette of *sense* material and enabling him to mold them with precision into an aesthetic unity as he never could before. And it is the invasion of such a relentlessly efficient and logical apparatus as the machine into the humane and heretofore romantic field of art that not only suggests but necessitates a clear, efficient *methodology* of art.

If at this point we scan back over our very brief history of art forms it becomes apparent that the first law of such an artistic methodology must be: "The nature of man's art is fundamentally rooted in his peculiar psychic apparatus and is limited by the material means at his disposal." Logically, then a proper science of art should be devoted to the revelation of the laws of his psyche and the invention of better means.

Although very little of it was conscious or intentional, nothing demonstrates this research and invention more dramatically than the cinema. The sense was vision—the material, light. The still camera had been invented but it could do no more than a skilled painter could do with time. But when in 1888 the Lumiere Brothers set up a little box before their factory and cranked away at it as a group of workers left, they did something no human being could ever do before. They captured visual movement in a form that could exactly reproduce the moving image as often as desired. Only after countless millenniums of existence had man learned how to do what his visual mechanism can do with no effort at all. With time every part of this new machine, from the lens to the film stock, was improved. Lenses were made faster and given wider angles. The iris became adjustable and the film finer-grained and faster. Always the criteria of invention were to reproduce as closely as possible man's miraculous mechanism of vision. The addition of color was inevitable. Man sees color, so must his mechanical eye. Now we have the so-called "revolutionary" 3-D and Wide Screen. The excitement and confusion are great but they need not be. First, 3-D was invented over 50 years ago and shown at the Paris Exposition. Financial, not technical, considerations held it up until 1953. The really exciting thing is that these new devices have clearly and dramatically revealed to everyone what painting, photography, and cinema have been semiconsciously trying to do all along—portray in its full glory the visual world of man as perceived by the human eye.

Side by side with the invention of means to freeze visual movement, machines were developed that could (this also for the first time in human history) freeze sound. But again, the public's deep and natural urge for more complete realism in its art had to wait on the wheel of finance until 1933. It is the addition of sound that represents the really great "revolution" in the history of cinema. For with the addition of sound, cinema stepped irrevocably out of the domain of the "pure arts" into the camp of the "combined arts." Rather than attempting to portray the whole through the part, it now began attempting to portray the whole directly. But with this tremendous difference from all other composite arts—it could do it without losing *control* or *permanence.* With the help of the machine two radically different sense materials, light and sound, could be dynamically combined into one work without losing any of the control, subtlety, or concreteness formerly attained only by the pure arts. Cinema was no longer just a visual art (notwithstanding the effort of some directors to keep it such by shooting visual films and pouring the sound track over it like some pleasant, superfluous goo), but had set itself the task of expressing in all its variety and vitality the full consciousness of man.

Instead of continuing to stumble along this road with the system of hit and miss, let us, according to our first law, deliberately turn to life and study the nature of man's consciousness.

Man's nervous system—sensory nerves, brain, and motor nerves—is the seat of his consciousness. The substance or component parts of this consciousness can be determined by the process of elimination. If a man lies still, or, due to some disease or drug, has his motor nerves blocked, his consciousness or wide-awakeness is not diminished in any way. If, however, he closes his eyes, it is. If he stops his ears, it is diminished further. If he pinches his nose and does not taste anything and avoids tactile impressions, his awakeness is diminished considerably. And if, as is done in anesthesia, all sensory nerves leading to the brain are blocked, he would lose consciousness completely. (Dreams and internal voices merely being sense impressions of former experiences stored away and served up later by memory.) Thus we can state our second law: "Consciousness is a composite of all the sense impressions conveyed to the brain by the sensory part of the nervous system which can be divided into the great receiving organs—the eyes, ears, nose, mouth, and skin."

By concentrating on one organ at a time, we can list the various elements affecting it. These are, for the eye, peripheral imagery—180° horizontal × 150° vertical, three dimensionality, color and movement; for the ear, pitch, volume, rhythm, sounds, words, and music; for the nose and mouth, odors and flavors; and for the skin, temperature, texture, and pressure. These divisions—although purely subjective and dependent on vocabulary and techniques of reproduction—are nonetheless useful for analysis.

These elements are the building bricks, which when united create the sensual form of man's consciousness, and the science of art must devote itself to inventing techniques for recording and projecting them in their entirety. Celluloid film is a very crude and primitive means of recording light and is already being replaced by a combination television camera and magnetic tape recorder. Similarly, sound recording on film or plastic records is being replaced by tape recording.

Odors will be reduced to basic qualities the way color is into primary colors. The intensity of these will be recorded on magnetic tape, which in turn will control the release from vials into the theatre's air conditioning system. In time all of the above elements will be recorded, mixed, and projected electronically—a reel of the cinema of the future being a roll of magnetic tape with a separate track for each sense material. With these problems solved it is easy to imagine the cinema of the future.

Open your eyes, listen, smell, and feel—sense the world in all its magnificent colors, depth, sounds, odors, and textures—this is the cinema of the future!

The screen will not fill only 5% of your visual field as the local movie screen does, or the mere 7.5% of Wide Screen, or 18% of the "miracle mirror" screen of Cinemascope, or the 25% of Cinerama—but 100%. The screen will curve past the spectator's ears on both sides and beyond his sphere of vision above and below. In all the praise about the marvels of "peripheral vision," no one paused to state that the human eye has a vertical span of 150°[1] as well as a horizontal one of 180°. The vertical field is difficult, but by no means impossible, to provide. Planetariums have vertical peripheral vision and the cinema of the future will provide it along similar lines as shown in the accompanying drawing. This 180° × 150° oval will be filled with true and not illusory depth. Why? Because as demonstrated above this is another essential element of man's consciousness. Glasses, however, will not be necessary. Electronic and optical means will be devised to create illusory depth without them.

Cinemascope, despite all the raving of its publicity men that it is the "crowning glory" of motion picture development, represents one small step forward, and one big one backward. Its increase of screen image from 5 to 18% of man's visual field is a definite improvement although there is still 82% to go. It has, however, regressed substantially in clarity. One reason that few critics noticed for Cinerama's excellent illusion of reality is its extraordinary clarity. The human eye is one of the most perfect in the animal kingdom. It is not spotty, out-of-focus, or jumpy the way average movie images are. The image it records is limpid, razor-sharp, and solid as a rock, and Cinerama, by using three film strips instead of one, and specially designed projectors, makes a great advancement towards this perfection. Cinemascope, on the other hand, by still using only one film strip to cover two and one-half the normal screen area, is also magnifying grain, and softening the focus two and one-half times, making clarity much worse than it is on the normal screen.[2] The electrically created image of tomorrow's film will be perfect in focus and stability—the grain and spots vanishing along with the film stock.

Stereophonic sound will be developed so that the spectator will be enclosed within a sphere, the walls of which will be saturated with dozens of speakers. Sounds will come from every direction—the sides, top, back and bottom—as they do in real life.

The large number of speakers will permit a much better identity of image and sound than is achieved now where the sound leaping from one distant speaker to another is either behind or ahead of the image. The air will be filled with odors and up to the point of discretion or aesthetic function we will feel changes of temperature and the texture of things. We will feel physically and mentally transported into a new world.

Yes, the cinema of the future will far surpass the "Feelies" of Aldous Huxley's *Brave New World.* And like many other things in this book that are nightmarish because superficially understood, it will be a great new power, surpassing conventional art forms like a Rocket Ship outspeeds the horse and whose ability to destroy or build men's souls will depend purely on the people behind it.

The mastery of so many sense materials pose another problem—selection. People already complain about the excess of realism in films and say the new inventions shall plunge us from bad to worse. Although the spirit of their complaint is valid, their use of the word "realism" is not. "Realism," or, in aesthetic terms, "experience," is that something which is created by the unity of the outer world with the inner. No matter how extensive the artist's means, he must use them to provoke more of the spectator's participation, not less. For without the active participation of the spectator there can be no transfer of *consciousness, no art.* Thus art is never "too" realistic. When either too much or too little is given, there just isn't any "realism." Poor use of cinema's remarkable new powers is no more of a case against them than daubing with oils is a case against painting.

It is estimated that each sense monopolizes man's attention in the following proportions:

Sight	70%
Hearing	20%
Smell	5%
Touch	4%
Taste	1%

Men can have their attention led for them as a bird will do by flying across an empty sky, or can willfully direct it as everyone does at the dinner table when singling one voice out of the maze of chatter. In each case the criterion is "what is the point of greatest interest and significance to me?" Thousands of sense impressions stimulate the sensory nerves every second of the day, but only one or a few are permitted to enter the *realm of higher consciousness at a time.* The organ that screens them out is the brain. The brain is the storehouse—the memory of the physical and spiritual needs of the individual, and through him of the human race, and it is according to this criteria—"what is beneficial for the development of the individual and racial life force?" that a decision is made. We can now state the third law of our methodology of art: "The brain of man shifts rapidly from element to element within each sense and from sense to sense in the approximate proportion of sight, 70%; hearing, 20%; smell, 5%; touch, 4%; and taste, 1%, selecting one impression at a time according to the needs of individual and racial development." These unite into the dynamic stream of sensations we call "consciousness." The cinema of the future will be the first direct, complete and conscious application of this law. Since the conventional movie screen fills only 5% of the spectator's field of vision, it automatically represents his point of visual attention and the director needs only to point his camera to control the point of attention. But with the invention of means to fill 100% of the spectator's field of vision with sharp imagery, he must solve the problem of visual attention another way or lose his main aesthetic power.

Every capable artist has been able to draw men into the realm of a new experience by making (either consciously or subconsciously) a profound study of the way their attention shifts. Like a magician he learns to lead man's attention with a line, a color, a gesture, or a sound. Many are the devices to control the spectator's attention at the opera, ballet, and theatre. But the inability to eliminate the unessential is what loosens their electrifying grip on the attention of a spectator and causes them to remain secondary arts.

The evolution of the aesthetic form of cinema can, in a way, be described as a continuation of the artist's struggle to master attention. Griffith began using the "close-up" to draw the spectator's attention to a significant visual detail. Lenses with narrow focus fields were devised to throw foreground and background out of focus, riveting the eye only on the sharp part of the image. Pudovkin developed the close-up in time by varying camera speeds to parallel the varying intensity of man's observations. Eisenstein proclaimed "montage" and Griffith discovered "parallel cutting," both magnificent weapons in the director's arsenal of attention. Shots and scenes could now be shifted with the same freedom and rapidity possible in man's natural observation or imagination. Sound arrived with undiminished intensity, but in time it too became refined in content, pitch, and volume, sometimes dominating the scene, sometimes leaving it completely, leading the ear as precisely as the eye. But like the search for an additional number of sense materials, the principle involved in this refinement of attention were mostly stumbled on by accident—rarely searched for deliberately, and never formulated consciously.

Again, the only place to search is in the mind of man. We must try to learn how man shifts his attention normally in any situation.

Suppose we are standing on a hilltop overlooking the countryside. First we are struck by the huge sweep of the view before us. Then we notice the vivid green of the fields and the sunshine. Then the silent expansion and rolling of a cumulus cloud entrances us. We feel a warm gust of wind and our nostrils dilate at the smell of new-mown hay. Suddenly, our ears sharpen as the shriek of a jet plane cuts the air. We cannot see it, but we linger on the way its high-tone lowers in pitch and fades away. Here is an example of how attention shifts from one element (space—color—then movement, in our example) within a single sense (the eye) and from one sense to another (the eye, the skin, the nose, the ear). In each moment it fixes itself, if for only an instant, on one sense element to the partial or complete exclusion of all others.

In life, only the object being observed is in focus. The area of focus is not necessarily rectangular, including everything in the same plane, as it is in today's films, but can be circular, triangular, vertical, or horizontal, depending on the shape of the objects of interest. Electrical and optical means will be developed to duplicate this flexibility, retaining the hazy frame of peripheral vision as the human eye does for added realism. Naturally, the great visual oval of the camera field will include, exclude, move closer, and recede as it does in life. This zone of focus will generally be at the center of the visual field, but it will be free to shift up and down, or around to the sides, leading the eye wherever it goes. The direction, quality, and intensity of all other sense elements will be controlled and pin-pointed in the same subtle manner.

Each basic sense will dominate the scene in roughly the same proportion we found them to have in man. That is, sight, 70%; sound, 20%; smell, 5%; touch, 4%; and taste, 1%. Nature turns them on and off without a whimper but filmmakers once in possession of a new power usually cling to it like a drowning man to a life raft. Eye irritating colors, ear deafening dialogue, and soul sickening music are loaded one on top of another just to "make sure the point gets across." The cinema of the future will turn any and all of it off, including the visual part, when the theme calls for it. For, and it cannot be stressed too strongly, the cinema of the future will no longer be a "visual art," but an "art of consciousness."

When a great many sense materials are presented in sharp focus simultaneously the spectator must do his own selecting. He is no longer being led along by a work of art, but must begin with great fatigue to create his own patterns. This situation is so life-like that it gives the spectator the sensation of being *physically* in the scene. For example, in Cinerama's famous roller coaster sequence, the spectator's body, not his soul, is riding the roller coaster. This is a tremendous faculty and will, I am sure, be used to great effect in the cinema of the future, but it must be used with great discretion. For aside from being very tiring, after one too many loops, the spectator may be so thoroughly convinced that he is shooting the chutes as to throw up on the lady in front of him. As stated before, art is a specific technique for living vicariously, of weeping without actually losing a loved one, of thrilling to the hunt without being mangled by a lion, in short of reaping the lessons and spiritual nourishment of experience without any loss. The solution of the problem of focus will invalidate the opinion that the wide screen is no good for "the intimate thing." If man can have intimate moments in life with his peripheral vision, stereophonic hearing, smell, and touch, so can his art.

It would seem from the preceding analysis that my conception of the function of the cinema of the future is to faithfully reproduce man's superficial and immediate perception of the world about him. Nothing could be further from the truth. The history of art demonstrates over and over again that some of the most valid experiences come from the inner and not the outer world. But the history, not only of art but any other human endeavor, also proves that the outer precedes the inner. The outer world supplies the raw materials of creation. Man cannot originate. He can only take the forces of nature and rearrange them into shapes more friendly to his own existence. Just as nature had to provide water, iron, fire, and the laws of thermodynamics before someone could invent the steam engine, so nature must supply man with raw impressions before he can fashion them into an imagery more meaningful and useful to himself. The first task of painting was to copy the world, and only when the camera relieved it of this mirror-like function was it really free to explore the full range of man's fantasy. At first, motion picture cameras and sound recorders could not even capture the simplest aspects of man's perception of the outer world. Now, though still far from matching some of these, they are far superior to others. Slow motion, fast motion, and infrared ray photography are able to "see" things no human eye can.

Supersensitive microphones are now able to "hear" sounds way beyond the range of human ears. Similarly, directors at first had to be content with what the natural scene about them offered. Then, in studios, they began to select and arrange what went before the lens. By building sets, and developing trick photography, they could set the world of history and fantasy before the lens. Then, by perfecting the technique of animation, they could do without bulky sets and intricate models entirely and give free reign to their wildest imagination.

Sound has followed a similar evolution—from the objective to the subjective world. First we recorded only natural sounds, or the sounds created by human voices and musical instruments. Then we invented a whole series of odd new sound-making instruments and set them about the microphone. Now people like Norman McLaren are dispensing with expensive instruments and microphones entirely and are creating sound never heard before by painting directly on the sound track.

These developments bring us to our fourth law: "In his creative process, man is imposed on by outer impressions. He learns the secrets of their basic principles through imitation and then subjects these to the needs of his own expression. He goes from reception to imitation to creation, i.e., from portraying the outer to portraying the inner world."

This law will inevitably hold true for the cinema of the future. While it still must learn to faithfully reproduce man's outer world as perceived in his consciousness, it will eventually learn to create totally new sense materials for each of the senses—shapes, movements, colors, sounds, smells, and tastes—they have never known before, and to arrange them into forms of consciousness never before experienced by man in his contact with the outer world. . . .

The theatre will provide for the full exercise of the social instincts. It will incorporate a promenade and cafe around the theatre proper. The film will not be presented as "entertainment" but as an evening of community culture. A speaker will review the personalities in and history of the film being viewed. After the performance the audience will criticize the film in a discussion facilitated by television relays and led by a moderator. The audience will be able to continue the discussion in the cafe-lounge or on the promenade where they can see, be seen, and enjoy the evening in a thoroughly social fashion. Thus, individually and collectively, by thoroughly applying the methodology of art, the cinema of the future will become the first art form to reveal the new scientific world to man in the full sensual vividness and dynamic vitality of his consciousness.

—Translated by Uri Feldman

Ivan Sutherland

<< 21 >> "The Ultimate Display" (1965)

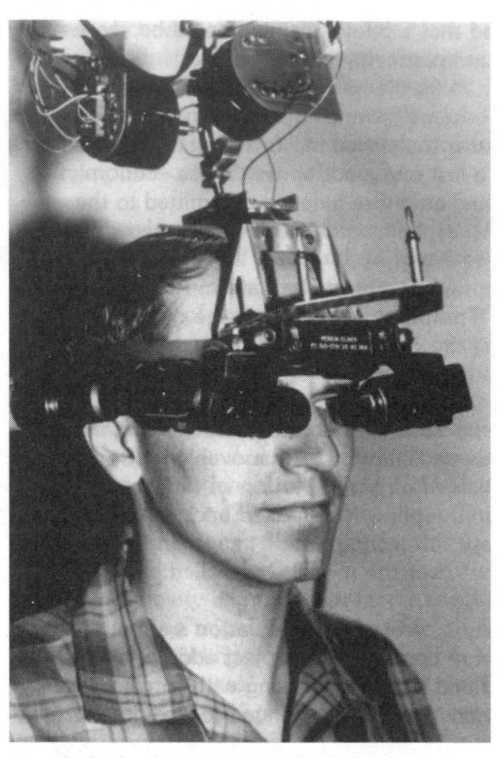

Ivan Sutherland, Head-Mounted Display (worn by Don Vickers). Courtesy of University of Utah.

"The ultimate display would, of course, be a room within which the computer can control the existence of matter. A chair displayed in such a room would be good enough to sit in. Handcuffs displayed in such a room would be confining, and a bullet displayed in such a room would be fatal. With appropriate programming such a display could literally be the Wonderland into which Alice walked."

<< Ivan Sutherland made his first significant contribution to computer science in 1963 as a twenty-two-year-old graduate student at MIT, where he pioneered the field of interactive computer graphics with his legendary Sketchpad system. Later, in 1965, while conducting his initial research in immersive technologies, Sutherland wrote "The Ultimate Display," in which he made the first advance toward marrying the computer to the design, construction, navigation, and habitation of virtual worlds. This essay has since become known by generations of scientists, computer artists, and media theorists as the seminal statement to define the field of computer-based virtual reality.

Morton Heilig provided the first glimpse of a simulated reality, but had no means to pursue his vision. Sutherland, on the other hand, had access to the abundant resources of ARPA. He predicted that advances in computer science would eventually make it possible to engineer virtual experiences that were convincing to the senses. Sutherland understood that pure information can be expressed in any number of forms, through an interface of the engineer's choosing. That interface might be the lines of type on a screen, or, he reasoned, it could be a fully realized three-dimensional environment. While this speculation seems influenced more by science fiction than scientific fact, Sutherland believed in the ineffable potential of computers to transform the abstract nature of mathematical constructions into habitable, expressive worlds in the spirit of Lewis Carroll's *Alice in Wonderland.*

In this essay, Sutherland presents a survey of the hardware and software that would someday constitute the computer-driven "reality" engine. Although it was several years before the invention of the personal computer, Sutherland had already conceived an interactive, immersive system that went well beyond the keyboard, light pen, and joystick, which were the cutting-edge technologies of the day. In 1966, Sutherland took a crucial step toward the implementation of his vision by inventing the head-mounted display—a helmet-shaped apparatus designed to immerse the viewer in a visually simulated 3-D environment. He completed the prototype in 1970 at the University of Utah, and fondly nicknamed it the "Sword of Damocles," because the mass of hardware threatened to decapitate the immersant. This invention led to the development of the more sophisticated and visually arresting representations of reality that became possible with advances in computing power and 3-D graphics. >>

We live in a physical world whose properties we have come to know well through long familiarity. We sense an involvement with this physical world which gives us the ability to predict its properties well. For example, we can predict where objects will fall, how well-known shapes look from other angles, and how much force is required to push objects against friction. We lack corresponding familiarity with the forces on charged particles, forces in nonuniform fields, the effects of nonprojective geometric transformations, and high-inertia, low friction motion. A display connected to a digital computer gives us a chance to gain familiarity with concepts not realizable in the physical world. It is a looking glass into a mathematical wonderland.

Computer displays today cover a variety of capabilities. Some have only the fundamental ability to plot dots. Displays being sold now generally have built in line-drawing capability. An ability to draw simple curves would be useful. Some available displays are able to plot very short line segments in arbitrary directions, to form characters or more complex curves. Each of these abilities has a history and a known utility.

It is equally possible for a computer to construct a picture made up of colored areas. Knowlton's movie language, BEFLIX,[1] is an excellent example of how computers can produce area-filling pictures. No display available commercially today has the ability to present such area-filling pictures for direct human use. It is likely that new display equipment will have area-filling capability. We have much to learn about how to make good use of this new ability.

The most common direct computer input today is the typewriter keyboard. Typewriters are inexpensive, reliable, and produce easily transmitted signals. As more and more on-line systems are used, it is likely that many more typewriter consoles will come into use. Tomorrow's computer [user] will interact with a computer through a typewriter. He ought to know how to touch type.

A variety of other manual-input devices are possible. The light pen or RAND Tablet stylus serve a very useful function in pointing to displayed items and in drawing or printing for input to the computer. The possibilities for very smooth interaction with the computer through these devices is only just beginning to be exploited. RAND Corporation has in operation today a debugging tool which recognizes printed changes of register contents, and simple pointing and moving motions for format relocation. Using RAND's techniques you can change a digit printed on the screen by merely writing what you want on top of it. If you want to move the contents of one displayed register into another, merely point to the first and "drag" it over to the second. The facility with which such an interaction system lets its user interact with the computer is remarkable.

Knobs and joysticks of various kinds serve a useful function in adjusting parameters of some computation going on. For example, adjustment of the viewing angle of a perspective view is conveniently handled through a three-rotation joystick. Push-buttons with lights are often useful. Syllable voice input should not be ignored.

In many cases the computer program needs to know which part of a picture the man is pointing at. The two-dimensional nature of pictures makes it impossible to order the parts of a picture by neighborhood. Converting from display coordinates to find the object pointed at is, therefore, a time-consuming process. A light pen can interrupt at the time that the display circuits transfer the item being pointed at, thus automatically indicating its address and coordinates. Special circuits on the RAND Tablet or other position input device can make it serve the same function.

What the program actually needs to know is where in memory is the structure which the man is pointing to. In a display with its own memory, a light pen return tells where in the display file the thing pointed to is, but not necessarily where in main memory. Worse yet, the program really needs to know which sub part of which part the man is pointing to. No existing display equipment computes the depths of recursions that are needed. New displays with analog memories may well loose the pointing ability altogether.

Other Types of Display

If the task of the display is to serve as a looking glass into the mathematical won derland constructed in computer memory, it should serve as many senses as possible. So far as I know, no one seriously proposes computer displays of smell, or taste. Excellent audio displays exist, but unfortunately we have little ability to have the computer produce meaningful sounds. I want to describe for you a kinesthetic display.

The force required to move a joystick could be computer controlled, just as the actuation force on the controls of a Link Trainer are changed to give the feel of a real airplane. With such a display, a computer model of particles in an electric field could combine manual control of the position, of a moving charge, replete with the sensation of forces on the charge, with visual presentation of the charge's position. Quite complicated "joysticks" with force feedback capability exist. For example, the controls on the General Electric "handyman" are nothing but joysticks with nearly as many degrees of freedom as the human arm. By use of such an input/output device, we can add a force display to our sight and sound capability.

The computer can easily sense the positions of almost any of our body muscles. So far only the muscles of the hands and arms have been used for computer control. There is no reason why these should be the only ones, although our dexterity with them is so high that they are a natural choice. Our eye dexterity is very high also. Machines to sense and interpret eye motion data can and will be built. It remains to be seen if we can use a language of glances to control a computer. An interesting experiment will be to make the display presentation depend on where we look.

For instance, imagine a triangle so built that whichever corner of it you look at becomes rounded. What would such a triangle look like? Such experiments will lead not only to new methods of controlling machines but also to interesting understandings of the mechanisms of vision.

There is no reason why the objects displayed by a computer have to follow the ordinary rules of physical reality with which we are familiar. The kinesthetic display might be used to simulate the motions of a negative mass. The user of one of today's visual displays can easily make solid objects transparent—he can "see through matter!" Concepts which never before had any visual representation can be shown, for example the "constraints" in Sketchpad.[2] By working with such displays of mathematical phenomena we can learn to know them as well as we know our own natural world. Such knowledge is the major promise of computer displays.

The ultimate display would, of course, be a room within which the computer can control the existence of matter. A chair displayed in such a room would be good enough to sit in. Handcuffs displayed in such a room would be confining, and a bullet displayed in such a room would be fatal. With appropriate programming such a display could literally be the Wonderland into which Alice walked.

Scott Fisher

“Virtual Interface Environments” (1989)

<< 22 >>

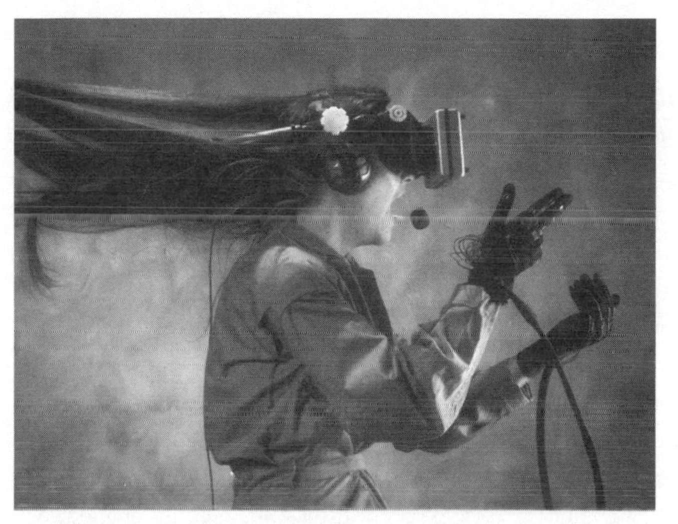

Virtual Interface Environmental Workstation. *Photo by W. Fisher and S. Fisher. Courtesy of Scott Fisher, NASA Ames Research Center.*

"The possibilities of virtual realities, it appears, are as limitless as the possibilities of reality. They can provide a human interface that disappears—a doorway to other worlds."

<< Scott Fisher is one of the new breed of artist-engineers who have emerged in recent years from interdisciplinary academic institutions such as MIT's Center for Advanced Visual Study. While a student in the 1970s, Fisher worked on the groundbreaking Aspen Movie Map project, in which the viewer-participant navigates through an interactive video-disc representation of Aspen, Colorado. This form of environmental simulation, dubbed "surrogate travel," was a strong influence on Fisher's subsequent work, which integrates scientific research in virtual reality with the artistic application of immersive environments. Inspired by experiments in simulation such as the nineteenth-century zoetrope, the nickelodeon, Cinerama, Morton Heilig's *Sensorama,* and Ivan Sutherland's head-mounted display, Fisher's work, which combines these influences, defines our contemporary notion of virtual reality.

This article documents Fisher's seminal research conducted in the late 1980s at the NASA-Ames Research Center in Mountain View, California, where he worked on the Virtual Environment Workstation (VIEW) project. Intending to extend the possibilities of the virtual environment beyond the "surrogate travel" of the Aspen project, Fisher set out to develop an interface that would engage all the senses, thrusting the viewer into a realm of full sensory immersion. The NASA system included an updated version of the head-mounted display, with stereoscopic images that provided depth of field, a major advancement over the monoscopic vision of Sutherland's earlier device. Fisher added headphones for surround-sound, a microphone for speech recognition, and, in collaboration with scientist Tom Zimmerman, adapted the "dataglove"—the wired glove worn by the user that makes it possible to grasp virtual objects in cyberspace.

This multisensory interaction with cybernetic devices created the powerful illusion of entering a digitized landscape. By pursuing Heilig's concept of experience theater, Fisher made a significant advance toward what he termed "telepresence"—the projection of the self into a remote location or virtual world. This experience could range from simulations of undersea exploration to flights through abstract, imaginary space. Influenced by the work of artist Myron Krueger, Fisher's research into the use of collaborative telepresent environments has led to such diverse applications as telesurgery, architectural walkthroughs, on-line communities, and multi-user games. >>

On the morning of July 29, 2001, Helen and Graham Lesh commuted to work in their accustomed manner. They got up from the breakfast table, kissed each other goodbye for the morning, and walked down the hall to their respective workspaces, where they slipped into their datasuits and donned their virtual environment headsets.

An outside observer would see Helen standing almost immobile in the center of her dimly lit space, making a series of fine movements with her hands. An outside observer would not be seeing the same environment that Helen saw, however. Helen, a plastic microsurgeon, found herself, as expected, in full sensory contact with her virtual operating theater: The bright white lights, the blue-garbed OR crew, and the phalanxes of equipment that surrounded them, looked and sounded exactly as they would in a real operation. Reaching for a scalpel, murmuring commands to her interface, Helen prepared for an hour of intensive rehearsal for the operation that would take place in the real, life-and-death operating theater at University Hospital the following day.

While Helen was zooming her view in on facial capillaries modeled on those of tomorrow's patient and etching virtual tissue a few hundred molecules at a time with a micromanipulated laser scalpel, Graham was in his workspace, trying to create a new pharmaceutical. An outside observer would think that Graham was miming a wrestling match with himself, or recapitulating his infancy, for Graham was on all fours, his head aimed at a point in space near one corner of the workspace. In Graham's reality, he was crawling around a complex molecule, reaching for other molecules and trying to fit them together, like a kind of dynamic Tinkertoy.

For the time being, Graham and Helen are fictitious, but they are using perfectly plausible technological descendants of the virtual environment display systems that have been developed over the past four years at NASA's Ames Research Center in the Aerospace Human Factors Research Division [Fisher et al., 1986, 1987, 1988].[1]

Here, in fact, there is a graphic simulation of a hemoglobin molecule that a person can fly through and move around in physically, as if reduced to molecular size. A head-mounted, wide-angle, stereoscopic display system controlled by operator position, voice, and gesture has been developed for use as a multipurpose, multimodal interface environment.[1] The system provides a multisensory, interactive display environment in which a user can virtually explore a 360-degree synthesized or remotely sensed environment and can viscerally interact with its components. Our objective has been to develop a new kind of interface that would be very closely matched to human sensory and cognitive capabilities. The effect of immersing one's sensoria in even the crudest prototypes of these interfaces has led many who have experienced it to refer to the experience as a kind of *virtual reality*. We prefer to use the term *virtual environment* to emphasize the ability to completely immerse a subject in a simulated space with its attendant realities.

A Bit of History: The Evolution of Personal Simulation and Telepresence Environments

Matching visual display technology as closely as possible to human cognitive and sensory capabilities in order to better represent "real" experience has been a major ambition for research and industry for decades. One example is the development of stereoscopic movies in the early 1950s, in which a perception of depth was created by presenting a slightly different image to each eye of the viewer. In competition with that during the same era was Cinerama, which involved three different projectors presenting a wide field of view display to the audience. Because the size of the projected image was extended, the viewer's peripheral field of view was also engaged. More recently, along the same lines, the Omnimax projection system situates the audience under a huge hemispherical dome, and the film image is predistorted and projected on the dome; the audience is now almost immersed in an image surround.

The idea of sitting inside an image has been used in the field of aerospace simulation for many decades to train pilots and astronauts to safely control complex, expensive vehicles through simulated mission environments. More recently, this technology has been adapted for entertainment and educational use. "Tour of the Universe" in Toronto and "Star Tours" at Disneyland are among the first entertainment applications of simulation technology and virtual display environments; about 40 people sit in a room on top of a motion platform that moves in synch with a computer-generated and model-based image display of a ride through a simulated universe.

The technology has been moving gradually toward lower-cost personal simulation environments in which viewers are also able to control their own viewpoint or motion through a virtual environment. An early example of this is the

Aspen Movie Map, done by the M.I.T. Architecture Machine Group in the late 1970s [Lippman, 1980]. Imagery of the town of Aspen, Colorado, was shot with a special camera system mounted on top of a car, filming down every street and around every corner in town, combined with shots above town from cranes, helicopters, and airplanes and also with shots inside buildings. The Movie Map gave the operators the capability of sitting in front of a touch-sensitive display screen and driving through the town of Aspen at their own rate, taking any route they chose, by touching the screen, indicating what turns they wanted to make and what buildings they wanted to enter. In one configuration, this was set up so that the operator was surrounded by front, back, and side-looking camera imagery so that, again, they were completely immersed in a virtual environment.

Another notable virtual environment developed for entertainment applications was "Sensorama." This was an elegant prototype of an arcade game designed by Morton Heilig in the mid-1960s and one of the first examples of a multisensory simulation environment that provided more than just visual input. When you put your head up to a binocular viewing optics system, you saw a first-person-viewpoint, stereo-film-loop of a motorcycle ride through New York City and you heard three-dimensional binaural sound that gave you sounds of the city of New York and of the motorcycle moving through it. As you leaned your arms on the handlebar platform built into the prototype and sat in the seat, simulated vibration cues were presented. The prototype also had a fan for wind simulation that combined with a chemical smell bank to blow simulated smells in your face.

Conceptual versions of virtual environments have been described by science fiction writers for many decades. One concept has been called *telepresence,* a technology that would allow remotely situated operators to receive enough sensory feedback to feel like they are really at a remote location and are able to do different kinds of tasks. Arthur Clarke has described "personalized television safaris" in which the operator could virtually explore remote environments without danger or discomfort. Heinlein's "waldoes" were similar, but were able to exaggerate certain sensory capabilities so that the operator could, for example, control a huge robot. Since 1950, technology has gradually been developed to make telepresence a reality. Research continues at other laboratories, such as the Naval Ocean Systems Center in Hawaii and MITI's Tele-existence Project in Tsukuba, Japan, to develop improved systems to help humans operate safely and effectively in hazardous environments such as undersea or outerspace.

One of the first attempts at developing a telepresence visual system was done by the Philco Corporation in 1958. With this system, an operator could see an image from a remote camera on a CRT mounted on his head in front of his eyes and could control the camera's viewpoint by moving his head [Comeau, 1961]. A variation of the head-mounted display concept was done by Ivan Sutherland in the late 1960s, first at M.I.T.'s Draper Lab in Cambridge, Massachusetts, and then later at the University of Utah [Sutherland, 1968]. This helmet-mounted display had a see-through capability so that computer-generated graphics could be viewed superimposed onto the real environment. As the viewer moved around, the computer-generated objects would appear to be stable within that real environment and could be manipulated with various input devices.

The Virtual Environment Workstation project at NASA Ames (VIEW) is based upon yet another kind of helmet-mounted display technology and, in addition, provides auditory, speech, and gesture interaction within a virtual environment.

The Ames Virtual Environment Workstation

The first virtual environment display system developed at Ames was literally a motorcycle helmet with a visor attachment containing two small liquid crystal display screens of 100-by-100-pixel resolution. Several iterations of this first prototype have resulted in a much lighter, less claustrophobic headset, with a long-term goal of designing a visor-like display package worn on the head. We are currently assembling five versions of this third prototype for use in various research environments. Included on the headset is a microphone for connected speech recognition, earphones for 3-D sound cueing, and a head-tracking device, as well as the package to hold the LCD display screens and viewing optics for each eye.

The liquid crystal display screens are inexpensive, flat panel displays that draw very little power and that have been used specifically for safety reasons and for their lightness. The current resolution is 640 pixels by 220 pixels. These transmissive displays accept a standard NTSC video signal and require some form of backlighting. In the first version, an LED backlight array was used, and, currently, very bright, miniature fluorescent tubes are being used. Another key part of the display is the very wide-angle optics through which the liquid crystal displays are viewed. The objective is to present a wide-angle field of view that closely matches human binocular visual capabilities rather than just a small window into this 3-D world. These optics allow the displays to completely fill your visual field of view, thus helping to achieve a sense of presence in the 3-D virtual environment.

On the helmet is a tracking device that tells the host computer where you're looking within the three-dimensional virtual environment. We're now using an electromagnetic device that measures where the head is within a magnetic field emitted by a source. Azimuth elevation and roll information of the head is combined with XYZ position at a resolution of .03 inch and .10 degree accuracy. When this information is received by the host computer, it's sent to the graphics system or remote camera system so that, as you turn your head, the new position is recorded and a new image that matches that position of regard is displayed. As you turn your head, new images are being drawn so quickly that you feel like you're completely immersed in the virtual environment. Another sensing technology that we're investigating may provide the ability to track eye position in order to display a three-dimensional cursor within this environment. Eye-tracking will tell us where your eyes are converged in three-dimensional space so that, for example, you'll be able to trigger menus with your eyes; we'll also be able to present different kinds of depth-of-field information, which is yet another depth cue that has been lacking in typical simulation display systems to date.

Virtual environments in the Ames system are synthesized with 3-D computer-generated imagery, or are remotely sensed by user-controlled, stereoscopic video camera configurations. The computer image system enables high-performance, real-time 3-D graphics presentation at resolutions of 640 by 480 and 1,000 by 1,000 pixels. This imagery is generated at rates up to 30 frames per second as required to update image viewpoints in coordination with head and limb motion. Dual independent, synchronized display channels are used to present disparate imagery to each eye of the viewer for true stereoscopic depth cues. For real-time video input of remote environments, two miniature CCD video cameras are used to provide stereoscopic imagery. Development and evaluation of several head-coupled, remote camera platform and gimbal prototypes is in progress to determine optimal hardware and control configurations for remotely controlled, free-flying, or telerobot-mounted camera systems.

As this display technology evolved to give a feeling of being surrounded by virtual objects or being inside a virtual environment, we also began to develop means of interacting with that virtual world—of literally reaching in and touching the virtual objects, picking them up, interacting with virtual control panels, etc. Earlier research in this area had been done for various applications that required manipulation of remote objects with reasonable dexterity and feedback to the operator. For example, in 1954, Ralph Mosher at General Electric developed the "Handyman" system—a complicated exoskeleton around the operator's arm that was used to control a remote robotic arm device. Although operationally useful, this control system was rather large and unportable.

We were interested in getting away from such an invasive device and in 1985 contracted VPL Research, Inc., to develop a "dataglove"—a lightweight, flexible glove instrumented with flex-sensing devices to measure the amount of bend of each joint of the fingers and the amount of abduction, the splay between fingers [Fisher, 1986].[2]

Information from the glove is transmitted to the host computer to represent what the hand is doing at any moment. In the current, commercially available DataGlove, the amount of bend is measured by light attenuation within fiber optic bundles covering the fingers, and magnetic tracking devices give absolute position and orientation of the hand within the 3-D space. Position and orientation combined with the finger bend is used to control a graphic model of the hand in the virtual environment or to control a remote robot hand. With this capability the user can pick up and manipulate virtual objects that appear in the surrounding virtual environment [Foley, 1987]. We've also programmed the dataglove so that gestures can be recognized, similar to American sign language. The operator can make a particular gesture and the host will recognize it as user input to trigger a particular command or subroutine. For example, pointing with one finger moves your viewpoint through the computer-generated environment as if you were flying through that space, with the distance between your finger and your body determining your velocity; making a fist lets you grab different objects; and using a three-finger point invokes a menu floating in visual space that you can then use to choose other subroutines or information displays. In actual use, operators can design their own set of gestures as preferred.

Additional work in progress for the glove will provide the capability for tactile feedback. An array of very small solenoid actuators has been assembled that will present a sense of texture as you touch a virtual object and, for example, if you're touching an edge of a virtual cube, as you intersect the edge of the cube, one line of the array of solenoids is triggered to present some sense of edgeness. In future scenarios, detailed virtual environment databases or tactile sensors on remote robot end-effectors will transmit tactile information to arrays such as this integrated in the DataGlove. A further requirement for tactile feedback to the operator is force reflection, through which some sense of solidity and surface boundaries of virtual objects is communicated back to the operator. Now, if the hand is closed on a virtual object, the fingers will pass right through the object.

Interim solutions include the use of auditory feedback to the operator to indicate contact with and forces applied to virtual objects. In the longer term, technology is required that will constrain the user's fingers so that they will close only to the outside boundary of that virtual object; eventually, full arm and body interaction will need to be enabled.

The auditory channel is another major part of our sensory input and display research. The auditory display in the VIEW project is capable of presenting a wide variety of binaural sounds to the user via headphones using sound-synthesis technology developed for music synthesizers. The primary function of this display is to provide both discrete and dynamic auditory cues to augment or supply information missing from the visual or gestural displays. For example, discrete sound cues can signal contact between telerobot end-effectors and target objects; alert the operator to attend to information in a data window that is currently out of the field of view; or indicate successful recognition of a hand gesture. Similarly, auditory parameters can be dynamically modulated and coordinated with the other subsystem displays to represent, for example, the relative positions of other objects moving within the virtual environment. In the real world, we also have a very good sense of where sounds are coming from around us. Additional research at Ames has developed an auditory display prototype called the Convolvotron that is capable of synthetically generating three-dimensional sound cues in real-time. These cues are presented via headphones and are perceived outside of the user's head at a discrete distance and direction in the 3-D space surrounding the user. When it is integrated into the VIEW system, the position of the operator can be monitored in real-time and the information used to maintain up to four localized sound cues in fixed positions or in motion trajectories relative to the user. This capability will further aid the operator's situational awareness by augmenting spatial information from the visual display and providing navigational and cueing aids from outside the field of view [Wenzel et al., 1988].

The VIEW system also includes commercially available, speaker-dependent, connected speech-recognition technology that allows the user to give voice input in a natural, conversational format that cannot be achieved with highly constrained discrete word recognition systems or through keyboard input. Typical speech-mediated interactions are requests for display/report of system status, instructions for supervisory control tasks, and verbal commands to change interface mode or configuration. Taken together, the capabilities of the Virtual Environment Display system described enable a wide range of applications. . . .

Telecollaboration

A major near-term goal for the Virtual Environment Workstation Project is to connect at least two of the current prototype interface systems to a common virtual environment database. The two users will participate and interact in a shared virtual environment, but each will view it from their relative, spatially disparate viewpoints. The objective is to provide a collaborative workspace in which remotely located participants can virtually interact with some of the nuances of face-to-face meetings while also having access to their personal dataspace facility. This could enable valuable interaction between scientists collaborating from different locations across the country or even between astronauts on a space station and research labs on Earth. With full body tracking capability, it would also be possible for users to be represented in this space by life-size virtual representations of themselves in whatever form they choose—a kind of electronic persona. For interactive theater or interactive fantasy applications, these styles might range from fantasy figures to inanimate objects, or different figures to different people. Eventually, telecommunication networks may develop that will be configured with virtual environment servers for users to dial into remotely in order to interact with other virtual proxies.

Although the current prototype of the Virtual Environment Workstation has been developed primarily to be used as a laboratory facility, the components have been designed to be easily replicable for relatively low cost. As the processing power and graphics frame rate on microcomputers quickly increases, portable, personal virtual environment systems will also become available. The possibilities of virtual realities, it appears, are as limitless as the possibilities of reality. They can provide a human interface that disappears—a doorway to other worlds.

William Gibson

“Academy Leader” (1991)

<< 23 >>

William Gibson. Courtesy of Karen Moskowitz.

> *"A year here and he still dreamed of cyberspace, hope fading nightly. All the speed he took, all the turns he'd taken and the corners he'd cut in Night City, and still he'd see the matrix in his sleep, bright lattices of logic unfolding across that colorless void."*
>
> —FROM *NEUROMANCER,* BY WILLIAM GIBSON

<< In his science fiction novels, William Gibson's hallucinatory account of cyberspace provided the first social and spatial blueprint for the digital frontier. In his 1984 novel *Neuromancer*—a colorful, disturbing account of our emerging information society—he added the word *cyberspace* to our vocabulary. His writings explore the implications of a wired, digital culture, and have had tremendous influence on the scientists, researchers, theorists, and artists working with virtual reality. Gibson's notion of an inhabitable, immersive terrain that exists in the connections between computer networks, a fluid, architectural space that could expand endlessly—an invitation to "jack in" to the "digital matrix"—has opened the door to a new genre of literary and artistic forms, and has shaped our expectations of what is possible in virtual environments.

In this essay, Gibson explains that the word *cyberspace* initially came to him as a term without meaning, that it "preceded any concept whatever." He tips his hat to the "dangerous old literary gentleman" from East St. Louis, William S. Burroughs, whose acidic depictions of twentieth-century life inspired Gibson's own fictions. He goes on to suggest that in coining the term *cyberspace* he was applying Burroughs's experimental cut-up technique. This entire essay is, in fact, written in the manner of the Burroughs cut-up—with its sudden leaps from one idea to another. It's an appropriate homage to the Beat writer, whose novels influenced not only Gibson but also an entire generation of dystopian writers, filmmakers, and artists.

In *Neuromancer,* as well as in his later novels *Count Zero* (1987) and *Mona Lisa Overdrive* (1988), Gibson's vision of cyberspace, with its antiheroes who reside in the void between the physical world and the network, helped spark an age of the posthuman. The cyborgian redefinition of self is staged in such immersive cyberhabitats as MUDs, virtual communities, and on-line chat spaces, where identity has become malleable and interchangeable. Gibson's strange, menacing virtual world meshed perfectly with the detached, ironic stance of late-twentieth-century culture. >>

"Ride music beams back to base."

He phases out on a vector of train whistles and the one particular steel-engraved slant of winter sun these manifestations favor, leaving the faintest tang of Players Navy Cut and opening piano bars of East St. Louis, this dangerous old literary gentleman who sent so many of us out, under sealed orders, years ago . . .

Inspector Lee taught a new angle—

Frequencies of silence; blank walls at street level. In the flat field. We became field operators. Decoding the lattices. Patrolling the deep faults. Under the lights. Machine Dreams. The crowds, swept with con . . . Shibuya Times Square Picadilly. A parked car, an arena of grass, a fountain filled with earth. In the hour of the halogen wolves . . . The hour remembered. In radio silence . . .

Just a chance operator in the gasoline crack of history, officer . . .

Assembled word *cyberspace* from small and readily available components of language. Neologic spasm: the primal act of pop poetics. Preceded any concept whatever. Slick and hollow—awaiting received meaning.

All I did: folded words as taught. Now other words accrete in the interstices.

"Gentlemen, that is not now nor will it ever be *my* concern . . ."

Not what I do.

I work the angle of transit. Vectors of neon plaza, licensed consumers, acts primal and undreamed of . . .

The architecture of virtual reality imagined as an accretion of dreams: tattoo parlors, shooting galleries, pinball arcades, dimly lit stalls stacked with damp-stained years of men's magazines, chili joints, premises of unlicensed denturists, of fireworks and cut bait, betting shops, sushi bars, purveyors of sexual appliances, pawnbrokers, wonton counters, love hotels, hotdog stands, tortilla factories, Chinese greengrocers, liquor stores, herbalists, chiropractors, barbers, bars.

These are dreams of commerce. Above them rise intricate barrios, zones of more private fantasy . . .

Angle of transit sets us down in front of this dusty cardtable in an underground mall in the Darwin Free Trade Zone, muzak-buzz of seroanalysis averages for California-Oregon, factoids on EBV mutation rates and specific translocations at the breakpoint near the c-myc oncogene . . .

Kelsey's second week in Australia and her brother is keeping stubbornly in-condo, doing television, looping *Gladiator Skull* and a new Japanese game called *Torture Garden*. She walks miles of mall that could as easily be Santa Barbara again or Singapore, buying British fashion magazines, shoplifting Italian eyeshadow; only the stars at night are different, Southern Cross, and the Chinese boys skim the plazas on carbon-fiber skateboards trimmed with neon.

She pauses in front of the unlicensed vendor, his face notched with pale scars of sun-cancer. He has a dozen cassettes laid out for sale, their plastic cases scratched and dusty. "Whole city in there," he says, "Kyoto, yours for a twenty." She sees the security man, tall and broad, Kevlar-vested, blue-eyed, homing in to throw the old man out, as she tosses the coin on impulse and snatches the thing up, whatever it is, and turns, smiling blankly, to swan past the guard. She's a licensed consumer, untouchable, and looking back she sees the vendor squinting, grinning his defiance, no sign of the $20 coin . . .

No sign of her brother when she returns to the condo. She puts on the glasses and the gloves and slots virtual Kyoto . . .

Once perfected, communication technologies rarely die out entirely; rather, they shrink to fit particular niches in the global info-structure. Crystal radios have been proposed as a means of conveying optimal seed-planting times to isolated agrarian tribes. The mimeograph, one of many recent dinosaurs of the urban office-place, still shines with undiminished *samisdat* potential in the century's backwaters, the Late Victorian answer to desktop publishing. Banks in uncounted Third World villages still crank the day's totals on black Burroughs adding machines, spooling out yards of faint indigo figures on long, oddly festive curls of paper, while the Soviet Union, not yet sold on throw-away new-tech fun, has become the last reliable source of vacuum tubes. The eight-track tape format survives in the truckstops of the Deep South, as a medium for country music and spoken-word pornography.

The Street finds its own uses for things—uses the manufacturers never imagined. The micro-tape recorder, originally intended for on-the-jump executive dictation, becomes the revolutionary medium of *magnetisdat,* allowing the covert spread of banned political speeches in Poland and China. The beeper and the cellular phone become economic tools in an increasingly competitive market in illicit drugs. Other technological artifacts unexpectedly become means of communication . . . The aerosol can gives birth to the urban graffitti-matrix. Soviet rockers press homemade flexidisks out of used chest x-rays . . .

Fifteen stones against white sand.

The sandals of a giant who was defeated by a dwarf.

A pavilion of gold, another of silver.

A waterfall where people pray . . .

Her mother removes the glasses. Her mother looks at the timer. Three hours. "But you don't like games, Kelsey . . ."

"It's not a game," tears in her eyes. "It's a city." Her mother puts on the glasses, moves her head from side to side, removes the glasses.

"I want to go there," Kelsey says.

"It's different now. Everything changes."

"I want to go there," Kelsey insists. She puts the glasses back on because the look in her mother's eyes frightens her.

The stones, the white sand: cloud-shrouded peaks, islands in the stream . . .

She wants to go there . . .

"The targeted numerals of the ACADEMY LEADER were hypnogogic sigils preceding the dreamstate of film."

Marcos Novak

<< 24 >> "Liquid Architectures in Cyberspace" (1991)

Marcos Novak, Liquid Architectures. *© Marcos Novak. Courtesy of Marcos Novak.*

"Liquid architecture is an architecture that breathes, pulses, leaps as one form and lands as another. Liquid architecture is an architecture whose form is contingent on the interests of the beholder; it is an architecture that opens to welcome me and closes to defend me; it is an architecture without doors and hallways, where the next room is always where I need it to be and what I need it to be."

<< Marcos Novak describes himself as a "trans-architect," due to his work with computer-generated architectural designs, conceived specifically for the virtual domain, that do not exist in the physical world. His immersive, three-dimensional creations are responsive to the viewer, transformable though user interaction. Exploring the potential of abstract and mathematically conceived forms, Novak has invented a set of conceptual tools for thinking about and constructing territories in cyberspace.

In this essay, Novak introduces the concept of "liquid architecture," a fluid, imaginary landscape that exists only in the digital domain. Novak suggests a type of architecture cut loose from the expectations of logic, perspective, and the laws of gravity, one that does not conform to the rational constraints of Euclidean geometries. He views trans-architecture as an expression of the "fourth dimension," which incorporates time alongside space among its primary elements. Novak's liquid architecture bends, rotates, and mutates in interaction with the person who inhabits it. In liquid architecture, "science and art, the worldly and the spiritual, the contingent and the permanent," converge in a poetics of space made possible by emerging, virtual reality technologies.

Novak describes his work as a process of metamorphosis, a "symphony of space," in which 3-D constructions have the properties of music, an experience he has since referred to as "navigable music." His is a poetic language that attempts to describe the indescribable—"cyberspace as habitat of the imagination, a habitat for the imagination." >>

Introduction

What is cyberspace? Here is one composite definition:

> Cyberspace is a completely spatialized visualization of all information in global information processing systems, along pathways provided by present and future communications networks, enabling full copresence and interaction of multiple users, allowing input and output from and to the full human sensorium, permitting simulations of real and virtual realities, remote data collection and control through telepresence, and total integration and intercommunication with a full range of intelligent products and environments in real space.[1]

Cyberspace involves a reversal of the current mode of interaction with computerized information. At present such information is external to us. The idea of cyberspace subverts that relation; we are now within information. In order to do so we ourselves must be reduced to bits, represented in the system, and in the process become information anew.

Cyberspace offers the opportunity of maximizing the benefits of separating *data, information,* and *form,* a separation made possible by digital technology. By reducing selves, objects, and processes to the same underlying ground-zero representation as binary streams, cyberspace permits us to uncover previously invisible relations simply by modifying the normal mapping from data to representation.

To the composite definition above I add the following: Cyberspace is a habitat for the imagination. Our interaction with computers so far has primarily been one of clear, linear thinking. Poetic thinking is of an entirely different order. To locate the difference in terms related to computers: poetic thinking is to linear thinking as random access memory is to sequential access memory. Everything that can be stored one way can be stored the other; but in the case of sequential storage the time required for retrieval makes all but the most predictable strategies for extracting information prohibitively expensive.

Cyberspace is a habitat of the imagination, a habitat for the imagination. Cyberspace is the place where conscious dreaming meets subconscious dreaming, a landscape of rational magic, of mystical reason, the locus and triumph of poetry over poverty, of "it-can-be-so" over "it-should-be-so."

> The greater task will not be to impose science on poetry, but to restore poetry to science.

This chapter is an investigation of the issues that arise when we consider cyberspace as an inevitable development in the interaction of humans with computers. To the extent that this development inverts the present relationship of human to information, placing the human within the information space, it is an architectural problem; but, beyond this, cyberspace has an architecture of its own and, furthermore, can contain architecture. To repeat: cyberspace *is* architecture; cyberspace *has* an architecture; and cyberspace *contains* architecture.

Cyberspace relies on a mix of technologies, some available, some still imaginary. This chapter will not dwell on technology. Still, one brief comment is appropriate here. A great number of devices are being developed and tested that promise to allow us to enter cyberspace with our bodies. As intriguing as this may sound, it flies in the face of the most ancient dream of all: magic, or the desire to will the world into action. Cyberspace will no doubt have physical aspects; the visceral has genuine power over us. And though one of the major themes of this essay has to do with the increasing recognition of the physicality of the mind, I find it unlikely that once inside we will tolerate such heavy devices for long. Gloves and helmets and suits and vehicles are all mechanocybernetic inventions that still rely on the major motor systems of the body, and therefore on coarse motor coordination, and more importantly, low nerve-ending density. The course of invention has been to follow the course of desire, with its access to the parts of our bodies that have the most nerve endings. When we enter cyberspace we will expect to feel the mass of our bodies, the reluctance of our skeleton; but we will choose to *control* with our eyes, fingertips, lips, and tongues, even genitals.

The trajectory of Western thought has been one moving from the concrete to the abstract, from the body to the mind; recent thought, however, has been pressing upon us the frailty of that Cartesian distinction. The mind is a property of the body, and lives and dies with it. Everywhere we turn we see signs of this recognition, and cyberspace, in its literal placement of the body in spaces invented entirely by the mind, is located directly upon this blurring boundary, this fault.

At the same time as we are becoming convinced of the embodiment of the mind, we are witnessing the acknowledgment of the inseparability of the two in another way: the mind affects what we perceive as real. Objective reality itself seems to be a construct of our mind, and thus becomes subjective.

The "reality" that remains seems to be the reality of fiction. This is the reality of what can be expressed, of how meaning emerges. The trajectory of thought seems to be from concrete to abstract to concrete again, but the new concreteness is not that of Truth, but of *embodied* fiction.

The difference between embodied fiction and Truth is that we are the authors of fiction. Fiction is there to serve our purposes, serious or playful, and to the extent that our purposes change as we change, its embodiment also changes. Thus, while we reassert the body, we grant it the freedom to change at whim, to become liquid.

It is in this spirit that the term *liquid architecture* is offered. Liquid architecture of cyberspace; liquid architecture in cyberspace.

Part One: Cyberspace

Poetics and Cyberspace

> Well then, before reading poems aloud to so many people, the first thing one must do is invoke the *duende*. This is the only way all of you will succeed at the hard

> task of understanding metaphors as soon as they arise, without depending on intelligence or the critical apparatus, and be able to capture, as fast as it is read, the rhythmic design of the poem. For the quality of a poem can never be judged on just one reading, especially not poems like these which are full of what I call "poetic facts" that respond to a purely poetic logic and follow the constructs of emotion and of poetic architecture. Poems like these are not likely to be understood without the cordial help of the *duende.*
>
> —Federico García Lorca, *Poet in New York*

The *duende* is a spirit, a demon, invoked to make comprehensible a "poetic fact," an "hecho poético." An "hecho poético," in turn, is a poetic image that is not based on analogy and bears no direct, logical explanation (Lorca, 1989). This freeing of language from one-to-one correspondence, and the parallel invocation of a "demon" that permits access to meanings that are beyond ordinary language, permits Lorca to produce some of the most powerful and surprising poetry ever written. It is this power that we need to harness in order to be able to contend with what William Gibson called the "unimaginable complexity" of cyberspace.

How does this poetry operate?

Concepts, like subatomic particles, can be thought to have world lines in space-time. We can draw Feynman diagrams for everything that we can name, tracing the trajectories from our first encounter with an idea to its latest incarnation. In the realm of prose, the world lines of similar concepts are not permitted to overlap, as that would imply that during that time we would be unable to distinguish one concept from another. In poetry, however, as in the realm of quantum mechanics, world lines may overlap, split, divide, blink out of existence, and spontaneously reemerge.[2] Meanings overlap, but in doing so call forth associations inaccessible to prose. Metaphor moves mountains. Visualization reconciles contradiction by a surreal and permissive blending of the disparate and far removed. Everything can modify everything: *"Green oh how I love you green,"* writes Lorca, *"Giant stars of frost / come with the shadow-fish / that leads the way for dawn. / The fig tree chafes its wind / with its sandpapered branches, / and the mountain, untamed cat, / bristles sour maguey spears."*

If cyberspace holds an immense fascination, it is not simply the fascination of the new. Cyberspace stands to thought as flight stands to crawling. The root of this fascination is the promise of control over the world by the power of the will. In other words, it is the ancient dream of magic that finally nears awakening into some kind of reality. But since it is technology that promises to deliver this dream, the question of "how" must be confronted. Simply stated, the question is, What is the technology of magic? For the answer we must turn not only to computer science but to the most ancient of arts, perhaps the only art: poetry. It is in poetry that we find a developed understanding of the workings of magic, and not only that, but a wise and powerful knowledge of its purposes and potentials. Cyberspace is poetry inhabited, and to navigate through it is to become a leaf on the wind of a dream.

Tools of poets: image and rhythm, meter and accent, alliteration and rhyme, tautology, simile, analogy, metaphor, strophe and antistrophe, antithesis, balance and caesura, enjambment and closure, assonance and consonance, elision and inflection, hyperbole, lift, onomatopoeia, prosody, trope, tension, ellipsis . . . poetic devices that allow an inflection of language to produce an inflection of meaning. By push and pull applied to both syntax and symbol, we navigate through a space of meaning that is sensitive to the most minute variations in articulation. Poetry is liquid language.

As difficult as it may sound, it is with operations such as these that we need to contend in cyberspace. Nothing less can suffice.

> I am in cyberspace. I once again resort to a freer writing, a writing more fluid and random. I need to purge a mountain of brown thoughts whose decay blocks my way. I seek the color of being in a place where information flies and glitters, connections hiss and rattle, my thought is my arrow. I combine words and occupy places that are the consequence of those words. Every medium has its own words, every mixture of words has a potential for meaning. Poets have always known this. Now I can mix the words of different media and watch the meaning become navigable, enter it, watch magic and music merge. . . .

Underlying Considerations

Minimal Restriction and Maximal Binding The key metaphor for cyberspace is "being there," where both the "being" and the "there" are user-controlled variables, and the primary principle is that of *minimal restriction,* that is, that it is not only desirable but necessary to impose as few restrictions as possible on the definition of cyberspace, this in order to allow both ease of implementation and richness of experience. In addition, *maximal binding* implies that in cyberspace anything can be combined with anything and made to "adhere," and that it is the responsibility of the user to discern what the implications of the combination are for any given circumstance. Of course, defaults are given to get things started, but the full wealth of opportunity will only be harvested by those willing and able to customize their universe. Cyberspace is thus a user-driven, self-organizing system.

Multiple Representations Cyberspace is an invented world; as a world it requires "physics," "subjects" and "objects," "processes," a full ecology. But since it is an invented world, an embodied fiction, one built on a fundamental representation of our own devising, it permits us to redirect data streams into different representations: selves become multiple, physics become variable, cognition becomes extensible. The boundaries between subject and object are conventional and utilitarian; at any given time the data representing a user may be combined with the data representing an object to produce . . . what?

Digital technology has brought a dissociation between data, information, form, and appearance. Form is now governed by representation, data is a binary stream, and information is pattern perceived in the data after the data has been seen through the expectations of a representation scheme or code. A stream of bits, initially formless, is given form by a representation scheme, and information emerges through the interaction of the data with the representation; different representations allow different correlations to become apparent within the same body of data. Appearance is a late aftereffect, simply a consequence of many sunken layers of patterns acting upon patterns, some patterns acting as data, some as codes. This leads to an interesting question: what is the information conveyed by the representation that goes beyond that which is in the data itself? If a body of data seen one way conveys different information than the same body of data seen another way, what is the additional information provided by one form that is not provided by the other? Clearly, the answer is *pattern,* that is, *perceived* structure. And if different representations provide different perceived information, how do we choose representations? Not only do different representations provide different information, but in the comparison *between* representations new information may become apparent. We can thus distinguish two kinds of *emergent* information: *intrarepresentational* and *interrepresentational.*

> I substitute the characters on this and the next page of this text with grey scale values; two images emerge, pleasingly rhythmical. The gray tone of the letter e stands out, forming snakes along the pages; I apply spline curves to the snakes, and, in another space, my text itself is changing—what will it say? Now I combine the two pages, and convert the result into a landscape, using the grey values to represent height. What snakes were left after the combination of the two pages

become Serpent Mounds. The Others who were reading my text with me a while ago are now flying over this landscape with me, but only I can command it to change. Today....

From Poetics to Architecture We have examined various aspects of cyberspace from a viewpoint that stresses the power of poetic language over ordinary, reductive language. Poetic language is language in the process of making and is best studied by close examination of poetic artifacts, or, better yet, by making poetic artifacts.

The transition from real space to cyberspace, from prose to poetry, from fact to fiction, from static to dynamic, from passive to active, from the fixed in all its forms to the fluid in its everchanging countenance, is best understood by examining that human effort that combines science and art, the worldly and the spiritual, the contingent and the permanent: architecture.

Even as cyberspace represents the acceptance of the body in the realm of the mind, it attempts to escape the mortal plane by allowing everything to be converted to a common currency of exchange. Architecture, especially visionary architecture, the architecture of the excess of possibility, represents the manifestation of the mind in the realm of the body, but it also attempts to escape the confines of a limiting reality. The story of both these efforts is illuminating, and in both directions. Cyberspace, as a world of our creation, makes us contemplate the possibility that the reality we exist in is already a sort of "cyberspace," and the difficulties we would have in understanding what is real if such were the case. Architecture, in its strategies for dealing with a constraining reality, suggests ways in which the limitations of a fictional reality may be surmounted.

Architecture, most fundamentally, is the art of space. There are three fundamental requirements for the perception of space: reference, delimitation, and modulation. If any one is absent, space is indistinguishable from nonspace, being from nothingness. This, of course, is the fundamental observation of categorical relativity. This suggests that cyberspace does not exist until a distance can be perceived between subject and boundary, that is to say, until it is delimited and modulated.

A space modulated so as to allow a subject to observe it but not to inhabit it is usually called sculpture. A space modulated in a way that allows a subject to enter and inhabit it is called architecture. Clearly, these categories overlap a great deal: architecture is sculptural, and sculpture can be inhabited.

We can now draw an association between sculpture and the manner in which we are accustomed to interacting with computers. The interface is a modulated information space that remains external to us, though we may create elaborate spatial visualizations of its inner structure in our minds. Cyberspace, on the other hand, is intrinsically about a space that we enter. To the extent that this space is wholly artificial, even if it occasionally looks "natural," it is a modulated space, an architectural space. But more than asserting that there is architecture within cyberspace, it is more appropriate to say that cyberspace cannot exist without architecture, cyberspace is architecture, albeit of a new kind, itself long dreamed of.

Part Two: Liquid Architecture in Cyberspace

> . . . and we can in our Thought and Imagination contrive perfect Forms of Buildings entirely separate from Matter, by settling and regulating in a certain Order, the Disposition and Conjunction of the Lines and Angles.
>
> —Leon Battista Alberti, *The Ten Books of Architecture*

Visionary Architecture: The Excess of Possibility

Just as poetry differs from prose in its controlled intoxication with meanings to be found beyond the limits of ordinary language, so visionary architecture exceeds ordinary architecture in its search for the conceivable. Visionary architecture, like poetry, seeks an extreme, any extreme: beauty, awe, structure, or the lack of structure, enormous weight, lightness, expense, economy, detail, complexity, universality, uniqueness. In this search for that which is beyond the immediate, it proposes embodiments of ideas that are both powerful and concise. More often than not these proposals are well beyond what can be built. This is not a weakness: in this precisely is to be found the poignancy of vision.

The Space of Art In imagining how information is to be "spatialized" in cyberspace, it is easy to be overwhelmed by the idea of "entering" the computer in the first place, and to only consider relatively mundane depictions of space: perspectival space, graph space, the space of various simple projection systems. Humanity's library of depictions of space is far richer than that: synchronically and diachronically, across the globe and through time, artists have invented a wealth of spatial systems. What would it be like to be inside a cubist universe? a hieroglyphic universe? a universe of cave drawings or Magritte paintings? Just as alternative renditions of the same reality by different artists, each with a particular style, can bring to our attention otherwise invisible aspects of that reality, so too can different modes of cyberspace provide new ways of interrogating the world.

The development of abstract art by Malevich, Kandinsky, Klee, Mondrian, and other early modern artists prefigures cyberspace in that it is an explicit turning away from representing known nature. Perhaps artists have always invented worlds—one is reminded how varied the representations used in the arts have been—from Chinese watercolors, to Byzantine icons, to the strange, conflicting backgrounds of Leonardo—but those worlds usually made reference to some familiar reality. Even when that reality was of a cosmic or mystical nature, we find an assumption of similarity to the everyday world. Modern artists took on the task of inventing entire worlds without explicit reference to external reality. Malevich is pertinent: his *architectones* are architectural studies imagined to exist in a world beyond gravity, *against gravity*. Created as an architecture without functional program or physical constraint, they are also studies for an *absolute architecture,* in the same sense that we speak of absolute music, architecture for the sake of architecture. Paul Klee, in his dairies, speaks about being a "god" in a universe of his own making.

The paintings of Max Ernst or, even more so, Hieronymous Bosch come to mind, for their ability to create mysterious new worlds. In the works of these artists we not only see worlds fashioned out of unlikely combinations of a code consisting of familiar elements, but also meaningful crossings of expected conceptual and categorical boundaries.

As with cyberspace, the space of art *is* architecture, *has* an architecture and *contains* architecture. It *is* architecture in its ability to create a finely controlled sense of depth, even within depictions that are inherently two-dimensional; it *has* an architecture in its compositional structure; and, by depiction, it *contains* architecture. It can serve as a bridge between cyberspace and architecture. . . .

Cyberspace Architecture

Architecture has been earthbound, even though its aspirations have not. Buckminster Fuller remarked that he was surprised that, in spite of all the advances made in the technology of building, architecture remained rooted to the ground by the most mundane of its functions, plumbing. Rooted by waste matter, architecture has nevertheless attempted to fly in dreams and projects, follies and cathedrals.

Architecture has never suffered a lack of fertile dreams. Once, however, in times far less advanced technologically, the distance between vision and embodiment was smaller, even though the effort required for that embodiment was often crushing. Most "grand traditions" began with an experimental stage of danger and discovery and did not become fossilized until much later. Hard as it may be for us to fathom, a gothic cathedral was an extended experiment often lasting over a century, at the end of which there was the literal risk of collapse. The dream and the making were one. Curiously, the practice of architecture has become increasingly disengaged from those dreams. Cyberspace permits the schism that has emerged to be bridged once again.

Cyberspace alters the ways in which architecture is conceived and perceived. Beyond computer-aided design (CAD), design computing (DC), or the development of new formal means of describing, generating, and transforming architectural form, cyberspace encodes architectural knowledge in a way that indicates that our conception of architecture is becoming increasingly musical, that architecture is spatialized music. Computational composition, in turn, combines these new methods with higher-level compositional concepts of overall form subject to local and global constraints to transform an input pattern into a finished work. In principle, and with the proper architectural knowledge, any pattern can be made into a work of architecture, just as any pattern can be made into music. In order for the data pattern to qualify as music or architecture it is passed through compositional "filters," processes that select and massage the data according to the intentions of the architect and the perceptual capacity of the viewer. This "adaptive filtering," to use a neural net term, provides the beginning of the intelligence that constitutes a cyberspace and not a hypergraph. This, of course, means that any information, any data, can become architectonic and habitable, and that cyberspace and cyberspace architecture are one and the same.

A radical transformation of our conception of architecture and the public domain is implied by cyberspace. The notions of city, square, temple, institution, home, infrastructure become permanently extended. The city, traditionally the continuous city of physical proximity, becomes the discontinuous city of cultural and intellectual community. Architecture, normally understood in the context of the first, conventional city, shifts to the structure of relationships, connections and associations that are webbed over and around the simple world of appearances and accommodations of commonplace functions.

> I look to my left, and I am in one city; I look to my right, and I am in another. My friends in one can wave to my friends in the other, through my having brought them together.

It is possible to envision architecture nested within architecture. Cyberspace itself is architecture, but it also contains architecture, but now without constraint as to phenomenal size. Cities can exist within chambers as chambers may exist within cities. Since cyberspace signifies the classical object yielding to space and relation, all "landscape" is architecture, and the objects scattered upon the landscape are also architecture. Everything that was once closed now unfolds into a place, and everything invites one to enter the worlds within worlds it contains.

> I am in an empty park. I walk around a tree, and I find myself in a crowded chamber. The tree is gone. I call forth a window, and in the distance see the park, leaving.

Liquid Architecture

> That is why we can equally well reject the dualism of appearance and essence. The appearance does not hide the essence, it reveals it; it is the essence. The essence of an existent is no longer a property sunk in the cavity of this existent; it is the manifest law which presides over the succession of its appearances, it is the principle of the series.
>
> . . . But essence, as the principle of the series is definitely only the concatenation of appearances; that is, itself an appearance.
>
> . . . The reality of a cup is that it is there and that it is not me. We shall interpret this by saying that the series of its appearances is bound by a principle which does not depend on my whim.
>
> —Jean-Paul Sartre, *Being and Nothingness*

The relationship established between architecture and cyberspace so far is not yet complete. It is not enough to say that there is architecture in cyberspace, nor that *that* architecture is animistic or animated. Cyberspace calls us to consider the difference between animism and animation, and animation and metamorphosis. Animism suggests that entities have a "spirit" that guides their behavior. Animation adds the capability of change in *location,* through time. Metamorphosis is change in *form,* through time *or space.* More broadly, metamorphosis implies changes in one aspect of an entity as a function of other aspects, continuously or discontinuously. I use the term liquid to mean animistic, animated, metamorphic, as well as crossing categorical boundaries, applying the cognitively supercharged operations of poetic thinking.

> Cyberspace is liquid. Liquid cyberspace, liquid architecture, liquid cities. Liquid architecture is more than kinetic architecture, robotic architecture, an archi-

> tecture of fixed parts and variable links. Liquid architecture is an architecture that breathes, pulses, leaps as one form and lands as another. Liquid architecture is an architecture whose form is contingent on the interests of the beholder; it is an architecture that opens to welcome me and closes to defend me; it is an architecture without doors and hallways, where the next room is always where I need it to be and what I need it to be. Liquid architecture makes liquid cities, cities that change at the shift of a value, where visitors with different backgrounds see different landmarks, where neighborhoods vary with ideas held in common, and evolve as the ideas mature or dissolve.

The locus of the concept "architecture" in an architecture that fluctuates is drastically shifted: Any particular appearance of the architecture is devalued, and what gains importance is, in Sartre's terms, "the principle of the series." For architecture this is an immense transformation: for the first time in history the architect is called upon to design not the object but the principles by which the object is generated *and varied* in time. For a liquid architecture requires more than just "variations on a theme," it requires the invention of something equivalent to a "grand tradition" of architecture at each step. A work of liquid architecture is no longer a single edifice, but a continuum of edifices, smoothly or rhythmically evolving in both space and time. Judgments of a building's *performance* become akin to the evaluation of dance and theater.

If we described liquid architecture as a symphony in space, this description would still fall short of the promise. A symphony, though it varies within its duration, is still a fixed object and can be repeated. At its fullest expression a liquid architecture is more than that. It is a symphony in space, but a symphony that never repeats and continues to develop. If architecture is an extension of our bodies, shelter and actor for the fragile self, a liquid architecture is that self in the act of becoming its own changing shelter. Like us, it has an identity; but this identity is only revealed fully during the course of its lifetime.

Conclusion

A liquid architecture in cyberspace is clearly a dematerialized architecture. It is an architecture that is no longer satisfied with only space and form and light and all the aspects of the real world. It is an architecture of fluctuating relations between abstract elements. It is an architecture that tends to music.

Music and architecture have followed opposite paths. Music was once the most ephemeral of the arts, surviving only in the memory of the audience and the performers. Architecture was once the most lasting of the arts, reaching as it did into the caverns of the earth, changing only as slowly as the planet itself changes. Symbolic notation, analog recording, and, currently, digital sampling and quantization, and computational composition, have enabled music to become, arguably, the most permanent of the arts. By contrast, the life span of architecture is decreasing rapidly. In many ways architecture has become the least durable of the arts. The dematerialized, dancing, difficult architecture of cyberspace, fluctuating, ethereal, temperamental, transmissible to all parts of the world simultaneously but only indirectly tangible, may also become the most enduring architecture ever conceived.

> I am in a familiar place. Have I been here before? I feel I know this place, yet even as I turn something appears to have changed. It is still the same place, but not quite identical to what it was just a moment ago. Like a new performance of an old symphony, its intonation is different, and in the difference between its present and past incarnations something new has been said in a language too subtle for words. Objects and situations that were once thought to have a fixed identity, a generic "self," now possess personality, flaw and flavor. All permanent categories are defeated as the richness of the particular impresses upon me that in this landscape, if I am to benefit fully, attention is both required and rewarding. Those of us who have felt the difference nod to each other in silent acknowledgment, knowing that at the end of specificity lies silence, and what is made speaks for itself, not in words, but in presences, ever changing, liquid. . . .

Daniel Sandin, Thomas DeFanti, and Carolina Cruz-Neira

<< 25 >> "A Room with a View" (1993)

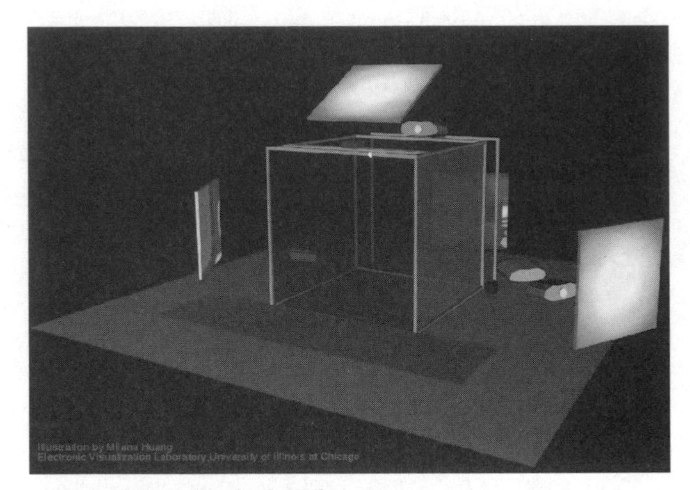

CAVE System. Illustration by Milana Huang. Courtesy of Daniel Sandin.

"More than one user will, however, have to interact in the same environment if virtual reality is to become an effective visualization tool."

<< Media artist Daniel Sandin and engineer Thomas DeFanti joined the Electronic Visualization Laboratory at the University of Illinois, Chicago, during the 1970s, where they combined their talents in computer graphics and video imaging. Assisted by graduate student Carolina Cruz-Neira, their research in electronic visualization culminated in 1991 with the design and construction of the CAVE (Cave Automatic Virtual Environment). This paper describes their achievement in advancing the medium of VR by creating a virtual environment that joins interactive, computer-generated imagery and 3-D audio with physical space—freeing the user from the confinement of the head-mounted display and dataglove.

The suspension of disbelief, so critical to the overall effect of virtual reality, is enhanced by the specific qualities of the CAVE's interface, which is, in fact, a small room of about three cubic meters. After entering the room, the user finds himself surrounded by projected images that are seamlessly synchronized on three walls, as well as on the floor. It is like stepping onto the stage of a virtual theater. The participant wears a pair of stereoscopic viewing glasses with built-in head-tracking technology, and carries a "magic wand" that provides the means of manipulating real-time imagery with the subtle movement of the wrist. The immersive experience of the CAVE was intended as an allusion to Plato's cave; its multiple screens and surround-sound audio evoke the metaphor of a shadowy representation of reality, suggesting how perception is always filtered through the mind's veil of illusion.

Unlike other systems of virtual reality, the properties of the CAVE are enhanced by the interplay between the real and the virtual. Influenced by Myron Krueger's installations, the CAVE immersant does not experience dislocation and disembodiment, but rather is viscerally aware of his or her physical presence "on stage" amid the animated imagery and orchestrated sound. Also influenced by Krueger, the multiparticipant capability of the CAVE has been extended by networking separate CAVE systems together to explore new forms of telematic artworks. >>

The media's confidence that virtual reality has already arrived is a little premature (they seem to believe that a set of videogame goggles and gloves is all that is needed). But as recent experience at the Electronic Visualization Laboratory (EVL) of the University of Illinois at Chicago has shown, the technology still needs a good deal of work.

In pursuit of a practical virtual reality (VR) system, researchers at EVL had to develop sophisticated software applications, and real-time networks to link advanced high-speed computers with a variety of high-tech peripherals (such as sound synthesizers and location trackers). In so doing, they had to solve technical problems that had limited the usefulness of such systems.

Real Work The benefits make the development worth the effort. When perfected, virtual reality systems may enhance how people work and play, contriving comfortable cybernetic environments that enliven and accelerate education and scientific modeling, in addition to devising new forms of recreation. However, VR is not child's play, as is shown by the Chicago laboratory's struggle to get the discrete parts of such systems to function flawlessly together in real time.

The laboratory's virtual reality installation is mostly configured from commercially available, state-of-the-art equipment. The installation is called the Cave, which is a recursive acronym (that is, an acronym that doubles as one of its own elements) that stands for Cave Automatic Virtual Environment.

While the name suggests the system's physical appearance, it is intended more strongly as an allusion to the Allegory of the Cave found in Plato's *Republic.* There the Greek philosopher explored the ideas of perception, reality, and illusion, using the analogy of a person facing the back of a cave alive with shadows that are his only basis for his ideas of what real objects are.

The laboratory's Cave is a new model for the design of virtual reality systems, one that offers several advantages over existing models. Unlike users of the video-arcade type of virtual reality system, Cave "dwellers" do not need to wear helmets, which would limit their view of and mobility in the real world, nor don bulky gloves and heavy electronics packs, nor be pushed about by movement-restricting platforms, to experience virtual reality. Instead, they put on a pair of lightweight "glasses" and walk into the Cave, a 27-cubic-meter room with an open side and no ceiling.

The Cave is in fact a partial cube, with the top and one vertical side missing. The other three vertical sides are 3-by-3-meter rear-projection screens facing the viewer, while the floor is a fourth screen—a front-projection one. The screens are wrapped around hard-to-detect cables that form seams at their edges. Behind each screen (or, in the case of the floor screen, above it) is a video projector that shines stereo images onto it as the viewer stands inside the cube.

The glasses trick a user's mind into seeing the screen images as three-dimensional objects. Each of the glasses' lenses is a shutter made of a liquid-crystal material; an electric pulse rapidly turns the material from clear to opaque or opaque to clear. In this way, each of the viewer's eyes can be momentarily shielded.

Every sixtieth of a second, a pair of stereoscopically different images is projected onto the screen; each of the images that make up the pair (a left-eye view and a right-eye view) is displayed for 1/120 second. The crystal shutters are synchronized so that one clears and closes and the other closes and clears 120 times per second. Thus each eye sees a slightly different image, so as to create the illusion of three dimensions.

Note that while the glasses alter what the wearer sees on the screens, they do not change the way he or she sees real objects (or other people) in the environment. Thus, when a computer-generated image of an object is projected into the environment, the viewer may be tempted to ask, "Is it live?" (Well, one just has to try to grasp the object; for now, at any rate, one's hand will pass through a projected image.)

The stereo glasses also have an electromagnetic transmitter attached to them that can be tracked by a location sensor. When the viewer moves, the computers that generate the images can determine that a change in viewing perspective is needed, and adjust the images appropriately.

A continuous image across several surfaces can be produced when each projector is controlled by a sophisticated computer graphics workstation. Each workstation calculates the perspective from each eye position, and sends the data to the projectors.

A hand-held tracking wand with its own location trackers lets the viewer manipulate the images projected onto the screen. This the person does by locating an imaginary position in virtual reality and interacting with it when the wand "touches" an image. For example, a user can move an object from one place to another, select various options from a menu, or strike images of bells and cause them to "ring."

An important ingredient in a virtual environment is sound. To generate the right kind of environmental sounds for whatever application is being run, the Cave employs a synthesizer with a musical-instrument digital interface (MIDI) and up to eight speakers located at the eight corners of the room.

Masterminds The Cave's virtual world is presided over by four Crimson VGX workstations from Silicon Graphics Inc., Mountain View, CA. Each workstation, with 256M-byte memory and two 1.6-gigabyte disks, is connected to an Electrohome Marquee 8000 projection display. A Silicon Graphics Personal Iris serves as a master controller for the system.

All workstations communicate via a ScramNet optical-fiber network from

Systran Corp., Dayton, OH. The network transfers data between these systems within the 8-ms window needed to synchronize images displayed on the Cave's four screens. It can accurately and synchronously broadcast data in real time at the high speeds necessary.

The workstations display stereo images by dividing the VGX frame buffer into two fields, each of which contains data for the image to be seen by the left or right eye; each field equals half the vertical resolution (1280 by 512 pixels). The liquid-crystal-shutter glasses (CrystalEyes from StereoGraphics Corp., San Rafael, CA) alternately allow vision in one eye at the field rate of the displays—60 Hz for each eye or 120 Hz combined. The glasses are synchronized by infrared signals generated by the CrystalEyes controller.

Without the glasses, the wall screens show double images; with them, the viewer gains a striking sense of visual depth and three-dimensional movement. Although the illusion of motion begins, albeit jerkily, in animation at approximately 10 Hz (or 10 frames per second), highly disparate areas flickered noticeably even at 30 Hz (60 Hz for both eyes). To avoid that effect, the update rate was doubled to 60 Hz (120 Hz).

Sound is also provided to enhance the illusion of depth. The use of a sound board and MIDI synthesizers in conjunction with the display allows the computer to generate echoes, Doppler effects, and other sounds associated with real three-dimensional environments and objects, and to direct them to the appropriate speakers. Thus the system persuades the listener the sound "comes from" the proper spot on the horizon.

For the display and sound to create the best illusion of reality, the system should be alert to the viewer's location and direction of gaze. Accordingly, an electromagnetic transponder mounted on the glasses (a Flock of Birds location sensor from Ascension Technology Corp., Burlington, VT) monitors where the lead viewer's head is inside the Cave and in what direction it is turned. Information on viewer location is sent to the controlling workstation, which does the computations for image generation and sends the data to the appropriate projectors. The hand-held tracking wand also contains a tracking receiver so the system can determine where the user is holding and pointing it. . . .

Some Advantages While it is much larger than some other approaches, the Cave immerses the user in the virtual reality world. Its high-resolution and distortion-free wide-angle views are lacking in other systems. Its horizontal resolution of 3840 two-by-six-millimeter pixels across three screens is roughly twice that of High-Definition Television (HDTV). The visual acuity is about 20/110, which is four times better than current head-mounted displays and on a par with binocular omni-oriented monitors (Booms).

The Cave is relatively immune to errors due to head rotation and nodding. With head-mounted displays, the projection plane must move as the viewer's head moves to look left or right, up or down. Therefore, in head-mounted displays, the viewer must wait for the computer to adjust the monitors accordingly after turning, a lag that may cause a loss of equilibrium and even nausea. In the Cave, the projection plane is stable; a glance to the left inside the Cave shows the image already on-screen.

Also, the Cave is a less intrusive interface. A head-mounted display can be uncomfortable and disorienting, because the viewer is cut off from the real visual world. Someone wearing it is blind to the real-world risks of, say, walking into walls or tripping over the cables needed to connect the headset back to the computer. A viewer in the Cave can move around at will, because the outside world is still visible through the liquid-crystal glasses. Better still, because he or she can still see the outside world, the Cave frees the viewer to collaborate more fully.

More than one user will, however, have to interact in the same environment if virtual reality is to become an effective visualization tool in fields such as research and education. While the head-mounted display and the Boom interfaces do allow multiple users to share an environment, it is only at the high cost of duplicating the interface hardware. The participants must themselves be represented with virtual selves in each other's simulated environments, which adds to a system's location-tracking and computational workloads. In other words, in order for me to see you in a head-mounted display environment, the computer must calculate in virtual terms physically in relation to me and render your image accordingly on my headset monitors.

In the Cave, two or more users may look around as they wish. They need not be represented virtually, because they represent themselves physically. At present, though, the screen images are shown from the perspective of only one person, the so-called lead viewer, who is being tracked.

Continuing Challenges Viewer location tracking remains a challenge. A lag between changes in viewer position and the updating of the screen was sometimes noticeable. While less troublesome than the lag experienced with head-mounted displays, it still shows up if the viewer moves too quickly. The lab is looking into extrapolation techniques to better predict user motion and reduce interactive delays to sensor input.

More advanced technology may be capable of maintaining several perspectives separately on-screen, so that each viewer enjoys an individual perspective. However, this technological step is not trivial; with each additional viewer, the display update rate increases, and more brightness is required from the projector.

Despite these difficulties, the Cave has drawn long lines of eager would-be visitors at industry shows since it was first demonstrated at the showcase event at Siggraph '92, held in Chicago. The applications demonstrated have included practical and educational programs, such as molecular biology models, superconductor design, fractal mathematics, weather mapping, and environmental management. Some programs were simply entertaining. For example, some visitors had the opportunity to play music with the hand-held wand by ringing bells floating around them.

Other useful applications of the Cave model of virtual reality include scientific visualization, medicine, architecture, and art. In 1994, the Electronic Visualization Laboratory plans to return to Siggraph with an event called Vroom (Virtual Reality Room). The intent is to display the latest in head-mounted displays, binocular omni-oriented monitor, and Cave technologies. The lab is currently asking others for applications to display in the Cave.

At the same time, EVL is also working on the creation of a "road-show" Cave, which would be easier to transport. While this one-of-a-kind working model costs about $600,000 in all, it could eventually become the prototype for home- or business-based Caves that might cost no more than a "home theater" does today.

<< part V >>

Narrativity

William Burroughs

"The Future of the Novel" (1964)

<< 26 >>

William Burroughs and Brion Gysin. Copyright © 1978 William S. Burroughs and Brion Gysin, reprinted with the permission of The Wylie Agency, Inc.

"Certainly if writing is to have a future it must at least catch up with the past and learn to use techniques that have been used for some time past in painting, music and film."

<< William Burroughs, appropriately the grandson of the inventor of the adding machine, was a catalyst in instigating the literary movement surrounding the emerging digital culture in the 1980s through his surreal mythologies of alien influence and space-age futurisms. Yet it was Burroughs's preoccupation with the deconstruction of words and language, most notably through the cut-up and fold-in techniques that he developed with artist Brion Gysin beginning in 1959, which constitutes his most significant contribution to the fragmentary, nonlinear approach to contemporary narrative—from the cyberpunk novel to the electronically mediated hypertext.

In this essay Burroughs describes the cut-up technique he developed after completing his novel *Naked Lunch* in the late fifties, and later implemented in his famous trilogy *The Soft Machine, The Ticket That Exploded,* and *Nova Express.* His intent was to find a way, through writing, to explore nonlinear perceptions of space and time. Burroughs viewed the cut-up as a way for the writer to discover new connections that extended and ultimately dissolved the rational associations of conventional narrative. Borrowing from the collage technique of visual artists, most notably the surrealists, his method links fragments of texts in surprising juxtapositions, offering unexpected leaps into uncharted territories. For this reason, Burroughs refers to himself as "a map maker, an explorer of psychic areas."

Burroughs's writing experiments of the early 1960s express the essential narrative strategies of computer-based multimedia storytelling long before their time. He regards the reading experience as one of entering into a multidirectional web of different voices, ideas, perceptions, and periods of time. This kind of navigation echoes the work of Vannevar Bush, Douglas Engelbart, and Ted Nelson, who used personal association as a guiding principle for the organization of information, and proposed hypermedia systems that would make private, associative thinking accessible to others. For Burroughs, narrative operates as a vast, multithreaded network that reflects the associative tendencies of the mind. >>

In my writing I am acting as a map maker, an explorer of psychic areas, to use the phrase of Mr Alexander Trocchi, as a cosmonaut of inner space, and I see no point in exploring areas that have already been thoroughly surveyed—A Russian scientist has said: "We will travel not only in space but in time as well—" That is to travel in space is to travel in time—If writers are to travel in space time and explore areas opened by the space age, I think they must develop techniques quite as new and definite as the techniques of physical space travel—Certainly if writing is to have a future it must at least catch up with the past and learn to use techniques that have been used for some time past in painting, music and film—Mr Lawrence Durrell has led the way in developing a new form of writing with time and space shifts as we see events from different viewpoints and realize that so seen they are literally not the same events, and that the old concepts of time and reality are no longer valid—Brion Gysin, an American painter living in Paris, has used what he calls "the cut-up method" to place at the disposal of writers the collage used in painting for fifty years—Pages of text are cut and rearranged to form new combinations of word and image—In writing my last two novels, *Nova Express* and *The Ticket That Exploded,* I have used an extension of the cut-up method I call "the fold-in method"—A page of text—my own or someone else's—is folded down the middle and placed on another page—The composite text is then read across half one text and half the other—The fold-in method extends to writing the flashback used in films, enabling the writer to move backward and forward on his time track—For example I take page one and fold it into page one hundred—I insert the resulting composite as page ten—When the reader reads page ten he is flashing forward in time to page one hundred and back in time to page one—the *déjà vu* phenomenon can so be produced to order—This method is of course used in music, where we are continually moved backward and forward on the time track by repetition and rearrangements of musical themes—

In using the fold-in method I edit, delete and rearrange as in any other method of composition—I have frequently had the experience of writing some pages of straight narrative text which were then folded in with other pages and found that the fold-ins were clearer and more comprehensible than the original texts—Perfectly clear narrative prose can be produced using the fold-in method—Best results are usually obtained by placing pages dealing with similar subjects in juxtaposition—

What does any writer do but choose, edit and rearrange material at his disposal?—The fold-in method gives the writer literally infinite extension of choice—Take for example a page of Rimbaud folded into a page of St John Perse—(two poets who have much in common)—From two pages an infinite number of combinations and images are possible—The method could also lead to a collaboration between writers on an unprecedented scale to produce works that were the composite effort of any number of writers living and dead—This happens in fact as soon as any writer starts using the fold-in method—I have made and used fold-ins from Shakespeare, Rimbaud, from newspapers, magazines, conversations and letters so that the novels I have written using this method are in fact composites of many writers—

I would like to emphasize that this is a technique and like any technique will, of course, be useful to some writers and not to others—In any case a matter for experimentation not argument—The conferring writers have been accused by the press of not paying sufficient attention to the question of human survival—In *Nova Express* (reference is to an exploding planet) and *The Ticket That Exploded,* I am primarily concerned with the question of survival—with nova conspiracies, nova criminals, and nova police—A new mythology is possible in the space age where we will again have heroes and villains with respect to intentions toward this planet—

Allan Kaprow

"Untitled Guidelines for Happenings" (1965)

<< 27 >>

Allan Kaprow, Household. *Photo by Sol Goldberg.*
© Sol Goldberg.

"*The images in each situation can be quite disparate: a kitchen in Hoboken, a* pissoir *in Paris, a taxi garage in Leopoldville, and a bed in some small town in Turkey. . . . None of these planned ties are absolutely required, for preknowledge of the Happening's cluster of events by all participants will allow each one to make his own connections.*"

<< Allan Kaprow is best known as a prime mover behind the Happenings. He coined the term in the late 1950s, and led the movement into the bright lights of popular culture that characterized the 1960s. Happenings are notoriously difficult to describe, in part because each was a unique event shaped by the actions of the audience that participated in any given performance. Simply put, Happenings were held in physical environments—loft spaces, abandoned factories, buses, parks, etc.—and brought people, objects, and events in surprising juxtaposition to one another. With the Happenings, Kaprow rejected the authority and primacy of the artwork as an idealized creation and holy act. Rather, he views art as a vehicle for expanding our awareness of life by prompting unexpected, provocative interactions. For Kaprow, art is a continual work-in-progress, with an unfolding narrative that is realized through the active participation of the audience.

In this essay, Kaprow sets down guidelines for creating Happenings. He begins by dissolving the boundaries between art and everyday life, distributing responsibility for the execution of the performance to all of its participants. John Cage's influence was important to Kaprow's work, particularly the composer's interest in indeterminacy, chance operations, and audience involvement. Cage's experimental performances at Black Mountain College in the late 1940s are regarded as proto-Happening events. Kaprow's approach actually attempted to eliminate the distinction between audience and performer all together.

Kaprow developed techniques to prompt a creative response from the audience, encouraging audience members to make their own connections between ideas and events. These narrative strategies relied on a nonlinear sequencing of events, and the use of indeterminacy to shape the course of the Happening. Kaprow describes how a Happening might take place over an extended period, across vast distances, or in many places at once. His techniques reordered past, present, and future within a single work in a remarkable variety of arrangements. The decentralization of authorship, location, and narrative—here united by the intent of the artist and the imagination of the participating audience members—foreshadows nonlinear forms in digital media that make use of interactive and networked technology to expand the boundaries of space and time. >>

(A) *The line between art and life should be kept as fluid, and perhaps indistinct, as possible.* The reciprocity between the man-made and the ready-made will be at its maximum potential this way. Something will always happen at this juncture, which, if it is not revelatory, will not be merely bad art—for no one can easily compare it with this or that accepted masterpiece. I would judge this a foundation upon which may be built the specific criteria of the Happenings.

(B) *Therefore, the source of themes, materials, actions, and the relationships between them are to be derived from any place or period* except *from the arts, their derivatives, and their milieu.* When innovations are taking place it often becomes necessary for those involved to treat their tasks with considerable severity. In order to keep their eyes fixed solely upon the essential problem, they will decide that there are certain "don'ts" which, as self-imposed rules, they will obey unswervingly. Arnold Schoenberg felt he had to abolish tonality in music composition and, for him at least, this was made possible by evolving the twelve-tone series technique. Later on his more academic followers showed that it was very easy to write traditional harmonies with that technique. But still later, John Cage could permit a C major triad to exist next to the sound of a buzz saw, because by then the triad was thought of differently—not as a musical necessity but as a sound as interesting as any other sound. This sort of freedom to accept all kinds of subject matter will probably be possible in the Happenings of the future, but I think not for now. Artistic attachments are still so many window dressings, unconsciously held on to to legitimize an art that otherwise might go unrecognized.

Thus it is not that the known arts are "bad" that causes me to say "Don't get near them": it is that they contain highly sophisticated habits. By avoiding the artistic modes there is the good chance that a new language will develop that has its own standards. The Happening is conceived as an art, certainly, but this is for lack of a better word, or one that would not cause endless discussion. I, personally, would not care if it were called a sport. But if it is going to be thought of in the context of art and artists, then let it be a distinct art which finds its way into the art category by realizing its species outside of "culture." A United States Marine Corps manual on jungle-fighting tactics, a tour of a laboratory where polyethylene kidneys are made, the daily traffic jams on the Long Island Expressway, are more useful than Beethoven, Racine, or Michelangelo.

(C) *The performance of a Happening should take place over several widely spaced, sometimes moving and changing locales.* A single performance space tends toward the static and, more significantly, resembles conventional theater practice. It is also like painting, for safety's sake, only in the center of a canvas. Later on, when we are used to a fluid space as painting has been for almost a century, we can return to concentrated areas, because then they will not be considered exclusive. It is presently advantageous to experiment by gradually widening the distances between the events within a Happening. First along several points on a heavily trafficked avenue; then in several rooms and floors of an apartment house where some of the activities are out of touch with each other; then on more than one street; then in different but proximate cities; finally all around the globe. On the one hand, this will increase the tension between the parts, as a poet might by stretching the rhyme from two lines to ten. On the other, it permits the parts to exist more on their own, without the necessity of intensive coordination. Relationships cannot help being made and perceived in any human action, and here they may be of a new kind if tried-and-true methods are given up.

Even greater flexibility can be gotten by moving the locale itself. A Happening could be composed for a jetliner going from New York to Luxembourg with stopovers at Gander, Newfoundland, and Reykjavik, Iceland. Another Happening would take place up and down the elevators of five tall buildings in midtown Chicago.

The images in each situation can be quite disparate: a kitchen in Hoboken, a *pissoir* in Paris, a taxi garage in Leopoldville, and a bed in some small town in Turkey. Isolated points of contact may be maintained by telephone and letters, by a meeting on a highway, or by watching a certain television program at an appointed hour. Other parts of the work need only be related by theme, as when all locales perform an identical action which is disjoined in timing and space. But none of these planned ties are absolutely required, for preknowledge of the Happening's cluster of events by all participants will allow each one to make his own connections. This, however, is more the topic of form, and I shall speak further of this shortly.

(D) *Time, which follows closely on space considerations, should be variable and discontinuous.* It is only natural that if there are multiple spaces in which occurrences are scheduled, in sequence or even at random, time or "pacing" will acquire an order that is determined more by the character of movements within environments than by a fixed concept of regular development and conclusion. There need be no rhythmic coordination between the several parts of a Happening unless it is suggested by the event itself: such as when two persons must meet at a train departing at 5:47 P.M.

Above all, this is "real" or "experienced" time as distinct from conceptual time. If it conforms to the clock used in the Happening, as above, that is legitimate, but if it does not because a clock is not needed, that is equally legitimate. All of us know how, when we are busy, time accelerates, and how, conversely, when we are bored it can drag almost to a standstill. Real time is always connected with doing something, with an event of some kind, and so is bound up with things and spaces.

Imagine some evening when one has sat talking with friends, how as the conversation became reflective the pace slowed, pauses became longer, and the speakers "felt" not only heavier but their distances from one another increased proportionately, as though each were surrounded by great areas commensurate with the voyaging of his mind. Time retarded as space extended. Suddenly, from out on the street, through the open window a police car, siren whining, was heard speeding by, *its* space moving as the source of sound moved from somewhere to the right of the window to somewhere farther to the left. Yet it also came spilling into the slowly spreading vastness of the talkers' space, invading the transformed room, partly shattering it, sliding shockingly in and about its envelope, nearly displacing it. And as in those cases where sirens are only sounded at crowded street corners to warn pedestrians, the police car and its noise at once ceased and the capsule of time and space it had become vanished as abruptly as it made itself felt. Once more the protracted picking of one's way through the extended reaches of mind resumed as the group of friends continued speaking.

Feeling this, why shouldn't an artist program a Happening over the course of several days, months, or years, slipping it in and out of the performers' daily lives. There is nothing esoteric in such a proposition, and it may have the distinct advantage of bringing into focus those things one ordinarily does every day without paying attention—like brushing one's teeth.

On the other hand, leaving taste and preference aside and relying solely on chance operations, a completely unforeseen schedule of events could result, not merely in the preparation but in the actual performance; or a simultaneously performed single moment; or none at all. (As for the last, the act of finding this out would become, by default, the "Happening.")

But an endless activity could also be decided upon, which would apparently transcend palpable time—such as the slow decomposition of a mountain of sandstone. . . . In this spirit some artists are earnestly proposing a lifetime Happening equivalent to Clarence Schmidt's lifetime Environment.

The common function of these alternatives is to release an artist from conventional notions of a detached, closed arrangement of time-space. A picture, a piece of music, a poem, a drama, each confined within its respective frame, fixed number of measures, stanzas, and stages, however great they may be in their own right, simply will not allow for breaking the barrier between art and life. And this is what the objective is.

(E) *Happenings should be performed once only.* At least for the time being, this restriction hardly needs emphasis, since it is in most cases the only course possible. Whether due to chance, or to the lifespan of the materials (especially the perishable ones), or to the changeableness of the events, it is highly unlikely that a Happening of the type I am outlining could ever be repeated. Yet many of the Happenings have, in fact, been given four or five times, ostensibly to accommodate larger attendances, but this, I believe, was only a rationalization of the wish to hold on to theatrical customs. In my experience, I found the practice inadequate because I was always forced to do that which *could be repeated,* and had to discard countless situations which I felt were marvelous but performable only once. Aside from the fact that repetition is boring to a generation brought up on ideas of spontaneity and originality, to repeat a Happening at this time is to accede to a far more serious matter: compromise of the whole concept of Change. When the practical requirements of a situation serve only to kill what an artist has set out to do, then this is not a practical problem at all; one would be very practical to leave it for something else more liberating.

Nevertheless, there is a special instance of where more than one performance is entirely justified. This is the score or scenario which is designed to make every performance significantly different from the previous one. Superficially this has been true for the Happenings all along. Parts have been so roughly scored that there was bound to be some margin of imprecision from performance to performance. And, occasionally, sections of a work were left open for accidentals or improvisations. But since people are creatures of habit, performers always tended to fall into set patterns and stick to these no matter what leeway was given them in the original plan.

In the near future, plans may be developed which take their cue from games and athletics, where the regulations provide for a variety of moves that make the outcome always uncertain. A score might be written, so general in its instructions that it could be adapted to basic types of terrain such as oceans, woods, cities, farms; and to basic kinds of performers such as teenagers, old people, children, matrons, and so on, including insects, animals, and the weather. This could be printed and mail-ordered for use by anyone who wanted it. George Brecht has been interested in such possibilities for some time now. His sparse scores read like this:

DIRECTION

Arrange to observe a sign
indicating direction of travel.

- travel in the indicated direction
- travel in another direction

But so far they have been distributed to friends, who perform them at their discretion and without ceremony. Certainly they are aware of the philosophic allusions to Zen Buddhism, of the subtle wit and childlike simplicity of the activities indicated. Most of all, they are aware of the responsibility it places on the performer to make something of the situation or not.

This implication is the most radical potential in all of the work discussed here. Beyond a small group of initiates, there are few who could appreciate the moral dignity of such scores, and fewer still who could derive pleasure from going ahead and doing them without self-consciousness. In the case of those Happenings with more detailed instructions or more expanded action, the artist must be present at every moment, directing and participating, for the tradition is too young for the complete stranger to know what to do with such plans if he got them.

(F) *It follows that audiences should be eliminated entirely.* All the elements—people, space, the particular materials and character of the environment, time—can in this way be integrated. And the last shred of theatrical convention disappears. For anyone once involved in the painter's problem of unifying a field of divergent phenomena, a group of inactive people in the space of a Happening is just dead space. It is no different from a dead area of red paint on a canvas. Movements call up movements in response, whether on a canvas or in a Happening. A Happening with only an empathic response on the part of a seated audience is not a Happening but stage theater.

Then, on a human plane, to assemble people unprepared for an event and say that they are "participating" if apples are thrown at them or they are herded about is to ask very little of the whole notion of participation. Most of the time the response of such an audience is half-hearted or even reluctant, and sometimes the reaction is vicious and therefore destructive to the work (though I suspect that in numerous instances of violent reaction to such treatment it was caused by the latent sadism in the action, which they quite rightly resented). After a few years, in any case, "audience response" proves to be so predictably pure cliché that anyone serious about the problem should not tolerate it, any more than the painter should continue the use of dripped paint as a stamp of modernity when it has been adopted by every lampshade and Formica manufacturer in the country.

I think that it is a mark of mutual respect that all persons involved in a Happening be willing and committed participants who have a clear idea what they are to do. This is simply accomplished by writing out the scenario or score for all and discussing it thoroughly with them beforehand. In this respect it is not different from the preparations for a parade, a football match, a wedding, or religious service. It is not even different from a play. The one big difference is that while knowledge of the scheme is necessary, professional talent is not; the situations in a Happening are lifelike or, if they are unusual, are so rudimentary that professionalism is actually uncalled for. Actors are stage-trained and bring over habits from their art that are hard to shake off; the same is true of any other kind of showman or trained athlete. The best participants have been persons not normally engaged in art or performance, but who are moved to take part in an activity that is at once meaningful to them in its ideas yet natural in its methods.

There is an exception, however, to restricting the Happening to participants only. When a work is performed on a busy avenue, passers-by will ordinarily stop and watch, just as they might watch the demolition of a building. These are not theater-goers and their attention is only temporarily caught in the course of their normal affairs. They might stay, perhaps become involved in some unexpected way, or they will more likely move on after a few minutes. Such persons are authentic parts of the environment.

A variant of this is the person who is engaged unwittingly with a performer in some planned action: a butcher will sell certain meats to a customer-performer without realizing that he is a part of a piece having to do with purchasing, cooking, and eating meat.

Finally, there is this additional exception to the rule. A Happening may be scored for *just watching.* Persons will do nothing else. They will watch things, each other, possibly actions not performed by themselves, such as a bus stopping to pick up commuters. This would not take place in a theater or arena, but anywhere else. It could be an extremely meditative occupation when done devotedly; just "cute" when done indifferently. In a more physical mood, the idea of called-for watching could be contrasted with periods of action. Both normal tendencies to observe and act would now be engaged in a responsible way. At those moments of relative quiet the observer would hardly be a passive member of an audience: he would be closer to the role of a Greek chorus, without its specific meaning necessarily, but with its required place in the overall scheme. At other moments the active and observing roles would be exchanged, so that by reciprocation the whole meaning of watching would be altered, away from something like spoon-feeding, toward something purposive, possibly intense. . . .

Bill Viola

"Will There Be Condominiums in Data Space?" (1982)

<< 28 >>

Bill Viola, Room for St. John of the Cross. © *Bill Viola. Courtesy of Bill Viola.*

"We can see the seeds of what some have described as the ultimate recording technology: total spatial storage, with the viewer wandering through some three-dimensional, possibly life-sized field of prerecorded or simulated scenes and events evolving in time."

<< Since he began producing video art in the early 1970s, Bill Viola has explored ways to manipulate and restructure our perception of time and space through electronic media. After several years of creating single-channel works for videotape, in which he explored the restructuring of time through techniques like extreme slow motion, jarring edits, and surreal superimpositions, Viola began to project video onto surfaces, which introduced a spatial element to his installations. In 1982, the same year that Lynn Hershman released the first art videodisc, *Lorna,* Viola wrote this article about the narrative potential of interactive media. Here he discusses the concept of *dataspace,* a territory of information in which all data exists in a continual present, outside the traditional definitions of time and space, available for use in endless juxtapositions.

Viola arrives at the notion of dataspace by considering the spaces that have been constructed over the ages to record cultural history in architectural form, from Greek temples to Gothic cathedrals. He compares these "memory palaces" to the personal computer, with its capacity for storage, instant access, and information retrieval. Cathedrals, Viola says, communicate through their symbolic ornamentation, illustrative paintings, and stained-glass narratives. Similarly, the branching pathways and forking paths of computer-controlled video works can be explored by the viewer searching for story and meaning. As Viola explains, this ability to enter a nonlinear information matrix at any point, to navigate its contents freely in any direction and at any speed, introduces a kind of "idea space" where the only constraint is the limit of the human imagination. The computer has introduced the "next evolutionary step," Viola claims, in which ancient models of memory and artistic expression are reborn through the fluid processes of information technologies. >>

Will There Be Condominiums in Data Space?

Possibly the most startling thing about our individual existence is that it is continuous. It is an unbroken thread—we have been living this same moment ever since we were conceived. It is memory, and to some extent sleep, that gives us the impression of a life of discrete parts, periods, or sections, of certain times or "highlights." Hollywood movies and the media, of course, reinforce this perception.

If things are perceived as discrete parts or elements, they can be rearranged.

Gaps become most interesting as places of shadow, open to projection. Memory can be regarded as a filter (as are the five senses)—it is a device implanted for our survival. The curse of the mnemonist is the flood of images that are constantly replaying in his brain. He may be able to demonstrate extraordinary feats of recall, but the rest of the banal and the mundane is playing back in there too, endlessly. The result can be lack of sleep, psychosis, and even willful death, driving some to seek professional psychiatric help (and thus become history on the pages of medical journals and books).[1] This reincarnates one of the curses of early video art—"record everything," the saturation-bombing approach to life which made so many early video shows so boring and impossible to sit through. Life without editing, it seems, is just not that interesting.

It is only very recently that the ability to forget has become a prized skill. In the age of "information overload," we have reached a critical mass that has accelerated the perfection of recording technologies, an evolution that leads back to ancient times. Artificial memory systems have been around for centuries. The early Greeks had their walks through temple,[2] and successive cultures have refined and developed so-called "mnemo-technics"—Thomas Aquinas described an elaborate memory scheme of projecting images and ideas on places; in 1482 Jacobus Publicius wrote of using the spheres of the universe as a memory system; Giulio Camillo created a "Memory Theater" in Italy in the early 1500s; and Giordano Bruno diagrammed his system of artificial memory in his work *Shadows,* published in 1582. Frances Yates describes this entire remarkable area in her brilliant book *The Art of Memory* (University of Chicago Press, 1966).

When I was in Japan in 1981, I visited a festival of the dead at one of the most sacred places in the country, Osoresan Mountain. There I saw blind female shamen called *itako* calling back the spirits of the dead for inquiring relatives, a centuries-old practice. Until that time I had felt that the large Japanese electronics companies were way ahead in the development of communications technology. After witnessing the *itako,* however, I realized they were way behind. Right in their own backyard were people who, without the aid of wires or hardware of any sort, have been for ages regularly communicating through time and space with ancestors long gone. An interesting place at the temple site (which was perched in the surreal landscape of an extinct volcanic crater) was a special walk for the visiting pilgrims to take along a prescribed trail. The way led from the temple through a volcanic wasteland of rockpiles and smoking fissures to the shores of a crater lake. It was called "the walk through Hell." The path through the landscape and the points along it all had special significance. The *itako,* to call up the dead, took this "walk through Hell" in their minds, bringing the spirits in along the familiar path, and when they were through, sent them back the same way.

The interesting thing about idea spaces and memory systems is that they presuppose the existence of some sort of place, either real or graphic, which has its own structure and architecture. There is always a whole space, which already exists *in its entirety,* onto which ideas and images can be mapped, using only that portion of the space needed.

In addition to the familiar model of pre-recorded time unfolding along a linear path (as evidenced by many things, from our writing system to the thread of magnetic tape playing in a videotape recorder), there is another parallel to be linked with modern technology. "Data space" is a term we hear in connection with computers. Information must be entered into a computer's memory to create a set of parameters, defining some sort of ground, or field, where future calculations and binary events will occur. In three-dimensional computer graphics, this field exists as an imaginary but real chunk of space, a conceptual geometry, theoretically infinite, within which various forms may be created, manipulated, extended, and destroyed. The graphics display screen becomes that mysterious third point of view looking in on this space (we often call it our "mind's eye"), which can be moved about and relocated from any angle at will. The catch is that the space must exist in the computer first, so that there is a reference system within which to locate the various coordinates of points and lines called into being by the operator. In our brain, constantly flickering pulses of neuron firings create a steady-state field onto which disturbances and perturbations are registered as percepts and thought forms. This is the notion that something is already "on" before you approach it, like the universe, or like a video camera which always needs to be "video-ing" even if there is only a blank raster ("nothing") to see. Turn it off, and it's not video anymore.

When I had my first experience with computer videotape editing in 1976, one demand this new way of working impressed upon me has remained significant. It is the idea of holism. I saw then that my piece was actually finished and in existence *before* it was executed on the VTRs. Digital computers and software technologies are holistic; they think in terms of whole structures. Wordprocessors allow one to write out, correct, and rearrange the *whole* letter before typing it. Data space is fluid and temporal, hardcopy is for real—an object is born and becomes fixed in time. Chiseling in stone may be the ultimate hard copy.

When I edited a tape with the computer, for the first time in my life I saw that my video piece had a "score," a structure, a pattern that could be written out on paper. We view video and film in the present tense—we "see" one frame at a time passing before us in this moment. We don't see what is before it and what is after it—we only see the narrow slit of "now." Later, when the lights come on, it's gone. The pattern does exist, of course, but only in our memory. Notation systems have been around since the beginning of history, since what we call history *is* notation of events in time, i.e., historical "records." With speech we have graphic writing systems; with music we have the score. They are both symbolic coded systems for the recording and later playback of information events in time. Poetry has always had a level that video or film cannot approach (at least not yet): the existence of the words on paper (how the poem looks, how the words are placed on the page, the gaps, the spacing, etc.). The whole poem is there before us, and, starting at the top of the page, we can see the end before we actually get there.

Our cultural concept of education and knowledge is based upon the idea of building something up from a ground, from zero, and starting piece by piece to put things together, to construct edifices. It is additive. If we approach this process from the other direction, considering it to be backwards, or subtractive, all sorts of things start to happen. Scientists always marvel at nature, at how it seems to be some grand code, with a built-in sense of purpose. Discoveries are made which reveal that more and more things are related, connected. Everything appears to be aware of itself and everything else, all fitting into an interlocking whole. We quite literally carve out our own realities. If you want to make a jigsaw puzzle, you must first start with the whole image, and *then* cut it up. The observer, working backwards into the system, has the point of view that he or she is building things up, putting it together piece by piece. The prophet Mohamed has said, "All knowledge is but a single point—it is the ignorant who have multiplied it."

The Whole Is the Sum of Its Parts

A friend of mine is an ethnomusicologist who spent several years studying the gamelan music of Central Java. He was trained in Western music in the States, and spent many years working on his own compositions and performing with other musicians. One of the most frustrating things about his studies in Java, he told me, was trying to work on specific parts of songs with the gamelan musicians. Once they were at a rehearsal, and after running through a piece, he asked

> them to play only a section from the middle so that he could make sure he got all the notes right. This proved to be an impossible request. After a lot of hemming and hawing, excuses, and several false starts, he realized that the group just could not do it. They insisted on playing the entire piece over again, from beginning to end. In Java, the music was learned by rote, from many years of observation and imitation, not from written notation. The idea of taking a small part out of context, or playing just a few bars, simply did not exist. The music was learned and conceived as a whole in the minds of the musicians.

Giulio Paolini, the contemporary Italian artist, made a little-known but far-reaching videotape in the mid-seventies. It was his first and only tape. Working at an experimental video studio in Florence in the cradle of Western art, he, like many other European artists who visited the art/tapes/22 studio, had his first encounter with video. Instead of simply re-translating into video what he had already been doing before, as most other artists had done, Paolini intuitively recognized the great power underlying the recording media. He took the slides of all his work, most of the pieces he had ever made, and recorded them one at a time on each frame of video. Playing back this tape, the viewer sees 15 years of Paolini's art, his life's work, go by in less than a minute. Poof! It's gone.

It is slowly becoming clear that structuralism, currently out of fashion in the fashion-conscious, ever shifting spotlight of the art world, must be reconsidered. It is vital. However, this new structuralism is not the same as the often over-intellectualized, didactic, structuralism-for-structuralism's-sake that took center stage in the art scene over a decade ago (most visibly through the work of experimental filmmakers). In retrospect, however, the core ideas being expressed then certainly remain important, and perhaps could only have emerged in the way they did given that particular place and moment in cultural time. Furthermore, the anti-content messages that have been espoused in various fields of art in the twentieth century also continue to merit attention. We have all been made aware that, since the Renaissance, Western eyes have been drawn to the visual, to the surface appearance of the world. "Realism" came to mean how something appeared to the eye alone. Looking at the Gothic art before it, along with Asian and so-called Primitive or Tribal Art, it is clear that something fundamental is missing. However, from our viewpoint today, it is also clear that pure structuralism alone is no answer either.

> Decadent art is simply an art which is no longer felt or energized, but merely denotes, in which there exists no longer any real correspondence between the formal and pictorial elements, its meaning, as it were, negated by the weakness or

> incongruity of the pictorial element; but it is often . . . *far less* conventional than are the primitive or classic stages of the same sequence. True art, pure art, never enters into competition with the unattainable perfection of the world.[3]
>
> —A. K. Coomaraswamy

Structure, or form, has always been the basis of the original pictorial art of both Europe and the East, but the Middle Ages were the last time when both Europe and Asia met on common artistic ground.

> In Western art, the picture is generally conceived as seen in a frame or through a window, and so brought towards the spectator; but the Oriental image really exists only in our mind and heart and thence is projected or reflected into space.
>
> The Indian, or Far Eastern icon, carved or painted, is neither a memory image nor an idealization, but a visual symbolism, ideal in the mathematical sense. . . . Where European art naturally depicts a moment of time, an arrested action, or an effect of light, Oriental art represents a continuous condition. In traditional European terms, we should express this by saying that modern European art endeavors to represent things as they are in themselves, Asiatic and Christian art to represent things more nearly as they are in God, or nearer their source.
>
> —A. K. Coomaraswamy

The idea of art as a kind of diagram has for the most part not made it down from the Middle Ages into modern European consciousness. The Renaissance was the turning point, and the subsequent history of Western art can be viewed as the progressive distancing of the arts away from the sacred and towards the profane. The original structural aspect of art, and the idea of a "data space" *was* preserved through the Renaissance, however, in the continued relation between the image and architecture. Painting became an architectural, spatial form, which the viewer experienced by physically walking through it. The older concept of an idea and an image architecture, a memory "place" like the mnemonic temples of the Greeks, is carried through in the great European cathedrals and palaces, as is the relation between memory, spatial movement, and the storage (recording) of ideas.

Something extraordinary is occurring today, in the 1980s, which ties together all these threads. The computer is merging with video. The potential offspring of this marriage is only beginning to be realized. Leaping directly into the farther future for a moment, we can see the seeds of what some have described as the ultimate recording technology: total spatial storage, with the viewer wandering through some three-dimensional, possibly life-sized field of prerecorded or simulated scenes and events evolving in time. At present, the interactive video discs currently on the market have already begun to address some of these possibilities. Making a program for interactive video disc involves the ordering and structuring (i.e., editing) of much more information than will actually be seen by an individual when he or she sits down to play the program. All possible pathways, or branches, that a viewer ("participant" is a better word) may take through the material must already exist at some place on the disc. Entire prerecorded sections of video may never be encountered by a given observer.

Soon, the way we approach making films and videotapes will drastically change. The notion of a "master" edit and "original" footage will disappear. Editing will become the writing of a software program that will tell the computer how to arrange (i.e., shot order, cuts, dissolves, wipes, etc.) the information on the disc, playing it back in the specified sequence in real time or allowing the viewer to intervene. Nothing needs to be physically "cut" or re-recorded at all. Playback speed, the cardinal 30 frames a second, will become intelligently variable and thus malleable, becoming, as in electronic music practice, merely one fundamental frequency among many which can be modulated, shifted up or down, superimposed, or interrupted according to the parameters of electronic wave theory. Different sections can be assigned to play back at specific speeds or reversed; and individual frames can be held still on the screen for predetermined durations. Other sections can be repeated over and over. Different priorities rule how and in what order one lays material down on the "master" (disc). New talents and skills are needed in making programs—this is not editing as we know it. It was Nikola Tesla, the original uncredited inventor of the radio, who called it "transmission of intelligence." He saw something there that others didn't. After all these years, video is finally getting "intelligence," the eye is being reattached to the brain. As with everything else, however, we will find that the limitations emerging lie more with the abilities and imaginations of the producers and users, rather than in the tools themselves.

As in the figure/ground shifts described in Gestalt psychology, we are in the process of a shift away from the temporal, piece-by-piece approach of *constructing* a program (symbolized by the camera and its monocular, narrow, tunnel-of-vision, single point of view), and towards a spatial, total-field approach of *carving out* potentially multiple programs (symbolized by the computer and its holistic software models, data spaces, and infinite points of view). We are proceeding from models of the eye and ear to models of thought processes and conceptual structures in the brain. "Conceptual Art" will take on a new meaning.

As we take the first steps into data space, we discover that there have been many previous occupants. Artists have been there before. Giulio Camillo's Memory Theater (which he actually constructed in wood, calling it a "constructed body and soul") is one example. Dante's *Divine Comedy* is another. Fascinating relationships between ancient and modern technologies become evident. A simple example can be found in the Indian Tantric doctrine of the three traditional expressions of the deity: the anthropomorphic, or visual, image; the yantra, or geometric "energy" diagram; and the mantra, or sonic representation through chanting and music. It is interesting to note that these are all considered to be equal—simply outward expressions of the same underlying thing. In form, this is not unlike the nature of electronic systems: the same electronic signal can be an image if fed into a video monitor, an energy diagram if fed into an oscilloscope, and a sequence of sounds if fed into an audio system.

Today, there are visual diagrams of data structures already being used to describe the patterns of information on the computer video disc. The most common one is called "branching," a term borrowed from computer science (figure 1). In this system, the viewer proceeds from top to bottom in time, and may either play the disc uninterrupted (arrow), or stop at predetermined branching points along the way and go off into related material at other areas on the disc for further study (like a form of "visual footnoting"). Examples of this system go something like—in a program on the desert, the viewer can stop at a point where plants are mentioned, and branch off to more detailed material on the various flora of the valley floor, etc. Although it is clear how this can enhance our current educational system, freeing students from boring and incompetent teachers so they can proceed at their own pace through information which now contains movement, dynamic action, and sound in addition to written words, artists know that there must be more out there than this. Even though the technology is interactive, this is still the same old linear logic system in a new bottle.

As a start, we can propose new diagrams, such as the "matrix" structure (Figure 1). This would be a non-linear array of information. The viewer could enter at any point, move in any direction, at any speed, pop in and out at any place. All directions are equal. Viewing becomes exploring a territory, traveling through a data space. Of course, it would not be the obviously literal one like the Aspen project.[4] We are moving into *idea* space here, into the world of thoughts and images as they exist in the brain, not on some city planner's drawing board. With the integration of images and video into the domain of computer logic, we are beginning the task of mapping the conceptual structures of our brain onto the technology. After the first TV camera with VTR gave us an eye connected to a

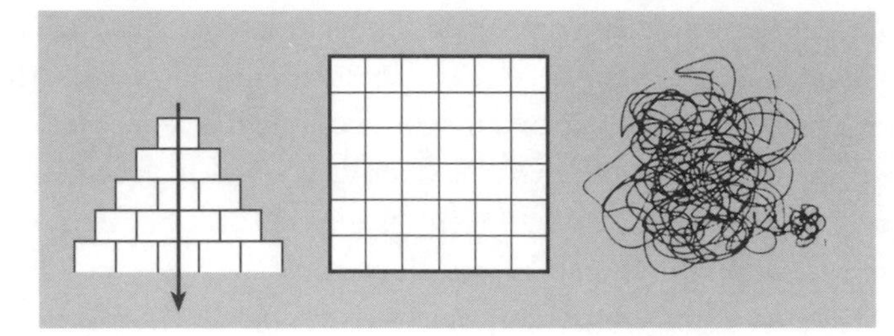

FIGURE 1. *Branching Structure, Matrix Structure, Schizo Structure.*

gross form of non-selective memory, we are now at the next evolutionary step—the area of intelligent perception and thought structures, albeit artificial.

Finally, we can envision other diagrams/models emerging as artists go deeper into the psychological and neurological depths in search of expressions for various thought processes and manifestations of consciousness. Eventually, certain forms of neurosis, so long the creative fuel of the tormented artist in the West, may be mapped into the computer disc. We may end up with the "schizo" or "spaghetti" model, in which not only are all directions equal, but all are not equal (Figure 1). Everything is irrelevant and significant at the same time. Viewers may become lost in this structure and never find their way out.

Worlds are waiting to be explored. It is to be hoped that artists will be given their share of access to experiment with this exciting new technology. I recently had a glimpse of some of the possibilities for art when I met a designer who had first encountered computers while working at a large French fashion design firm in New York. There, the graphics artist worked at computer terminals. With a light-pen, he could draw various designs, working with functions of computer memory and data manipulation. Furthermore, his terminal was linked to a large databank of fabric designs and images from around the world and throughout history. After completing a sketch, for example, he could call up a seventeenth-century Japanese kimono design, look at it or superimpose it with his own idea. Then he could call up a turn-of-the-century European dress pattern, combine that with his design or integrate it with the kimono, all the while storing the various stages in memory. When all of this was completed and the final design chosen, he could then tie into other offices in Europe and the Orient right on the same screen. Designers could compare notes, get availability data on his fabric from the mills (i.e., where is the best silk, who has stock, what is the order time, etc.). All phases of his work could occur on the same screen as digital information. He could travel in space (Europe, the Far East), as well as in time (art history), all in an instant and available either as written text or visual images.

Despite the anti-technology attitudes which still persist (some, it should be added, for very good reasons), the present generation of artists, filmmakers, and video-makers currently in school, and their instructors, who continue to ignore computer and video technology, will in the near future find that they have bypassed *the* primary medium, not only of their own fields, but of the entire culture as well. It is imperative that creative artists have a hand in the developments currently underway. Computer video discs are being marketed as a great new tool in training and education. At this moment, there are creative people experimenting with the technology, ensuring that innovative and unique applications will emerge; but for now, many of the examples return to the boring domain of linear logic in the school classroom. The Aspen city map project is perhaps one of the more interesting examples of new program formats. We are at the beginning, but even so, for the artist, standard educational logic structures are just not that interesting. Artists have been to different parts of the brain, and know quite well that things don't always work like they told you in school.

It is of paramount importance now, as we watch the same education system that brought us through school (and the same communications system that gave us the wonderful world of commercial TV and AM radio) being mapped onto these new technologies, that we go back and take a deeper look at some of the older systems described in these pages. Artists not shackled to the fad and fashion treadmill of the art world, especially the art world of the past few years, will begin to see the new meaning that art history is taking on. As I have begun to outline in this article, the relation between the image and architecture (as in Renaissance art), the structuralism of sacred art (Oriental, Early Christian, and Tribal art, with their mandalas, diagrams, icons, and other symbolic representations, including song, dance, poetry), and artificial memory systems (the first recording technologies from the time of the Greeks through the Middle Ages), are all areas that require further investigation.

As we continue to do our dance with technology, some of us more willingly than others, the importance of turning back towards ourselves, the prime mover of this technology, grows greater than the importance of any LSI circuit. The sacred art of the past has unified form, function, and aesthetics around this single ultimate aim. Today, development of self must precede development of the technology or we will go nowhere—there *will* be condominiums in data space (it has already begun with cable TV). Applications of tools are only reflections of the users—

chopsticks may be a simple eating utensil or a weapon, depending on who uses them.

The Porcupine and the Car

Late one night while driving down a narrow mountain highway, I came across a large porcupine crossing the road up ahead. Fortunately, I spotted him in time to bring the car to a stop a short distance from where he was standing. I watched him in the bright headlights, standing motionless, petrified at this "close encounter of the third kind." Then, after a few silent moments, he started to do a strange thing. Staying in his place, he began to move around in a circle, emitting a raspy hissing sound, with the quills rising up off his body. He didn't run away. I realized that this dance was actually a move of self-defence. I cut the car headlights to normal beams, but he still continued to move around, even more furiously, casting weird shadows on the trees behind. Finally, to avoid giving him a heart attack, and to get home, I cut the lights completely and turned off the engine. I watched him in the dim moonlight as he stopped his dance and moved off the road. Later, while driving off, I realized that he was probably walking proudly away, gloating over how he really gave it to that big blinding noisy thing that rushed toward him out of the night. I'm sure he was filled with confidence, so pleased with himself that he had won, his porcupine world-view grossly inflated as he headed home in the darkness.

Lynn Hershman

"The Fantasy Beyond Control" (1990)

<< 29 >>

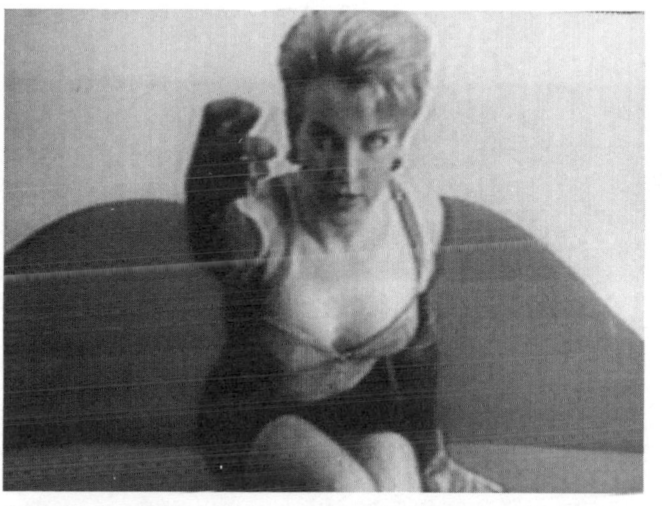

Lynn Hershman, Deep Contact. *© Lynn Hershman. Courtesy of Lynn Hershman*

"Traditional narratives are being restructured. As a result, people feel a greater need to personally participate in the discovery of values that affect and order their lives, to dissolve the division that separates them from control, freedom; replacing longing, nostalgia and emptiness with a sense of identity, purpose and hope."

<< Media artist Lynn Hershman divides her work into two categories: B.C. (Before Computers) and A.D. (After Digital). The line of demarcation occurred around 1980 as interactive technologies, including personal computers and laserdisc players, became commercially available. In her early performance works and site-specific installations (B.C.), Hershman had begun exploring themes that focused on issues of identity, alienation, and the blurring between reality and fiction. The introduction of computers into her work (A.D.) was a natural consequence of her pursuit of these themes. The confluence of these interests, the theatrical nature of her ideas, and the types of explorations suggested by emerging interactive technologies led her into groundbreaking creative terrain. She was one of the few artists to grasp the expressive possibilities of electronic media and to translate its potential into major interactive artworks for the medium.

In this essay, Hershman discusses two key works from the early 1980s that illustrate how she adopted electronic media to explore new narrative strategies that required and exploited viewer interaction. The first of these works was *Lorna* (1982), the seminal art videodisc: a labyrinthine journey through the mental landscape of an agoraphobic middle-aged woman. Trapped in the seclusion of her tiny apartment, the fictional character Lorna is obsessed with media, television, and the bizarre objects of her room. The viewer navigates through the branching structure of the videodisc to make decisions that ultimately determine the fate of Lorna's sordid existence. As Hershman points out in this essay, Lorna's passive relation to media and life is juxtaposed with the viewer's newfound agency to select and reassemble the narrative's branching themes, stories, interpretations, and conclusions.

Hershman describes her new media artworks as exploring "the interaction and intimacy of people despite technological separations." In *Deep Contact* (1984–1989), the other work discussed in this essay, Hershman uses a touchscreen interface to suggest that the viewer can reach through the work's glass surface, the computer's "fourth wall." The user's interactions are incorporated into the unfolding narrative to confront issues of voyeurism, eroticism, and narcissism. In *Deep Contact,* "touching" the protagonist Marion becomes an erotic gesture, the extension of a phantom limb into the illusionary world of virtual storyspace. As Hershman puts it, this type of interactivity constitutes a transgression of the screen, and transports the viewer into virtual reality. Interaction with virtual characters (including the viewer's own image, which is projected into the virtual space), encourages the viewer to break free from the traditional audience role of passive recipient by navigating through the work's many branching narrative paths. >>

> "A fascinating first. *Lorna* is a stream of consciousness collage that requires not only an interactive left brain, to deduce and make logical choices, but also an interactive right brain, to feel what the character's feeling and to understand her life. Whatever *Lorna*'s fate, Lynn Hershman has scored one for the history books."
>
> —*Video Magazine*

Despite some theories to the contrary, it is presumed that making art is active and viewing art is passive. Radical developments in communication technology, such as the marriage of image, sound, text, computers and interactivity, have challenged this assumption. The participants of *Lorna* have reported that they had the impression of being empowered because they could manipulate Lorna's life. The decision process was placed in their hands, literally. The media bath of transmitted, pre-structured and edited information that surrounds, and some say alienates, people is washed away, hosed down by viewer input. Altering the basis for the information exchange is subversive and encourages participation creating a different audience dynamic.

Interactive systems require viewers to react. Choices must be made. As technology expands, there will be more permutations available, not only between the viewer and the system, but between elements within the system itself. Computer systems will eventually reflect the personality of their users. However, there is a space between the system and player in which a link, a fusion or transplant occurs. Truth and fiction blur. According to Sigmund Freud, reality may be limited to perceptions that can be verified through words or visual codes. Therefore, perceptions are the drive to action that influence, if not control, real events. Perceptions become the key to reality.

Perhaps it was nostalgia that led me to search for an interactive video fantasy, a craving for control, a longing for real-time activities, a drive toward direct action. This chronic condition is reputedly a side effect, or for video artists an occupational hazard, that results from watching too much television. Television is a medium that is by its nature fragmentary, incomplete, distanced and unsatisfying, similar to platonic sex. A precondition of a video dialogue is that it does not talk back. Rather, it exists as a moving stasis, a one-sided discourse, a trick mirror that absorbs rather than reflects.

Lorna was developed as a research and development guide, but it is generally inaccessible as it was pressed in a limited edition of 20, of which only 14 now exist. *Lorna* is only occasionally installed in galleries or museums. Creating a truly interactive work demands that it exist on a mass scale, available and accessible to many people.

My path to interactive works began not with video, but with performance when, in 1971, I created an alternative identity named Roberta Breitmore. Her decisions were random, only very remotely controlled. Roberta's manipulated reality became a model for a private system of interactive performances. Instead of being kept on a disc or hardware, her records were stored as photographs and texts that could be viewed without predetermined sequences. This allowed viewers to become voyeurs into Roberta's history. Their interpretations shifted, depending on the perspective and order of the sequences.

Two years after Roberta's transformation, *Lorna,* the first interactive art videodisc, was completed. Unlike Roberta, who has many adventures directly in the environment, Lorna, a middle-aged, fearful agoraphobic, never leaves her tiny apartment. The premise was that the more she stays home and watches television, the more fearful she becomes, primarily because she absorbs the frightening messages of advertising and news broadcasts. Because she never leaves home, the objects in her room take on a magnificent proportion.

Every object in Lorna's room is numbered and becomes a chapter in her life that opens into branching sequences. The viewer, or participant, accesses information about her past, future and personal conflicts via these objects. Many images on the screen are of the remote control device Lorna uses to change television channels. Because the viewer uses a nearly identical unit to direct the disc action, a metaphoric link or point of identification is established between the viewer and Lorna. The viewer activates the live action and makes surrogate decisions for Lorna. Decisions are designed into a branching path. Although there are only 17 minutes of moving image on the disc, the 36 chapters can be sequenced differently and played over a period of time lasting several days. There are three separate endings to the disc, though the plot has multiple variations that include being caught in repeating dream sequences, or using multiple soundtracks, and can be seen backward, forward, at increased or decreased speeds, and from several points of view.

There is no hierarchy in the ordering of decisions. It should be noted that this idea is not new. It was explored by such artists as Stephen Mallarme, John Cage and Marcel Duchamp, particularly through his music. These artists pioneered ideas about random adventures and chance operations 50 years before the invention of the technology that would have allowed them to exploit their concepts more fully.

Lorna's passivity, caused by being controlled by the media, is a counterpoint to the direct action of the participants. As the branching path is deconstructed, players become aware of the subtle yet powerful effects of fear caused by the media, and become more empowered, more active. By acting on Lorna's behalf we travel through their own internal labyrinth to our innermost transgressions.

As interactive technology is increasingly visible in many areas of society the political impact is spectacular. Traditional narratives are being restructured. As a result, people feel a greater need to personally participate in the discovery of values that affect and order their lives, to dissolve the division that separates them from control, freedom; replacing longing, nostalgia and emptiness with a sense of identity, purpose and hope.

The second piece, *Deep Contact,* refers to the players' ability to travel the 57 different segments into the deepest part of the disc, determining, through their own intuition, the route to the centre, while simultaneously trying to find and to feel the deepest, most essential parts of themselves. Viewers choreograph their own encounters in the vista of voyeurism that is incorporated into *Deep Contact.*

This piece developed into a collaboration between many people. John Di Stefano had the difficult task of composing music that would work in modulated segments, as well as backward, forward or in slow-motion. Jiri Vsneska assisted immeasurably with the shooting and scanning of photographic images, and Marion Grabinsky, the leather-clad protagonist, gave the piece the erotic appeal so necessary for sexual transgressions. Toyoji Tomita played the Zen Master and Demon with equal charm, while the crew of camera operators, editors and production managers added tremendously to the success and joy of making this piece.

This touch-sensitive interactive videodisc installation compares intimacy with reproductive technology, and allows viewers to have adventures that change their sex, age and personality. Participants are invited to follow their instincts as they are instructed to actually touch their "guide" Marion on any part of her body. Adventures develop depending upon which body part is touched.

The sequence begins when Marion knocks on the projected video screen asking to be touched. She keeps asking until parts of her body, scanned and programmed to rotate onto the Microtouch screen, actually are touched. For instance, if you touch her head, you are given a choice of TV channels, some giving short, but humorous, analytic accounts of "reproductive technologies" and their effect on women's bodies while others show how women see themselves. The protagonist also talks about "extensions" into the screen that are similar to "phantom limbs," so that the screen becomes an extension of the participant's hand. Touching the screen encourages the sprouting of phantom limbs, virtual connections between viewer and image.

If Marion's torso is touched, the video image on the disc goes to a bar where the viewer can select one of three characters, Marion, a Demon or a Voyeur, to follow through interactive fiction that has a video component. If viewers touch Marion's legs, they enter a garden sequence in which they can follow Marion, a Zen Master, an Unknown Path or a Demon. Selections are made via images that have been photographically scanned onto the touchscreen. In the garden, for example, the image on the touchscreen is a hand that jumps forward depending upon selections, and that allows the viewer to follow the lines on the hands to different routes. The participant usually follows a character or a segment to a fork in the road. At this point, the disc automatically stops, requiring a selection to go left, to go to the right, to return to the first segment of the disc, or to repeat the segment just seen.

At certain instances viewers can see, close up, what they have just passed. For example, Marion runs past a bush that, examined closely, reveals a spider weaving a web. This allows new perceptions of the same scene, depending upon the speed at which it is seen. In some instances words are flashed on the screen for just three frames, forcing the viewer to go back slowly to see what was written. At other points lines are spoken backward, forcing the disc to be played in reverse. Whilst the Demon and Zen Master are played by the same actor, indicating different aspects of our personalities, suggesting that the same event can appear frightening or enlightening, depending upon its context.

A surveillance camera was programmed via a Fairlight to be switched "on" when a cameraman's shadow is seen. The viewer's image instantaneously appears on the screen, displacing and replacing the image. This suggests "transgressing the screen" being transported into "virtual reality."

Roy Ascott

"Is There Love in the Telematic Embrace?" (1990)

<< 30 >>

Roy Ascott, Aspects of Gaia: Digital Pathways Across the Whole Earth. *Courtesy of Roy Ascott.*

"Telematics . . . involves the technology of interactions among human beings and between the human mind and artificial systems of intelligence and perception. The individual user of networks is always potentially involved in a global net, and the world is always potentially in a state of interaction with the individual."

<< In the 1980s Roy Ascott presented a body of art based on telecommunications technologies, or, as he called it, "telematic art," at leading-edge European exhibitions of art and technology. These included a pivotal show at the 1986 Venice Biennale, "Art, Technology and Computer Science," and the 1989 debut of his *Aspects of Gaia: Digital Pathways across the Whole Earth* at Ars Electronica, in Austria. For the latter project, Ascott invited artists, musicians, and scientists from around the world to contribute digital representations of the earth (Gaia) by way of the global telecommunications network. Ascott's exploration of telematics prefigured the explosion of Internet art in the mid-1990s, shortly after the emergence of the World Wide Web.

With the title of this essay, Ascott begs the question of whether artistic work that incorporates telecommunications technologies is truly engaging and of quality. He observes how the global network is bringing about a "telematic culture" in which participants are "always potentially in a state of interaction." This dynamic communications environment, he notes, will lead to new opportunities for artistic expression. One consequence will be the emergence of the "integrated data work," which he defines as a digital artwork that derives its expressiveness from a "many-to-many" type of interaction. This collective interaction *between* viewers shares much with the Happenings of an earlier era.

Ascott concludes that telematic art encourages the artist to assume the role of facilitator, rather than be the sole arbiter of the artwork's creation. This empowers the participant and enhances the viewer's capacity for creative thought and action. In telematic art, he explains, the dynamic action between participants is determined by the work's interface. This approach is quite unlike the traditional view of the *objet d'art,* which concentrates on the static surface of the painting or sculpture as executed by the inspired *artiste.* As Ascott explains, "at the interface to telematic systems, content is created rather than received." By determining the interface, the artist shapes the participant's particular experience by using networked interaction to create a form of participatory narrative—a radical opportunity for personal expression unlike any established artistic genre. >>

The past decade has seen the two powerful technologies of computing and telecommunications converge into one field of operations that has drawn into its embrace other electronic media, including video, sound synthesis, remote sensing, and a variety of cybernetic systems. These phenomena are exerting enormous influence upon society and on individual behavior; they seem increasingly to be calling into question the very nature of what it is to be human, to be creative, to think and to perceive, and indeed our relationship to each other and to the planet as a whole. The "telematic culture" that accompanies the new developments consists of a set of behaviors, ideas, media, values, and objectives that are significantly unlike those that have shaped society since the Enlightenment. New cultural and scientific metaphors and paradigms are being generated, new models and representations of reality are being invented, new expressive means are being manufactured.

Telematics is a term used to designate computer-mediated communications networking involving telephone, cable, and satellite links between geographically dispersed individuals and institutions that are interfaced to data-processing systems, remote sensing devices, and capacious data-storage banks.[1] It involves the technology of interaction among human beings and between the human mind and artificial systems of intelligence and perception. The individual user of networks is always potentially involved in a global net, and the world is always potentially in a state of interaction with the individual. Thus, across the vast spread of telematic networks worldwide, the quantity of data processed and the density of information exchanged is incalculable. The ubiquitous efficacy of the telematic medium is not in doubt, but the question in human terms, from the point of view of culture and creativity, is: What is the content?

This question, which seems to be at the heart of many critiques of art involving computers and telecommunications, suggests deep-seated fears of the machine coming to dominate the human will and of a technological formalism erasing human content and values. Apart from all the particulars of personal histories, of dreams, desires, and anxieties that inform the content of art's rich repertoire, the question, in essence, is asking: Is there love in the telematic embrace?

In the attempt to extricate human content from technological form, the question is made more complicated by our increasing tendency as artists to bring together imaging, sound, and text systems into interactive environments that exploit state-of-the-art hypermedia and that engage the full sensorium, albeit by digital means. Out of this technological complexity, we can sense the emergence of a synthesis of the arts. The question of content must therefore be addressed to what might be called the *Gesamtdatenwerk*—the integrated data work—and to its capacity to engage the intellect, emotions, and sensibility of the observer.[2] Here, however, more problems arise, since the observer in an interactive telematic system is by definition a participator. In a telematic art, meaning is not something created by the artist, distributed through the network, and *received* by the observer. Meaning is the product of interaction between the observer and the system, the content of which is in a state of flux, of endless change and transformation. In this condition of uncertainty and instability, not simply because of the crisscrossing interactions of users of the network but because content is embodied in data that is itself immaterial, it is pure electronic *difference,* until it has been reconstituted at the interface as image, text, or sound. The sensory *output* may be differentiated further as existing on screen, as articulated structure or material, as architecture, as environment, or in virtual space.

Such a view is in line with a more general approach to art as residing in a cultural communications system rather than in the art object as a fixed semantic configuration—a system in which the viewer actively negotiates for meaning.[3] In this sense, telematic networking makes explicit in its technology and protocols what is implicit in all aesthetic experience where that experience is seen as being as much creative in the act of the viewer's perception as it is in the act of the artist's production.[4] Classical communications theory holds, however, that communication is a one-way dispatch, from sender to receiver, in which only contingent "noise" in the channel can modify the message (often further confused as the meaning) initiated at the source of transmission.[5] This is the model that has the artist as sender and therefore originator of meaning, the artist as creator and owner of images and ideas, the artist as controller of context and content. It is a model that requires, for its completion, the viewer as, at best, a skilled decoder or interpreter of the artist's "meaning" or, at worst, simply a passive receptacle of such meaning. It gives rise to the industry of criticism and exegesis in which those who "understand" this or that work of art explain it to those who are too stupid or uneducated to receive its meaning unaided. In this scenario, the artwork and its maker are viewed in the same way as the world and its creator. The beauty and truth of both art and the world are "out there" in the world and in the work of art. They are as fixed and immutable as the material universe appears to be. The canon of determinism decrees prefigured harmony and composition, regulated form and continuity of expression, with unity and clarity assured by a cultural consensus and a linguistic uniformity shared by artist and public alike.

The problem of content and meaning within a telematic culture gives added poignancy to the rubric "Issues of Content" under which this present writing on computers and art is developed: "issue" is open to a plurality of meanings, no one of which is satisfactory. The metaphor of a semantic sea endlessly ebbing and flowing, of meaning constantly in flux, of all words, utterances, gestures, and images in a state of undecidability, tossed to and fro into new collusions and conjunctions within a field of human interaction and negotiation, is found as much in new science—in quantum physics, second-order cybernetics,[6] or chaology,[7] for example—as in art employing telematic concepts or the new literary criticism that has absorbed philosophy and social theory into its practice. This sunrise of uncertainty, of a joyous dance of meaning between layers of genre and metaphoric systems, this unfolding tissue woven of a multiplicity of visual codes and cultural imaginations, was also the initial promise of the postmodern project before it disappeared into the domain of social theory, leaving only its frail corpus of pessimism and despair.

In the case of the physicists, the radical shift in metaphors about the world and our participation in its creation and redescription mean that science's picture window onto reality has been shattered by the very process of trying to measure it. John Wheeler uses this analogy succinctly:

> Nothing is more important about the quantum principle than this, that it destroys the concept of the world as "sitting out there," with the observer safely separated from it by a 20-centimeter slab of plate glass. Even to observe so minuscule an object as an electron, he must shatter the glass. He must reach in. He must install his chosen measuring equipment. It is up to him whether he shall measure position or momentum . . . the measurement changes the state of the electron. The universe will never afterwards be the same. To describe what has happened one has to cross out that old word "observer" and put in its place "participator." In some strange sense the universe is a participatory universe.[8]

In the context of telematic systems and the issue of content and meaning, the parallel shift in art of the status of "observer" to that of "participator" is demonstrated clearly if in accounts of the quantum principle we substitute "data" for "quanta." Indeed, finding such analogies between art and physics is more than just a pleasant game; the web of connections between new models of theory and practice in the arts and the sciences, over a wide domain, is so pervasive as to suggest a paradigm shift in our world view, a redescription of reality and a recontextualization of ourselves. We begin to understand that chance and change, chaos and indeterminacy, transcendence and transformation, the immaterial and the numinous, are terms at the center of our self-understanding and our new visions of reality. How, then, could there be a content—sets of meanings—contained within telematic art when every aspect of networking in dataspace is in a state of transformation and of becoming? The very technology of computer telecommunications extends the gaze, transcends the body, amplifies the mind into unpredictable configurations of thought and creativity. . . .

In the recent history of Western art, it was Marcel Duchamp who first took the metaphor of the glass, of the window onto the world, and turned it back on itself to reveal what is invisible. We see in the work known as *The Bride Stripped Bare by Her Bachelors, Even,* or *The Large Glass,* a field of vitreous reality in which energy and emotion are generated from the tension and interaction of male and female, natural and artificial, human and machine.[9] Its subject is attraction in Charles Fourier's sense,[10] or, we might even say, love. *The Large Glass,* in its transparent essence, always includes both its environment and the reflection of the observer. Love is contained in this total embrace; all that escapes is reason and certainty. By participating in the embrace, the viewer comes to be a progenitor of the semantic issue. The glass as "ground" has a function and status anticipating that of the computer monitor as a screen of operations—of transformations—and as the site of interaction and negotiation for meaning. But it is not only through the *Glass* that we can see Duchamp as prophetic of the telematic mode. The very metaphor of networking interaction in a field of uncertainty, in which the observer is creator and meaning is unstable, is implicit in all his work. Equally prophetic in the *Glass* is the horizontal bar that joins the upper and lower parts of the work and serves as a metaphor for the all-around viewing, the inclusive, all-embracing scope of its vision. This stands in opposition to the vertical, head-to-toe viewing of Renaissance space, embodied in the Western pictorial tradition, where the metaphor of verticality is employed insistently in its monuments and architecture—emblems often as not of aggression, competition, and dominance, always of a tunnel vision. The horizontal, on the other hand, is a metaphor for the bird's-eye view, the all-over, all-embracing, holistic systems view of structures, relationships, and events—viewing that can include the ironic, the fuzzy, and the ambiguous. This is precisely the condition of perception and insight to which telematic networking aspires. . . .

As communications networks increase, we will eventually reach a point where the billions of information exchanges, shuttling through the networks at any one time, can create coherence in the global brain, similar to those found in the human brain.[11]

This suggests equally the need for a redescription of human consciousness as it emerges from the developing symbiosis of the human mind and the artificial thought of parallel distributed processing (PDP).[12]

If one of the great rituals of emergence into a new world—that of the American Indian Hopi—is centered around the sacred sipapu that connects to the Underworld of power and transformation, so our emergence into the new world of telematic culture similarly calls for celebration at the interface to those PDP systems that can link us with superconnectivity, mind to mind, into a new planetary community. And just as the Hopi seek to exploit the full measure of their expressive means by joining image, music, chant, and dance into a holistic unity, so we too now seek a synthesis of digital modes—image, sound, text, and cybernetic structure—by which to recontextualize our own world, that numinous whole of all our separate realities.

The emerging new order of art is that of interactivity, of "dispersed authorship";[13] the canon is one of contingency and uncertainty. Telematic art encompasses a wide array of media: hypermedia, videotex, telefacsimile, interactive video, computer animation and simulation, teleconferencing, text exchange, image transfer, sound synthesis, telemetry and remote sensing, virtual space, cybernetic structures, and intelligent architecture. These are simply broad categories of technologies and methodologies that are constantly evolving—bifurcating, joining, hybridizing—at an accelerated rate.

At the same time, the status of the art object changes. The culturally dominant objet d'art as the sole focus (the uncommon carrier of uncommon content) is replaced by the interface. Instead of the artwork as a window onto a composed, resolved, and ordered reality, we have at the interface a doorway to undecidability, a dataspace of semantic and material potentiality. The focus of the aesthetic shifts from the observed object to the participating subject, from the analysis of observed systems to the (second-order) cybernetics of observing systems: the canon of the immaterial and participatory. Thus, at the interface to telematic systems, content is created rather than received. By the same token, content is disposed of at the interface by reinserting it, transformed by the process of interaction, back into the network for storage, distribution, and eventual transformation at the interface of other users, at other access nodes across the planet.

A telematic network is more than the sum of its parts, more than a computer communications web. The new order of perception it constitutes can be called "global vision," since its distributed sensorium and distributed intelligence—networked across the whole planet as well as reaching remotely into galactic space and deep into quantum levels of matter—together provide for a holistic, integrative viewing of structures, systems, and events that is global in its scope. This artificial extension of human intelligence and perception, which the neural nets of PDP and sophisticated remote-sensing systems provide,[14] not only amplifies human perception but is in the process of changing it. The transformation is entirely consistent with the overarching ambition of both art and science throughout this century: to make the invisible visible. Even now, it must be recognized that our human cognitive processes (whether involving linear or associative thought, imaging, remembering, computing, or hypothesizing) are rarely carried out without the computer being involved.[15] A great proportion of the time that we are involved in communicating, learning, or being entertained entails our interaction with telecommunications systems.[16] Similarly, with feeling and sensing, artificial, intelligent sensors of considerable subtlety are becoming integral to human interaction with the environment and to the monitoring of both internal and external ecologies. Human perception, understood as the product of active negotiation rather than passive reception, thus requires, within this evolving symbiosis of human/machine, telematic links of considerable complexity between the very diverse nodes of the worldwide artificial reticular sensorium.

Telematic culture means, in short, that we do not think, see, or feel in isolation. Creativity is shared, authorship is distributed, but not in a way that denies the individual her authenticity or power of self-creation, as rather crude models of collectivity might have done in the past. On the contrary, telematic culture amplifies the individual's capacity for creative thought and action, for more vivid and intense experience, for more informed perception, by enabling her to participate in the production of global vision through networked interaction with other minds, other sensibilities, other sensing and thinking systems across the planet—thought circulating in the medium of data through a multiplicity of different cultural, geographical, social, and personal layers. Networking supports endless redescription and recontextualization such that no language or visual code is final and no reality is ultimate. In the telematic culture, pluralism and relativism shape the configurations of ideas—of image, music, and text—that circulate in the system.

It is the computer that is at the heart of this circulation system, and, like the heart, it works best when least noticed—that is to say, when it becomes invisible. At present, the computer as a physical, material presence is too much with us; it dominates our inventory of tools, instruments, appliances, and apparatus as the ultimate machine. In our artistic and educational environments it is all too solidly there, a computational block to poetry and imagination. It is not transparent, nor is it yet fully understood as pure system, a universal transformative matrix. The computer is not primarily a thing, an object, but a set of behaviors, a system, actually a system of systems. Data constitute its lingua franca. It is the agent of the datafield, the constructor of dataspace. Where it is seen simply as a screen presenting the pages of an illuminated book, or as an internally lit painting, it is of no artistic value. Where its considerable speed of processing is used simply to simulate filmic or photographic representations, it becomes the agent of passive voyeurism. Where access to its transformative power is constrained by a typewriter keyboard, the user is forced into the posture of a clerk. The electronic palette, the light pen, and even the mouse bind us to past practices. The power of the computer's presence, particularly the power of the interface to shape language and thought, cannot be overestimated. It may not be an exaggeration to say that the "content" of a telematic art will depend in large measure on the nature of the interface; that is, the kind of configurations and assemblies of image, sound, and text, the kind of restructuring and articulation of environment that telematic interactivity might yield, will be determined by the freedoms and fluidity available at the interface.

The essence of the interface is its potential flexibility; it can accept and deliver images both fixed and in movement, sounds constructed, synthesized, or sampled, texts written and spoken. It can be heat sensitive, body responsive, environmentally aware. It can respond to the tapping of the feet, the dancer's arabesque, the direction of a viewer's gaze. It not only articulates a physical environment with movement, sound, or light; it is an environment, an arena of dataspace in which a distributed art of the human/computer symbiosis can be acted out, the issue of its cybernetic content. Each individual computer interface is an aspect of a telematic unity such that to be in or at any one interface is to be in the virtual presence of all the other interfaces throughout the network of which it is a part. This might be defined as the "holomatic" principle in networking. It is so because all the data flowing through any access node of the network are equally and at the same time held in the memory of that network: they can be accessed, through cable or satellite links, from any part of the planet at any time of day or night, by users of the network (who, in order to communicate with each other, do not need to be in the same place at the same time).

This holomatic principle was well demonstrated during the Venice Biennale of 1986, when the *Planetary Network* project had the effect of pulling the exhibition from its rather elite, centralized, and exclusive domain in Venice and stretching it out over the face of the globe;[17] the flow of creative data generated by the interaction of artists networking all over the world was accessible everywhere. Set within the interactive environment "Laboratory Ubiqua," a wide range of telematic media was involved, including electronic mail, videotex, digital-image exchange, slow-scan TV, and computer conferencing. Interactive videodiscs, remote sensing systems, and cybernetic structures were also included. . . .

A more elaborate and complex multimedia interface was created for the project *Aspects of Gaia: Digital Pathways across the Whole Earth* as part of the Ars Electronica Festival of Art and Technology in Linz, Austria, in 1989.[18] The transmission of digital image and sound by file transfer and the computer storage of telefacsimile material via modem, by this time economically feasible, invited the creation of a more dramatic and engaging environment for public participation. Invitations to participate in the project were emailed, faxed, or airmailed to "artists, scientists, poets, shamen [sic], musicians, architects, visionaries, aboriginal artists of Australia, native artists of the Americas." The subject of the project was the many aspects of the earth, Gaia, seen from a multiplicity of spiritual, scientific, cultural, and mythological perspectives. An energizing stream of integrated digital images, texts, and sounds (a *Gesamtdatenwerk*) would then constitute a kind of invisible cloak, a digital noosphere that might contribute to the harmonization of the planet. In accessing the meridians at various nodes, participants became involved in a form of global acupuncture, their interactions endlessly transforming and reconstituting the worldwide flow of creative data. . . .

To the objection that such a global vision of an emerging planetary art is uncritically euphoric, or that the prospectus of a telematic culture with its *Gesamtdatenwerk* of hypermediated virtual realities is too grandiose, we should perhaps remind ourselves of the essentially political, economic, and social sensibilities of those who laid the conceptual foundations of the field of interactive systems. This cultural prospectus implies a telematic politic, embodying the features of feedback, self-determination, interaction, and collaborative creativity not unlike the "science of government" for which, over 150 years ago, Andre Marie Ampere coined the term "cybernetics"—a term reinvigorated and humanized by Norbert Wiener in this century.[19] Contrary to the rather rigid determinism and positivism that have shaped society since the Enlightenment, however, these features will have to accommodate notions of uncertainty, chaos, autopoiesis, contingency, and the second-order cybernetics or fuzzy-systems view of a world in which the observer and observed, creator and viewer, are inextricably linked in the process of making reality—all our many separate realities interacting, colliding, re-forming, and resonating within the telematic-noosphere of the planet.

Within these separate realities, the status of the "real" in the phenomenology of the artwork also changes. Virtual space, virtual image, virtual reality—these are categories of experience that can be shared through telematic networks, allowing for movement through "cyberspace" and engagement with the virtual presence of others who are in their corporeal materiality at a distance, physically inaccessible or otherwise remote.[20] The adoption of a headset, DataGlove, or other data wear can make the personal connection to cyberspace—socialization in hyperreality—wherein interaction with others will undoubtedly be experienced as "real," and the feelings and perceptions so generated will also be "real." The passage from real to virtual will probably be seamless, just as social behavior derived from human-computer symbiosis is flowing unnoticed into our consciousness.

But the very ease of transition from "reality" to "virtuality" will cause confusion in culture, in values, and in matters of personal identity. It will be the role of the artist, in collaboration with scientists, to establish not only new creative praxes but also new value systems, new ordinances of human interaction and social communicability. The issue of content in the planetary art of this emerging telematic culture is therefore the issue of values, expressed as transient hypotheses rather than finalities, tested within the immaterial, virtual, hyperrealities of dataspace. Integrity of the work will not be judged by the old aesthetics; no antecedent criteria can be applied to network creativity since there is no previous canon to accommodate it. The telematic process, like the technology that embodies it, is the product of a profound human desire for transcendence: to be out of body, out of mind, beyond language. Virtual space and dataspace constitute the domain, previously provided by myth and religion, where imagination, desire, and will can reengage the forces of space, time, and matter in the battle for a new reality.

The digital matrix that brings all new electronic and optical media into its telematic embrace—being a connectionist model of hypermedia—calls for a "connective criticism." The personal computer yields to the interpersonal computer. Serial data processing becomes parallel distributed processing. Networks link memory bank to memory bank, intelligence to intelligence. Digital image and digital sound find their common ground, just as a synthesis of modes—visual, tactile, textual, acoustic, environmental—can be expected to "hypermediate" the networked sensibilities of a constellation of global cultures. The digital camera—gathering still and moving images from remote sensors deep in space, or directed by human or artificial intelligence on earth, seeking out what is unseen, imaging what is invisible—meets at a point between our own eyes and the reticular retina of worldwide networks, stretching perception laterally away from the tunnel vision, from the Cartesian sight lines of the old deterministic era. Our sensory experience becomes extrasensory, as our vision is enhanced by the extrasensory devices of telematic perception. The computer deals invisibly with the invisible. It processes those connections, collusions, systems, forces and fields, transformations and transferences, chaotic assemblies, and higher orders of organization that lie outside our vision, outside the gross level of material perception afforded by our natural senses. Totally invisible to our everyday unaided perception, for example, is the underlying fluidity of matter, the indeterminate dance of electrons, the "snap, crackle, and pop" of quanta, the tunneling and transpositions, nonlocal and superluminal, that the new physics presents. It is these patterns of events, these new exhilarating metaphors of existence—nonlinear, uncertain, layered, and discontinuous—that the computer can redescribe. With the computer, and brought together in the telematic embrace, we can hope to glimpse the unseeable, to grasp the ineffable chaos of becoming, the secret order of disorder. And as we come to see more, we shall see the computer less and less. It will become invisible in its immanence, but its presence will be palpable to the artist engaged telematically in the world process of autopoiesis, planetary self-creation.

The technology of computerized media and telematic systems is no longer to be viewed simply as a set of rather complicated tools extending the range of painting and sculpture, performed music, or published literature. It can now be seen to support a whole new field of creative endeavor that is as radically unlike each of those established artistic genres as they are unlike each other. A new vehicle of consciousness, of creativity and expression, has entered our repertoire of being. While it is concerned with both technology and poetry, the virtual and the immaterial as well as the palpable and concrete, the telematic may be categorized as neither art nor science, while being allied in many ways to the discourses of both. The further development of this field will clearly mean an interdependence of artistic, scientific, and technological competencies and aspirations and, urgently, on the formulation of a transdisciplinary education.

So, to link the ancient image-making process of Navajo sand painting to the digital imaging of modern supercomputers through common silicon, which serves them both as pigment and processor chip, is more than ironic whimsy. The holistic ambition of Native American culture is paralleled by the holistic potentiality of telematic art. More than a technological expedient for the interchange of information, networking provides the very infrastructure for spiritual interchange that could lead to the harmonization and creative development of the whole planet. With this prospectus, however naïvely optimistic and transcendental it may appear in our current fin-de-siècle gloom, the metaphor of love in the telematic embrace may not be entirely misplaced.

Pavel Curtis

"Mudding: Social Phenomena in Text-Based Virtual Realities" (1992)

<< 31 >>

Pavel Curtis. Photo by Anne Hamersky, © *Anne Hamersky,* Fast Company Magazine.

"*The emergence of MUDs has created a new kind of social sphere, both like and radically unlike the environments that have existed before. As they become more and more popular and more widely accessible, it appears likely that an increasingly significant proportion of the population will at least become familiar with mudding and perhaps become frequent participants in text-based virtual realities.*"

<< Pavel Curtis, a computer scientist at Xerox PARC (Palo Alto Research Center), created one of the first popular on-line role-playing environments, LambdaMOO, in 1991. Known as a MUD (the acronym stands for Multi-User Dungeons), LambdaMOO is a text-only fantasy realm that is descended from role-playing games from the 1970s such as "Dungeons and Dragons" (hence the dungeon allusion in MUD). After presiding over this experimental digital environment for one year as its creator, administrator, and self-proclaimed "wizard," Curtis documented his observations of the evolving social, cultural, and psychological effects of interaction in cyberspace. While not the first of its kind, LambdaMOO is perhaps the most famous text-based virtual environment, dissected and analyzed by media theorists, sociologists, and psychologists who see it as fertile breeding ground for a new hybrid form of literature, live performance, cinema, and interactive story making.

In his essay, Curtis directs our attention to LambdaMOO's narrative characteristics. Unlike conventional story structures, MUDs do not rely on defined plots with clear objectives. Rather, through the freewheeling dynamics of improvised dialogue and unrehearsed interactivity, participants lose themselves in their roles and collaborate in a form of collective authorship. Shielded (and even liberated) by the anonymity of their characters, players improvise their own conversations, story lines, props, and settings; they pursue their own adventures, and experiment with a myriad of alternate identities; sometimes they even switch gender and, occasionally, species. LambdaMOO has served as a model for thousands of MUDs that have since sprung from the Internet and the World Wide Web. MUDs are characterized by a tightly knit—though globally dispersed—community of characters engaged in an ongoing dialogue that combines the aimlessness of nomadic wandering with the focused creativity of world building. There is no omniscient narrator, no central author, in LambdaMOO, despite the shadowy presence of its "wizard," Pavel Curtis.

While Curtis emphasizes the richness of the text-based virtual world—referring to Marshall McLuhan's concept of the "cool" medium in which the viewer's imagination is engaged to "complete" the experience of the low-resolution environment—graphic, multimedia virtual worlds have supplanted their textual ancestors. Nonetheless, Curtis's observation of on-line role playing and virtual communities are fundamental to our understanding of the social implications of networked alternative realities, now popular in the form of chat rooms, teleconferencing, and email Listservs. >>

Abstract

A MUD (Multi-User Dungeon or, sometimes, Multi-User Dimension) is a network-accessible, multi-participant, user-extensible virtual reality whose user interface is entirely textual. Participants (usually called players) have the appearance of being situated in an artificially constructed place that also contains those other players who are connected at the same time. Players can communicate easily with each other in real time. This virtual gathering place has many of the social attributes of other places, and many of the usual social mechanisms operate there. Certain attributes of this virtual place, however, tend to have significant effects on social phenomena, leading to new mechanisms and modes of behavior not usually seen "IRL" (in real life). In this paper, I relate my experiences and observations from having created and maintained a MUD for over a year.

1 A Brief Introduction to Mudding

> The Machine did not transmit nuances of expression. It only gave a general idea of people—an idea that was good enough for all practical purposes.
>
> —E. M. Forster[1]

A MUD is a software program that accepts "connections" from multiple users across some kind of network (e.g., telephone lines or the Internet) and provides to each user access to a shared database of "rooms," "exits," and other objects. Each user browses and manipulates this database from "inside" one of those rooms, seeing only those objects that are in the same room and moving from room to room mostly via the exits that connect them. A MUD, therefore, is a kind of virtual reality, an electronically represented "place" that users can visit.

MUDs are not, however, like the kinds of virtual realities that one usually hears about, with fancy graphics and special hardware to sense the position and orientation of the user's real-world body. A MUD user's interface to the database is entirely text-based; all commands are typed in by the users and all feedback is printed as unformatted text on their terminal. The typical MUD user interface is most reminiscent of old computer games like Adventure and Zork;[2] a typical interaction is shown below.

```
>look
Corridor
The corridor from the west continues to the east here,
but the way is blocked by a purple-velvet rope
```

```
stretched across the hall. There are doorways leading
to the north and south.
You see a sign hanging from the middle of the rope here.
>read sign
This point marks the end of the currently-occupied
portion of the house. Guests proceed beyond this point
at their own risk.
—The residents
>go east
You step disdainfully over the velvet rope and enter
the dusty darkness of the unused portion of the house.
```

Three major factors distinguish a MUD from an Adventure-style computer game, though:

- A MUD is not goal-oriented; it has no beginning or end, no "score," and no notion of "winning" or "success." In short, even though users of MUDs are commonly called players, a MUD isn't really a game at all.
- A MUD is extensible from within; a user can add new objects to the database such as rooms, exits, "things," and notes. Certain MUDs, including the one I run, even support an embedded programming language in which a user can describe whole new kinds of behavior for the objects they create.
- A MUD generally has more than one user connected at a time. All of the connected users are browsing and manipulating the same database and can encounter the new objects created by others. The multiple users on a MUD can communicate with each other in real time.

This last factor has a profound effect on the ways in which users interact with the system; it transforms the activity from a solitary one into a social one.

Most inter-player communication on MUDs follows rules that fit within the framework of the virtual reality. If a player "says" something (using the say command), then every other player in the same room will "hear" them. For example, suppose that a player named Munchkin typed the command

say Can anyone hear me?

Then Munchkin would see the feedback

You say, "Can anyone hear me?"

and every other player in the same room would see

Munchkin says, "Can anyone hear me?"

Similarly, the emote command allows players to express various forms of "non-verbal" communication. If Munchkin types

> emote smiles.

then every player in the same room sees

> Munchkin smiles.

Most interplayer communication relies entirely on these two commands.[3]

There are two circumstances in which the realistic limitations of say and emote have proved sufficiently annoying that new mechanisms were developed. It sometimes happens that one player wishes to speak to another player in the same room, but without anyone else in the room being aware of the communication. If Munchkin uses the whisper command

> whisper "I wish he'd just go away . . ." to Frebble

then only Frebble will see

> Munchkin whispers, "I wish he'd just go away . . ."

The other players in the room see nothing of this at all.

Finally, if one player wishes to say something to another who is connected to the MUD but currently in a different and perhaps "remote" room, the page command is appropriate. It is invoked with a syntax very like that of the whisper command and the recipient sees output like this:

> You sense that Munchkin is looking for you in The Hall.
> He pages, "Come see this clock, it's tres cool!"

Aside from conversation, MUD players can most directly express themselves in three ways: by their choice of player name, by their choice of gender, and by their self-description.

When a player first connects to a MUD, they choose a name by which the other players will know them. This choice, like almost all others in MUDs, is not cast in stone; any player can rename themself at any time, though not to a name currently in use by some other player. Typically, MUD names are single words, in contrast to the longer "full" names used in real life.

Initially, MUD players appear to be neuter; automatically generated messages that refer to such a player use the family of pronouns including "it," "its," etc. Players can choose to appear as a different gender, though, and not only male or female. On many MUDs, players can also choose to be plural (appearing to be a kind of "colony" creature: "ChupChups leave the room, closing the door behind them"), or to use one of several sets of gender-neutral pronouns (e.g., "s/he," "him/her" and "his/her," or "e," "em" and "eir").

Every object in a MUD optionally has a textual description which players can view with the look command. For example, the description of a room is automatically shown to a player when they enter that room and can be seen again just by typing "look." To see another player's description, one might type "look Bert." Players can set or change their descriptions at any time. The lengths of player descriptions typically vary from short one-liners to dozen-line paragraphs.

Aside from direct communication and responses to player commands, messages are printed to players when other players enter or leave the same room, when others connect or disconnect and are already in the same room, and when objects in the virtual reality have asynchronous behavior (e.g., a cuckoo clock chiming the hours).

MUD players typically spend their connected time socializing with each other, exploring the various rooms and other objects in the database, and adding new such objects of their own design. They vary widely in the amount of time they spend connected on each visit, ranging from only a minute to several hours; some players stay connected (and almost always idle) for days at a time, only occasionally actively participating.

This very brief description of the technical aspects of mudding suffices for the purposes of this paper. It has been my experience, however, that it is quite difficult to properly convey the "sense" of the experience in words. Readers desiring more detailed information are advised to try mudding themselves, as described in the final section of this paper.

2 Social Phenomena Observed on One MUD

Man is the measure.[4]

—E. M. Forster

In October of 1990, I began running an Internet-accessible MUD server on my personal workstation here at PARC. Since then, it has been running continuously, with interruptions of only a few hours at most. In January of 1991, the existence of the MUD (called LambdaMOO[5]) was announced publicly, via the Usenet newsgroup rec.games.mud. As of this writing, well over 3,500 different players have connected to the server from over a dozen countries around the world and, at any given time, over 750 players have connected at least once in the last week. Recent statistics concerning the number of players connected at a given time of day (Pacific Standard Time) appear below.

```
4 a.m.   ************** 10-1/2
5 a.m.   ***************** 12/1/4
6 a.m.   ******************* 14
7 a.m.   ************************** 21-1/4
8 a.m.   ****************************** 21-1/4
9 a.m.   *********************************** 25-1/4
10 a.m.  *************************************** 28
11 a.m.  ********************************************* 32-1/4
noon     **************************************************** 37
1 p.m.   ******************************************************** 41-1/4
2 p.m.   ******************************************************** 39-3/4
3 p.m.   ************************************************* 35
4 p.m.   ******************************************************** 39-1/2
5 p.m.   ******************************************************** 40-3/4
6 p.m.   ******************************************************** 39-3/4
7 p.m.   ******************************************************* 40-1/2
8 p.m.   ***********************************************************
         42-1/2
9 p.m.   **************************************************************
         44-1/4
10 p.m.  **************************************************** 37-3/4
11 p.m.  ********************************************* 31
midnight*************************************** 26-3/4
1 a.m.   **************************************** 20-3/4
2 a.m.   ******************* 13-3/4
3 a.m.   ************** 10-3/4
4 a.m.   ************* 10 1/2
```

LambdaMOO is clearly a reasonably active place, with new and old players coming and going frequently throughout the day. This popularity has provided me with a position from which to observe the social patterns of a fairly large and diverse MUD clientele. I want to point out to the reader, however, that I have no formal training in sociology, anthropology, or psychology, so I cannot make any claims about methodology or even my own objectivity. What I relate below is merely my personal observations made over a year of mudding. In most cases, my discussions of the motivations and feelings of individual players is based upon in-MUD conversations with them; I have no means of checking the veracity of their statements concerning their real-life genders, identities, or (obviously) feelings. On the other hand, in most cases, I also have no reason to doubt them.

I have grouped my observations into three categories: phenomena related to the behavior and motivations of individual players, phenomena related to interactions between small groups of players (especially observations concerning MUD conversation), and phenomena related to the behavior of a MUD's community as a whole.

Cutting across all of these categories is a recurring theme to which I would like to draw the reader's attention in advance. Social behavior on MUDs is in some ways a direct mirror of behavior in real life, with mechanisms being drawn nearly unchanged from real-life, and in some ways very new and different, taking root in the new opportunities that MUDs provide over real life.

2.1 Observations about individuals

The Mudding Population The people who have an opportunity to connect to LambdaMOO are not a representative sample of the world population; they all read and write English with at least passable proficiency and they have access to the Internet. Based on the names of their network hosts, I believe that well over 90% of them are affiliated with colleges and universities, mostly as students and, to a lesser extent, mostly undergraduates. Because they have Internet access, it might be supposed that the vast majority of players are involved in the computing field, but I do not believe that this is the case. It appears to me that no more than half (and probably less) of them are so employed; the increasing general availability of computing resources on college campuses and in industry appears to be having an effect, allowing a broader community to participate.

In any case, it appears that the educational background of the mudding community is generally above average and it is likely that the economic background is similarly above the norm. Based on my conversations with people and on the names of those who have asked to join a mailing list about programming in LambdaMOO, I would guess that over 70% of the players are male; it is very difficult to give any firm justification for this number, however.

Player Presentation As described in the introduction to mudding, players have a number of choices about how to present themselves in the MUD; the first such decision is the name they will use. Some of the names used by players on LambdaMOO are shown below.

Toon	Gemba	Gary_Severn	Ford	Frand
li'ir	Maya	Rincewind	yduJ	funky
Grump	Foodslave	Arthur	EbbTide	Anathae
yrx	Satan	byte	Booga	tek

chupchups	waffle	Miranda	Gus	Merlin
Moonlight	MrNatural	Winger	Drazz'zt	Kendal
RedJack	Snooze	Shin	lostboy	foobar
Ted__Logan	Xephyr	King_Claudius	Bruce	Puff
Dirque	Coyote	Vastin	Player	Cool
Amy	Thorgeir	Cyberhuman	Gandalf	blip
Jayhirazan	Firefoot	JoeFeedback	ZZZzzz . . .	Lyssa
Avatar	zipo	Blackwinter	viz	Kilik
Maelstorm	Love	Terryann	Chrystal	arkanoiv

One can pick out a few common styles for names (e.g., names from or inspired by myth, fantasy, or other literature, common names from real life, names of concepts, animals, and everyday objects that have representative connotations, etc.), but it is clear that no such category includes a majority of the names. Note that a significant minority of the names are in lower case; this appears to be a stylistic choice (players with such names describe the practice as "cool") and not, as might be supposed, an indication of a depressed ego.

Players can be quite possessive about their names, resenting others who choose names that are similarly spelt or pronounced or even that are taken from the same mythology or work of literature. In one case, for example, a player names "ZigZag" complained to me about other players taking the names "ZigZag!" and "Zig."

The choice of a player's gender is, for some, one of great consequence and forethought; for others (mostly males), it is simple and without any questions. For all that this choice involves the fewest options for the player (unlike their name or description, which are limited only by their imagination), it is also the choice that can generate the greatest concern and interest on the part of other players.

As I've said before, it appears that the great majority of players are male and the vast majority of them choose to present themselves as such. Some males, however, taking advantage of the relative rarity of females in MUDs, present themselves as female and thus stand out to some degree. Some use this distinction just for the fun of deceiving others, some of these going so far as to try to entice male-presenting players into sexually explicit discussions and interactions. This is such a widely noticed phenomenon, in fact, that one is advised by the common wisdom to assume that any flirtatious female-presenting players are, in real life, males. Such players are often subject to ostracism based on this assumption.

Some MUD players have suggested to me that such transvestite flirts are perhaps acting out their own (latent or otherwise) homosexual urges or fantasies, taking advantage of the perfect safety of the MUD situation to see how it feels to approach other men. While I have had no personal experience talking to such players, let alone the opportunity to delve into their motivations, the idea strikes me as plausible given the other ways in which MUD anonymity seems to free people from their inhibitions. (I say more about anonymity later on.)

Other males present themselves as female more out of curiosity than as an attempt at deception; to some degree, they are interested in seeing "how the other half lives," what it feels like to be perceived as female in a community. From what I can tell, they can be quite successful at this.

Female-presenting players report a number of problems. Many of them have told me that they are frequently subject both to harassment and to special treatment. One reported seeing two newcomers arrive at the same time, one male-presenting and one female-presenting. The other players in the room struck up conversations with the putative female and offered to show her around but completely ignored the putative male, who was left to his own devices.

In addition, probably due mostly to the number of female-presenting males one hears about, many female players report that they are frequently (and sometimes quite aggressively) challenged to "prove" that they are, in fact, female. To the best of my knowledge, male-presenting players are rarely if ever so challenged.

Because of these problems, many players who are female in real life choose to present themselves otherwise, choosing either male, neuter, or gender-neutral pronouns. As one might expect, the neuter and gender-neutral presenters are still subject to demands that they divulge their real gender.

Some players apparently find it quite difficult to interact with those whose true gender has been called into question; since this phenomenon is rarely manifest in real life, they have grown dependent on "knowing where they stand," on knowing what gender roles are "appropriate." Some players (and not only males) also feel that it is dishonest to present oneself as being a different gender than in real life; they report feeling "mad" and "used" when they discover the deception.

While I can spare no more space for this topic, I enthusiastically encourage the interested reader to look up Van Gelder's fascinating article[6] for many more examples and insights, as well as the story of a remarkably successful deception via "electronic transvestism."

The final part of a player's self-presentation, and the only part involving prose, is the player's description. This is where players can, and often do, establish the details of a persona or role they wish to play in the virtual reality. It is also a significant factor in other players' first impressions, since new players are commonly looked at soon after entering a common room.

Some players use extremely short descriptions, either intending to be cryptic (e.g., "the possessor of the infinity gems") or straightforward (e.g., "an average-sized dark elf with lavender eyes") or, often, [they are] just insufficiently motivated to create a more complex description for themselves. Other players go to great efforts in writing their descriptions; one moderately long example appears below.

> You see a quiet, unassuming figure, wreathed in an oversized, dull-green Army jacket which is pulled up to nearly conceal his face. His long, unkempt blond hair blows back from his face as he tosses his head to meet your gaze. Small round gold-rimmed glasses, tinted slightly grey, rest on his nose. On a shoulder strap he carries an acoustic guitar and he lugs a backpack stuffed to overflowing with sheet music, sketches, and computer printouts. Under the coat are faded jeans and a T-shirt reading "Paranoid CyberPunks International." He meets your gaze and smiles faintly, but does not speak with you. As you surmise him, you notice a glint of red at the rims of his blue eyes, and realize that his canine teeth seem to protrude slightly. He recoils from your look of horror and recedes back into himself.

A large proportion of player descriptions contain a degree of wish fulfillment; I cannot count the number of "mysterious but unmistakably powerful" figures I have seen wandering around in LambdaMOO. Many players, it seems, are taking advantage of the MUD to emulate various attractive characters from fiction.

Given the detail and content of so many player descriptions, one might expect to find a significant amount of role-playing, players who adopt a coherent character with features distinct from their real-life personalities. Such is rarely the case, however. Most players appear to tire of such an effort quickly and simply interact with the others more-or-less straightforwardly, at least to the degree one does in normal discourse. One factor might be that the roles chosen by players are usually taken from a particular creative work and are not particularly viable as characters outside of the context of that work; in short, the roles don't make sense in the context of the MUD.

A notable exception to this rule is one particular MUD I've heard of, called "PernMUSH." This appears to be a rigidly maintained simulacrum of the world described in Ann McCaffrey's celebrated "Dragon" books. All players there have names that fit the style of the books and all places built there are consistent with what is shown in the series and in various fan materials devoted to it. PernMUSH apparently holds frequent "hatchings" and other social events, also derived in great detail from McCaffrey's works. This exception probably succeeds only because of its single-mindedness; with every player providing the correct context for every other, it is easier for everyone to stay more-or-less "in character."

Player Anonymity It seems to me that the most significant social factor in MUDs is the perfect anonymity provided to the players. There are no commands available to the players to discover the real-life identity of each other and, indeed, technical considerations make such commands either very difficult or impossible to implement.

It is this guarantee of privacy that makes players' self-presentation so important and, in a sense, successful. Players can only be known by what they explicitly project and are not "locked into" any factors beyond their easy control, such as personal appearance, race, etc. In the words of an old military recruiting commercial, MUD players can "be all that you can be."[7]

This also contributes to what might be called a "shipboard syndrome," the feeling that since one will likely never meet anyone from the MUD in real life, there is less social risk involved and inhibitions can safely be lowered.

For example, many players report that they are much more willing to strike up conversations with strangers they encounter in the MUD than in real life. One obvious factor is that MUD visitors are implicitly assumed to be interested in conversing, unlike in most real world contexts. Another deeper reason, though, is that players do not feel that very much is at risk. At worst, if they feel that they've made an utter fool of themself, they can always abandon the character and create a new one, losing only the name and the effort invested in socially establishing the old one. In effect, a "new lease on life" is always a ready option.

Players on most MUDs are also emboldened somewhat by the fact that they are immune from violence, both physical and virtual. The permissions systems of all MUDs (excepting those whose whole purpose revolves around adventuring and the slaying of monsters and other players) generally prevent any player from having any kind of permanent effect on any other player. Players can certainly annoy each other, but not in any lasting or even moderately long-lived manner.

This protective anonymity also encourages some players to behave irresponsibly, rudely, or even obnoxiously. We have had instances of severe and repeated sexual harassment, crudity, and deliberate offensiveness. In general, such cruelty seems to be supported by two causes: the offenders believe (usually correctly) that they cannot be held accountable for their actions in the real world, and the very same anonymity makes it easier for them to treat other players impersonally, as other than real people.

Wizards Usually, as I understand it, societies cope with offensive behavior by various group mechanisms, such as ostracism, and I discuss this kind of effect in detail in Section 2.3. In certain severe cases, however, it is left to the "authorities" or "police" of a society to take direct action, and MUDs are no different in this respect.

On MUDs, it is a special class of players, usually called wizards or (less frequently) gods, who fulfill both the "authority" and "police" roles. A wizard is a player who has special permissions and commands available, usually for the purpose of maintaining the MUD, much like a "system administrator" or "superuser" in real-life computing systems. Players can only be transformed into wizards by other wizards, with the maintainer of the actual MUD server computer program acting as the first such. . . .

2.2 Observations about small groups

MUD Conversation The majority of players spend the majority of their active time on MUDs in conversation with other players. The mechanisms by which those conversations get started generally mirror those that operate in real life, though sometimes in interesting ways.

Chance encounters between players exploring the same parts of the database are common and almost always cause for conversation. As mentioned above, the anonymity of MUDs tends to lower social barriers and to encourage players to be more outgoing than in real life. Strangers on MUDs greet each other with the same kinds of questions as in real life: "Are you new here? I don't think we've met." The very first greetings, however, are usually gestural rather than verbal: "Munchkin waves. Lorelei waves back."

The @who (or WHO) command on MUDs allows players to see who else is currently connected and, on some MUDs, where those people are. An example of the output of this command appears below.

PLAYER NAME	CONNECTED	IDLE TIME	LOCATION
Haakon (#2)	3 days	a second	Lambda's Den
Lynx (#8910)	a minute	2 seconds	Lynx' Abode
Garin (#23393)	an hour	2 seconds	Carnival Grounds

PLAYER NAME	CONNECTED	IDLE TIME	LOCATION
Gilmore (#19194)	an hour	10 seconds	Heart of Darkness
TamLin (#21864)	an hour	21 seconds	Heart of Darkness
Quimby (#23279)	3 minutes	2 minutes	Quimby's room
koosh (#24639)	50 minutes	5 minutes	Corridor
Nosredna (#2487)	7 hours	36 minutes	Nosredna's Hideaway
yduJ (#68)	7 hours	47 minutes	Hackers' Heaven
Zachary (#4670)	an hour	an hour	Zachary's Workshop
Woodlock (#2520)	2 hours	2 hours	Woodlock's Room

Total: 11 players, 6 of whom have been active recently.

This is, in a sense, the MUD analog of scanning the room in a real-life gathering to see who's present. . . .

Other Small-Group Interactions I would not like to give the impression that conversation is the only social activity on MUDs. Indeed, MUD society appears to have most of the same social activities as real life, albeit often in a modified form.

As mentioned before, PernMUSH holds large-scale, organized social gatherings such as "hatchings" and they are not alone. Most MUDs have at one time or another organized more or less elaborate parties, often to celebrate notable events in the MUD itself, such as an anniversary of its founding. We have so far had only one or two such parties on LambdaMOO, to celebrate the "opening" of some new area built by a player; if there were any other major parties, I certainly wasn't invited!

One of the more impressive examples of MUD social activity is the virtual wedding. There have been many of these on many different MUDs; we are in the process of planning our first on LambdaMOO, with me officiating in my role as archwizard.

I have never been present at such a ceremony, but I have read logs of the conversations at them. As I do not know any of the participants in the ceremonies I've read about, I cannot say much for certain about their emotional content. As in real life, they are usually very happy and celebratory occasions with an intriguing undercurrent of serious feelings. I do not know and cannot even speculate about whether or not the main participants in such ceremonies are usually serious or not, whether or not the MUD ceremony usually (or even ever) mirrors another ceremony in the real world, or even whether or not the bride and groom have ever met outside of virtual reality.

In the specific case of the upcoming LambdaMOO wedding, the participants first met on LambdaMOO, became quite friendly, and eventually decided to meet in real life. They have subsequently become romantically involved in the real world and are using the MUD wedding as a celebration of that fact. This phenomenon of couples meeting in virtual reality and then pursuing a real-life relationship is not uncommon; in one notable case, they did this even though one of them lived in Australia and the other in Pittsburgh!

It is interesting to note that the virtual reality wedding is not specific to the kinds of MUDs I've been discussing; Van Gelder[8] mentions an on-line reception on CompuServe and weddings are quite common on Habitat,[9] a half-graphical, half-textual virtual reality popular in Japan.

The very idea, however, brings up interesting and potentially important questions about the legal standing of commitments made only in virtual reality. Suppose, for example, that two people make a contract in virtual reality. Is the contract binding? Under which state's (or country's) laws? Is it a written or verbal contract? What constitutes proof of signature in such a context? I suspect that our real-world society will have to face and resolve these issues in the not-too-distant future.

Those who frequent MUDs tend also to be interested in games and puzzles, so it is no surprise that many real-world examples have been implemented inside MUDs. What may be surprising, however, is the extent to which this is so.

On LambdaMOO alone, we have machine-mediated Scrabble, Monopoly, Mastermind, Backgammon, Ghost, Chess, Go, and Reversi boards. These attract small groups of players on occasion, with the Go players being the most committed; in fact, there are a number of Go players who come to LambdaMOO only for that purpose. I say more about these more specialized uses of social virtual realities later on. In many ways, though, such games so far have little, if anything, to offer over their real-world counterparts except perhaps a better chance of finding an opponent.

Perhaps more interesting are the other kinds of games imported into MUDs from real life, the ones that might be far less feasible in a non-virtual reality. A player on LambdaMOO, for example, implemented a facility for holding food fights. Players throw food items at each other, attempt to duck oncoming items, and, if unsuccessful, are "splattered" with messes that cannot easily be removed. After a short interval, a semi-animate "Mr. Clean" arrives and one-by-one removes the messes from the participants, turning them back into the food items from which they came, ready for the next fight. Although the game was rather simple to implement, it has remained enormously popular nearly a year later.

Another player on LambdaMOO created a trainable Frisbee, which any player could teach to do tricks when they threw or caught it. Players who used the Frisbee seemed to take great pleasure in trying to out-do each other's trick descriptions. My catching description, for example, reads "Haakon stops the frisbee dead in the air in front of himself and then daintily plucks it, like a flower." I have also heard of MUD versions of paint-ball combat and fantastical games of Capture the Flag.

2.3 Observations about the MUD community as a whole

MUD communities tend to be very large in comparison to the number of players actually active at any given time. On LambdaMOO, for example, we have between 700 and 800 players connecting in any week but rarely more than 40 simultaneously. A good real-world analog might be a bar with a large number of "regulars," all of whom are transients without fixed schedules.

The continuity of MUD society is thus somewhat tenuous; many pairs of active players exist who have never met each other. In spite of this, MUDs do become true communities after a time. The participants slowly come to consensus about a common (private) language, about appropriate standards of behavior, and about the social roles of various public areas (e.g., where big discussions usually happen, where certain "crowds" can be found, etc.). . . .

It should be noted that different MUDs are truly different communities and have different societal agreements concerning appropriate behavior. There even exist a few MUDs where the only rule in the social contract is that there is no social contract. Such "anarchy" MUDs have appeared a few times in my experience and seem to be quite popular for a time before eventually fading away.

These are the main points of LambdaMOO manners:

- Be polite. Avoid being rude. The MOO is worth participating in because it is a pleasant place for people to be. When people are rude or nasty to one another, it stops being so pleasant.
- "Revenge is ours," sayeth the wizards. If someone is nasty to you, please either ignore it or tell a wizard about it. Please don't try to take revenge on the person; this just escalates the level of rudeness and makes the MOO a less pleasant place for everyone involved.
- Respect other players' sensibilities. The participants on the MOO come from a wide range of cultures and backgrounds. Your ideas about what constitutes offensive speech or descriptions are likely to differ from those of other players. Please keep the text that players can casually run across as free of potentially-offensive material as you can.
- Don't spoof. Spoofing is loosely defined as "causing misleading output to be printed to other players." For example, it would be spoofing for anyone

but Munchkin to print out a message like "Munchkin sticks out his tongue at Potrzebie." This makes it look like Munchkin is unhappy with Potrzebie even though that may not be the case at all.

- Don't shout. It is easy to write a MOO command that prints a message to every connected player. Please don't.
- Only teleport your own things. By default, most objects (including other players) allow themselves to be moved freely from place to place. This fact makes it easier to build certain useful objects. Unfortunately, it also makes it easy to annoy people by moving them or their objects around without their permission. Please don't.
- Don't teleport silently or obscurely. It is easy to write MOO commands that move you instantly from place to place. Please remember in such programs to print a clear, understandable message to all players in both the place you're leaving and the place you're going to.
- Don't hog the server. The server is carefully shared among all of the connected players so that everyone gets a chance to execute their commands. This sharing is, by necessity, somewhat approximate. Please don't abuse it with tasks that run for a long time without pausing.
- Don't waste object numbers. Some people, in a quest to own objects with "interesting" numbers (e.g., #17000, #18181, etc.) have written MOO programs that loop forever, creating and recycling objects until the "good" numbers come up. Please don't do this. . . .

3 The Prospects for Mudding in the Future

> The clumsy system of public gatherings had been long since abandoned; neither Vashti nor her audience stirred from their rooms. Seated in her arm-chair, she spoke, while they in their arm-chairs heard her, fairly well, and saw her, fairly well.
>
> —E. M. Forster[10]

It is substantially easier for players to give themselves vivid, detailed, and interesting descriptions (and to do the same for the descriptions and behavior of the new objects they create) in a text-based system than in a graphics-based one. In McLuhan's terminology,[11] this is because MUDs are a "cold" medium, while more graphically-based media are "hot"; that is, the sensorial parsimony of plain text tends to entice users into engaging their imaginations to fill in missing details while, comparatively speaking, the richness of stimuli in fancy virtual realities has an opposite tendency, pushing users' imaginations into a more passive role. I also find it difficult to believe that a graphics-based system will be able to compete with text for average users on the metric of believable detail per unit of effort expended; this is certainly the case now and I see little reason to believe it will change in the near future.

Finally, one of the great strengths of MUDs lies in the users' ability to customize them, to extend them, and to specialize them to the users' particular needs. The ease with which this can be done in MUDs is directly related to the fact that they are purely text-based; in a graphics-based system, the overhead of creating new moderate-quality graphics would put the task beyond the inclinations of the average user. Whereas, with MUDs, it is easy to imagine an almost arbitrarily small community investing in the creation of a virtual reality that was truly customized for that community, it seems very unlikely that any but the largest communities would invest the greatly-increased effort required for a fancier system.

4 Conclusions

> Vashti was seized with the terrors of direct experience. She shrank back into her room, and the wall closed up again.
>
> —E. M. Forster[12]

The emergence of MUDs has created a new kind of social sphere, both like and radically unlike the environments that have existed before. As they become more and more popular and more widely accessible, it appears likely that an increasingly significant proportion of the population will at least become familiar with mudding and perhaps become frequent participants in text-based virtual realities.

It thus behooves us to begin to try to understand these new societies, to make sense of these electronic places where we'll be spending increasing amounts of our time, both doing business and seeking pleasure. I would hope that social scientists will be at least intrigued by my amateur observations and perhaps inspired to more properly study MUDs and their players. In particular, as MUDs become more widespread, ever more people are likely to be susceptible to the kind of addiction I discuss; we must, as a society, begin to wrestle with the social and ethical issues brought out by such cases. . . .

Pierre Lévy

<< 32 >>

"The Art and Architecture of Cyberspace," *Collective Intelligence* (1994)

Pierre Lévy. Photo by Darcia Labrosse. Courtesy of Pierre Lévy.

"Rather than distribute a message to recipients who are outside the process of creation and invited to give meaning to a work of art belatedly, the artist now attempts to construct an environment, a system of communication and production, a collective event that implies its recipients, transforms interpreters into actors, enables interpretation to enter the loop with collective action."

<< French media theorist and philosopher Pierre Lévy brings a sociological perspective to the art and architecture of cyberspace and the emerging study of cyberculture. His books, *Collective Intelligence: Mankind's Emerging World in Cyberspace* (1994) and *Becoming Virtual: Reality in the Digital Age* (1997), have helped shape the dialogue about the aesthetic and social implications of multimedia, influencing artists and theorists alike. A counterpoint to the dystopic vision of Burroughs and Gibson, Lévy points to a digitally conceived utopian universe, a virtual world in which vast repositories of information, decentralized authorship, mutable identity, and telematic interaction form an "endless horizon" of evolving forms of art and communication.

Lévy identifies an active role for the recipient of the artwork in tandem with a dramatic dissolution of authorial control on the part of its creator. For Lévy, art is becoming a dynamic, fluid, changing environment, a "deterritorialized semiotic plane" in which "artist" and "recipient" unite in a consensual interplay in the formation, execution, and interpretation of art. This shift marks a final break from the notion of art as a sacred, discrete object, as indicated by the diminished authority of the artist's signature, an indelible aspect of art since its beginnings. In its place, he views the digital medium as a continuous and collaborative work-in-progress.

According to Lévy, the break from traditional notions of authorship is leading us toward cultural transformation. He envisions a collective society linked by electronic networks, with citizens actively engaged in the "continuous invention of the languages and signs of a community." Rather than approach the Internet, virtual reality, video games, and emerging media as disparate subjects, Lévy identifies the similarities between their narrative approaches, and treats them as a unified movement toward a global culture. Lévy furthers the case made by Douglas Englebart, and proposes that multimedia is a catalyst for social evolution. It is, he writes, "the architecture of the future"—or the language of the new era. >>

Cyberspace Under Construction

Communications networks and digital memories will soon incorporate nearly all forms of representation and messages in circulation. At this point I would like to consider the significance of the risks involved in developing these networks. Politics and aesthetics confront one another within the unbounded construction site of cyberspace. The perspective of collective intelligence is only one possible approach, however. Cyberspace might also presage, or even incarnate, the terrifying, often inhuman future revealed to us by science fiction: the cataloging of the individual, the processing of delocalized data, the anonymous exercise of power, implacable techno-financial empires, social implosion, the annihilation of memory, real-time warfare among maddened and out-of-control clones. Nevertheless, a virtual world of collective intelligence could just as easily be as replete with culture, beauty, intellect, and knowledge as a Greek temple, a Gothic cathedral, a Florentine palazzo, the *Encyclopédie* of Diderot and d'Alembert, or the Constitution of the United States. A site that harbors unimagined language galaxies, enables unknown social temporalities to blossom, reinvents the social bond, perfects democracy, and forges unknown paths of knowledge among men. But to do so we must fully inhabit this site; it must be designated, recognized as a potential for beauty, thought, and new forms of social regulation. I would like to end this first part of the book with a discussion of the aesthetic dimension associated with engineering the social bond, which primarily focuses on the design of cyberspace and the use of creative play within this new environment of communication and thought.

Cyberspace. Of American origin, the word was used for the first time in 1984 by the science-fiction writer William Gibson in his novel *Neuromancer.*[1] Cyberspace designates the universe of digital networks as a world of interaction and adventure, the site of global conflicts, a new economic and cultural frontier. There currently exists in the world a wide array of literary, musical, artistic, even political cultures, all claiming the title of "cyberculture." But cyberspace refers less to the new media of information transmission than to original modes of creation and navigation within knowledge, and the social relations they bring about. These would include, in no particular order: hyptertext, the World Wide Web, interactive multimedia, video games, simulations, virtual reality, telepresence, augmented reality (whereby our physical environment is enhanced with networks of sensors and intelligent modules), groupware (for collaborative activities), neuromimetic programs, artificial life, expert systems, etc. All of these tools are combined in exploiting the molecular character of digitized information. Various hybrids of the above technologies and conventional media (telephone, film, television, books, newspapers, museums) will come into existence in the near future. Cyberspace constitutes a vast, unlimited field, still partially indeterminate, which shouldn't be reduced to only one of its many components. It is designed to interconnect and provide an interface for the various methods of creation, recording, communication, and simulation.

While the true "great works" remain to be accomplished within the universe of digital information and at the new sites for the emergence of collective intelligence, we continue to encumber the landscape with cement, glass, and steel. We have built pyramids when we are in the process of again becoming nomads, when an architecture for a new exodus is needed. In the silence of thought, we will travel the digital avenues of cyberspace, inhabit weightless mansions that will now constitute our subjectivity. Cyberspace: urban nomad, software engineering, the liquid architecture of the knowledge space. It brings with it methods of collective perception, feeling, remembrance, working, playing, and being. It is an interior architecture, an unfinished system of intelligence hardware, a gyrating city with its rooftops of signs. The development of cyberspace, the quintessential medium of communication and thought, is one of the principal aesthetic and political challenges of the coming century.

Digital interactive multimedia, for example, explicitly poses the question of the end of logocentrism, the destitution of the supremacy of discourse over other modes of communication. It is likely that human language appeared simultaneously in several forms: oral, gestural, musical, iconic, plastic, each individual expression activating a given region of a semiotic continuum, bouncing back and forth from one language to another, from one meaning to another, following the rhizomes of signification, increasing the powers of mind as it traversed the body and its affects. The systems of domination founded on writing have isolated language, established its mastery over a semiotic territory that has been cut up, parceled out, and judged in terms of a sovereign *logos*. The appearance of hypermedia, however, sketches an interesting possibility (among others that are less interesting): A resurgence that lies well within the path opened by writing falls short of a triumphant logocentrism and moves toward the rediscovery of a deterritorialized semiotic plane. But such a resurgence will be enriched with the powers of the text; it will be based on instruments that were unknown during the Paleolithic and are capable of bringing signs to life. Rather than limiting ourselves to the facile opposition between reasonable text and fascinating image, shouldn't we attempt to explore the richer, subtler, more refined possibilities of thought and expression created by virtual worlds, multimodal simulations, dynamic writing media?

In evaluating these new technologies, should we limit ourselves to the concepts of the information highway, telecommuting, interactive CDs, and virtual reality gaming that are presented for public consumption by the media? In doing so we lose sight of the continuity between these spectacular phenomena and the invisible, day-to-day use of existing intellectual technologies. When such new technologies are presented as unrelated phenomena, as objects fallen from the sky, we lose sight of the open and dynamic system they form, their interconnection in cyberspace, their contentious insertion in ongoing cultural processes. We remain blind to the different possibilities they offer to human becoming, possibilities whose full scope is rarely perceived and which should be the subject of deliberation, choice, and judgments of taste, rather than the fiefdom of technical specialists. Even in terms of the apparatus of communication and thought, we are neglecting the dimension of interiority and collective subjectivity, ethics, and sensibility that even the most seemingly technical decisions imply.

From Design to Implementation

With respect to its relationship to future projects, cyberspace will assume the form of a cultural attractor, which we can summarize as follows.

1. Called, controlled, dismissed, distanced, combined, etc., no matter how they are orchestrated, messages, regardless of type, will now revolve around the individual receiver (the opposite of the situation represented by the mass media).
2. The distinctions between authors and readers, producers and spectators, creators and interpreters, will blend to form a reading-writing continuum, which will extend from machine and network designers to the ultimate recipient, each helping to sustain the activity of the others (dissolution of the signature).
3. The distinction between the message and the work of art, envisaged as a microterritory attributed to an author, is fading. Representation is now subject to sampling, mixing, and reuse. Depending on the emerging pragmatics of creation and communication, a nomadic distribution of information will fluctuate around an immense deterritorialized semiotic plane. It is therefore natural that creative effort be shifted from the message itself to the means, processes, languages, dynamic architectures, and environments used for its implementation.

Some of the questions that artists have been asking since the end of the nineteenth century will thus become more urgent with the emergence of cyberspace. These questions are directly concerned with the question of the frame: the limits of a work, its exhibition, reception, reproduction, distribution, interpretation, and the various forms of separation they imply. Under the present circumstances, however, no form of closure will be able to contain deterritorialization *in ex-*

tremis—a leap into a new space will be required. Mutation will occur in a sociotechnical environment in which works of art proliferate and are distributed. Yet, is it reasonable to even speak of a work of art in the context of cyberspace?

For the past several centuries in the West, artistic phenomena have been presented roughly as follows: a person (artist) signs an object or individual message (the work), which other persons (recipients, the public, critics) perceive, appreciate, read, interpret, evaluate. Regardless of the function of the work (religious, decorative, subversive, etc.) or its capacity to transcend function in search of the core of enigma and emotion that inhabits us, it is inscribed within a conventional pattern of communication. Transmitter and receiver are clearly differentiated and their roles uniquely assigned. The emerging technocultural environment, however, will encourage the development of new kinds of art, ignoring the separation between transmission and reception, composition and interpretation. Nevertheless, the ongoing mutation creates a realm of the possible that may never be realized or only incompletely. Our primary goal should be to prevent closure from occurring too quickly, before the possible has an opportunity to deploy the variety of its richness. With the disappearance of a traditional public, this new form of art will experiment with different modalities of communication and creation.

Rather than distribute a message to recipients who are outside the process of creation and invited to give meaning to a work of art belatedly, the artist now attempts to construct an environment, a system of communication and production, a collective event that implies its recipients, transforms interpreters into actors, enables interpretation to enter the loop with collective action. Clearly the "open work" prefigures such an arrangement. But it remains trapped in the hermeneutic paradigm. The recipients of the open work are invited to fill in the blanks, choose among possible meanings, confront the divergences among their interpretations. In all cases it involves the magnification and exploration of the possibilities of an unfinished monument, a succession of initials in a guest book signed by the artist. But the art of implication doesn't constitute a work of art at all, even one that is open or indefinite. It brings forth a process, attempts to open a career to autonomous lives, provides an introduction to the growth and habitation of a world. It places us within a creative cycle, a living environment of which we are always already the coauthors. Work in progress? The accent has now shifted from work to progress. Its embodiment is manifested in moments, places, collective dynamics, but no longer in individuals. It is an art without a signature.

The classic work of art is a gamble. The more it transmutes the language on which it rides, be it musical, plastic, verbal, or other, the more its author runs the risk of incomprehension and obscurity. But the larger the stake—the degree of change or fusion to which its language is subject—the greater the potential gain: the creation of an event in the history of a culture. Yet this game of language, this wager on incomprehension and recognition, is not restricted to artists alone. Each of us in our own way, as soon as we express ourselves, produces, reproduces, and alters language. From singular utterances to creative listening, languages emerge and drift along the stream of communication, borne by thousands of voices that call and respond to one another, take risks, provoke and deceive, hurling words, expressions, and new accents across the abyss of non-sense. In this way an artist can appropriate an expression inherited from earlier generations and help it evolve. This is one of the primary social functions of art: participation in the continuous invention of the languages and signs of a community. But the creator of language is always a community.

Radicalizing the classical function of the work of art, the art of implication creates tension and provides us with sign machines that will enable us to invent our languages. Critics may claim that we have been producing languages forever. True, but without our awareness. To avoid trembling in the face of our own audacity, to mask the void beneath our feet, or simply because this activity has been so slow that it has become invisible, or because it has had to encompass masses of people in constant motion, we have preferred the illusion of foundation. But the price of this illusion has been our sense of defeat. Powerless before the language of the absolute, overcome by the transcendence of the *logos*, exhausted in the presence of the artist's inspired effusions, castigated and corrected, bearing the weight of forgotten tongues, we falter beneath the exteriority of language. The art of implication, which can only give some idea of its true scope in cyberspace, by organizing cyberspace, is an art of therapy. It encourages us to experiment with the collective invention of a language that recognizes itself as such. And in so doing, it points to the very essence of artistic creation.

Having stepped out of the bath of life, far removed from their areas of competence, isolated from one another, individuals finally "have nothing to say." The difficulty lies in trying to comprehend them—in both the emotional and topological sense—as a group, in engaging them in an adventure in which they enjoy imagining, exploring, and constructing sentient environments together. Even if live and real-time technologies play a role in this undertaking, the time experienced by the imagining community overflows the staccato, accelerated, quasi-punctual temporality of "interactivity." The inadequacy of the immediate, of amnesiac channel hopping, no longer leads to lengthy sequences of interpretation, the infinite patience of tradition, which encompasses in a single sweep the ages of the living and the dead, and employs the quick currents of the present to erect a wall against time. In much the same way as madrepores erect coral reefs, commentary, strata upon strata, is always transformed into a subject for commentary. The rhythm of the imagining community resembles a very slow dance, a slow-motion choreography, in which gestures are slowly adjusted and respond with infinite precaution, in which the dancers gradually discover the secret *tempi* that will enable them to shift in and out of phase. Each learns from the others how to make their entrance in stately, slow, and complicated synchrony. Time in the intelligent community spreads itself out, blends with itself, and calmly gathers itself together like the constantly renewed outline of the delta of a great river. The imagining collective comes into being so that it may take the time to invent the ceremony by which it is introduced, which is at the same time a celebration of origin and origin itself, still undetermined.

Employing all the resources of cyberspace, the art of implication reveals the priority of music. But how can a symphony be created from the buzz of voices? Lacking a score, how can we progress from the murmur of the crowd to a chorus? The collective intellect continuously questions the social contract, maintains the group in a nascent state. Paradoxically, this requires time: the time to make sure people are involved, time to forge bonds, to bring objects into being, shared landscapes . . . the time to return. From the point of view of a watch or a calendar, the temporality of the imagining collective might seem displaced in time, interrupted, fragmented. But everything occurs within the obscure, invisible folds of the collective itself: the melodic line, the emotional tonality, the hidden intervals, the correspondences, the continuity that it weaves within the hearts of the individuals who compose it.

For an Architecture of Deterritorialization

The artists who explore such alternatives may be the pathfinders of the new architecture of cyberspace, which will undoubtedly become one of the major arts of the twenty-first century. The new architects could just as easily be engineers, network or interface designers, software programmers, international standards organizations, information lawyers, etc., as individuals with a background in traditional forms of art. In this field the most obviously "technical" choices will have considerable political, economic, and cultural impact. We know that traditional architects and urban planners have helped produce our material, practical, and even symbolic environment. In the same way, the sponsors, designers, and engineers of cyberspace will help produce the environments of thought (sign systems, intellectual technologies), action (telecommuting, remote operation), and communication (access rights, rate policies) that will, to a large extent, structure social and cultural developments.

To guide the construction of cyberspace, to help us choose among the different possible orientations or even imagine new ones, some criteria of ethical and political selection are needed, an organizing vision. Means that contribute to the production of a collective intelligence or imagination should be encouraged. In keeping with this general principle, I would suggest that we concentrate on the following:

1. Instruments that promote the development of the social bond through apprenticeship and the exchange of knowledge
2. Methods of communication that are predisposed to acknowledge, integrate, and restore diversity rather than simply reproduce traditional media-driven forms of distribution
3. Systems that promote the emergence of autonomous beings, regardless of the nature of the system (pedagogical, artistic, etc.) or the beings involved (individuals, groups, works of art, artificial creatures)
4. Semiotic engineering that will enable us to exploit and enhance, for the benefit of the greatest number, the veins of data, the capital of skills, and symbolic power accumulated by humanity

In terms of the creation and management of signs, the transmission of knowledge, the development of living and thinking spaces, the best propaedeutic is obviously supplied by literature, art, philosophy, and high culture in general. Barbarism is born of separation. Contrary to what they may think, technicians have a great deal to learn from humanists in this area. Likewise, those in the humanities must make an effort to employ the new tools, since they redefine the work of intelligence and sensation. Lacking such interaction, we will ultimately produce nothing more than a meaningless technology and a dead culture.

I am making a case here for an architecture without foundations, similar to a boat, with its system of practical oceanography and navigation. Not some benign symbolic structure, analogous to a static image of the body or mind, the reflection of a stable world. On the contrary, the architecture of the exodus will give rise to a nomadic cosmos that travels the universe of expanding signs; it will bring about endless metamorphoses of bodies; within the fissure of flesh and time, it will dispatch its fleets toward the inviolate archipelagos of memory. Far from engendering a theater of representation, the architecture of the future will assemble rafts of icons to help us cross the seas of chaos. Attentive to the voice of the collective brain, translating the thought of plurality, it will erect sonorous palaces, cities of voices and chants, instantaneous, luminous, and dancing like flames.

—Translated by Robert Bononno

<< notes and references >>

Richard Wagner, "Outlines of the Artwork of the Future"

1. The problem of the Theatrical edifice of the Future can in no wise be considered as solved by our modern stage buildings: for they are laid out in accord with traditional laws and canons which have nothing in common with the requirements of pure Art. Where speculation for gain, on the one side, joins forces with luxurious ostentation on the other, the absolute interests of Art must be cryingly affected; and thus no architect in the world will be able to raise our stratified and fenced-off auditoria—dictated by the parcelling of our public into the most diverse categories of class and civil station—to conformity with any law of beauty. If one imagines oneself, for a moment, within the walls of the common Theatre of the Future, one will recognise with little trouble, that an undreamt width of field lies therein open for invention.—R. WAGNER.
2. It can scarcely be indifferent to the modern landscape-painter to observe by how few his work is really understood to-day, and with what blear-eyed stupidity his nature-paintings are devoured by the Philistine world that pays for them; how the so-called "charming prospect" is purchased to assuage the idle, unintelligent, visual gluttony of those same *need*-less men whose sense of hearing is tickled by our modern, empty music-manufacture to that idiotic joy which is as repugnant a reward of his performance to the *artist* as it fully answers the intention of the *artisan*. Between the "charming prospect" and the "pretty tune" of our modern times there subsists a doleful affinity, whose bond of union is certainly not the musing calm of Thought, but that vulgar slipshod *sentimentality* which draws back in selfish horror from the sight of human suffering in its surroundings, to hire for itself a private heavenlet in the blue mists of Nature's generality. These sentimentals are willing enough to see and hear everything: only *not* the *actual, undistorted Man,* who lifts his warning finger on the threshold of their dreams. *But this is the very man whom we must set up in the forefront of our show!*—R. WAGNER.
3. It is a little difficult to quite unravel this part of the metaphor, for the same word "*Boden*" is used twice over. I have thought it best to translate it in the

first place as "loam," and in the second as "ground"; for it appears as though the idea were, in the former case, that of what agriculturists call a "top-dressing," and thus a substance which could break up the lower soil and make it fruitful. The "it" which occurs after the colon may refer either to the "feeling" or to the "orchestra," for both are neuter nouns.—TR.

4. The modern *Playwright* will feel little tempted to concede that Drama ought not to belong exclusively to *his* branch of art, the art of *Poesy;* above all will he not be able to constrain himself to share it with the Tone-poet,—to wit, as he understands us, allow the Play to be swallowed up by the Opera. Perfectly correct!—so long as Opera subsists, the Play must also stand, and, for the matter of that, the Pantomime too; so long as any dispute hereon is thinkable, the Drama of the Future must itself remain un-thinkable. If, however, the Poet's doubt lie deeper, and consist in this, that he cannot conceive how *Song* should be entitled to usurp entirely the place of spoken dialogue: then he must take for rejoinder, that in two several regards he has not as yet a clear idea of the character of the Art-work of the Future. Firstly, he does not reflect that Music has to occupy a very different position in this Art-work to what she takes in modern Opera: that only where her power is the *fittest,* has she to open out her full expanse; while, on the contrary, wherever another power, for instance that of dramatic Speech, is the most *necessary,* she has to subordinate herself to that; still, that Music possesses the peculiar faculty of, without entirely keeping silence, so imperceptibly linking herself to the thought-full element of Speech that she lets the latter seem to walk abroad alone, the while she still supports it. Should the poet acknowledge this, then he has to recognise in the second place, that thoughts and situations to which the lightest and most restrained accompaniment of Music should seem importunate and burdensome, can only be such as are borrowed from the spirit of our modern Play; which, from beginning to end, will find no inch of breathing-space within the Art-work of the Future. The Man who will portray himself in the Drama of the Future has done for ever with all the prosaic hurly-burly of fashionable manners or polite intrigue, which our modern "poets" have to tangle and to disentangle in their plays, with greatest circumstantiality. His nature-bidden action and his speech are: Yea, yea! and Nay, nay!—and all beyond is evil, *i.e.,* modern and superfluous.—R. WAGNER.

J.C.R. Licklider, "Man-Computer Symbiosis"

1. *Webster's New International Dictionary,* 2nd ed., G. and C. Merriam Co., Springfield, Mass., 1958, p. 2555.
2. J. D. North, "The Rational Behavior of Mechanically Extended Man," Boulton Paul Aircraft Ltd., Wolverhampton, Eng., September 1954.
3. H. Gelernter, "Realization of a Geometry Theorem Proving Machine," Unesco, NS, ICIP, 1.6.6, Internatl. Conf. on Information Processing, Paris, France, June 1959.
4. A. Bernstein and M. deV. Roberts, "Computer versus Chess-Player," *Scientific American,* vol. 198, June 1958, pp. 96–98. W. W. Bledsoe and I. Browning, "Pattern Recognition and Reading by Machine," presented at the Eastern Joint Computer Conf., Boston, Mass., December 1959. G. P. Dinneen, "Programming Pattern Recognition," *Proc. WJCC,* March 1955, pp. 94–100. B. G. Farley and W. A. Clark, "Simulation of Self-Organizing Systems by Digital Computers," *IRE Trans. on Information Theory,* vol. IT-4, September 1954, pp. 76–84. R. M. Friedberg, "A Learning Machine: Part I," *IBM J. Res & Dev.,* vol. 2, January 1958, pp. 2–13. P. C. Gilmore, "A Program for the Production of Proofs for Theorems Derivable Within the First Order Predicate Calculus from Axioms," Unesco, NS, ICIP, 1.6.14., Internatl. Conf. on Information Processing, Paris, France, June 1959. A. Newell, "The chess Machine: An Example of Dealing with a Complex Task by Adaptation," *Proc. WJCC,* March 1955, pp. 101–8. A. Newell and J. C. Shaw, "Programming the Logic Theory Machine," *Proc. WJCC,* March 1957, pp. 230–44. A. Newell, J. C. Shaw, and H. A. Simon, "Chess-Playing Programs and the Problem of Complexity," *IBM J. Res & Dev.,* vol. 2, October 1958, pp. 320–35. O. G. Selfridge, "Pandemonium, A Paradigm for Learning," *Proc. Symp. Mechanisation of Thought Processes,* Natl. Physical Lab., Teddington, Eng., November 1958. C. E. Shannon, "Programming a Computer for Playing Chess," *Phil. Mag.,* vol. 41, March 1950, pp. 256–75. H. Sherman, "A Quasi-Topological Method for Recognition of Line Patterns," Unesco, NS, ICIP, H.L.5, Internatl. Conf. on Information Processing, Paris, France, June 1959.
5. A. Newell, J. C. Shaw, and H. A. Simon, "Report on a General Problem-Solving Program," Unesco, NS, ICIP, 1.6.8, Internatl. Conf. On Information Processing, Paris, France, June 1959.

Douglas Engelbart, "Augmenting Human Intellect: A Conceptual Framework"

1. Kennedy and Putt bring out the importance of a conceptual framework to the process of research. They point out that new, multi-disciplinary research generally finds no such framework to fit within, that a framework of sorts would grow eventually, but that an explicit framework-search phase preceding the research is much to be preferred. J. L. Kennedy and G. H. Putt, "Administration of Research in a Research Corporation," *RAND Corporation Report P-847* (April 20, 1956).
2. Ross Ashby, *Design for a Brain,* New York: John Wiley & Sons, 1960. Ross Ashby, "Design for an Intelligence-Amplifier," in *Automata Studies,* C. E. Shannon and J. McCarthy, ed., Princeton, N.J.: Princeton University Press, 1956, pp. 215–34.

Roy Ascott, "Behaviourist Art and Cybernetic Vision"

1. J. Diebold, "The Economic and Social Effects of Automation," *Proc. 2nd International Congress on Cybernetics,* International Association for Cybernetics, Namur, 1958. F. H. George, *The Brain As a Computer,* London: Pergamon Press, 1961.
2. C. Chandessais, "Essai sur la formalisation dans les sciences du comportement," *Cybernetica, Journal of the International Association for Cybernetics,* Namur, 1964, no. 1.
3. W. Gordon, "Economic and Social Effects of Automation," *Cybernetica, Journal of the International Association for Cybernetics,* Namur, 1964, no. 4.
4. M. McLuhan, *Understanding Media,* London: Routledge and Kegan Paul, 1964.
5. J. Bronowski, *Science and Human Values,* London: Hutchinson, 1961.
6. L. Couffignal, "Que peut apporter la Cybernetique à la pédagogie," *Cybernetica, Journal of the International Association for Cybernetics,* Namur, 1964, no. 1.
7. P. de Latil, *La pensée artificielle,* Paris: Gallimard, 1952.
8. W. R. Ashby, "Design for an Intelligence Amplifier," in E. Shannon and J. McCarthy, eds., *Automata Studies,* Princeton, N.J.: Princeton University Press, 1956.
9. A. G. Pask, "The Growth Process inside the Cybernetic Machine," *Proc. 2nd*

Congress on Cybernetics, International Association for Cybernetics, Namur, 1958.

10. George, op. cit.

Alan Kay, "User Interface: A Personal View"

1. Bruner, Jerome. *Towards a Theory of Instruction.* New York: W. W. Norton and Company, 1966.
2. Dawkins, Richard. *The Blind Watchmaker.* New York: Penguin Books, 1986.
3. Gallwey, Tim. *The Inner Game of Tennis.* New York: Random House, 1974.
4. Haber, R. "How We Remember What We See," *Scientific American,* vol. 222, 1970, pp. 104–12.
5. Hadamard, Jacques. *An Essay on the Psychology of Invention in the Mathematical Field.* New York: Dover, 1945.
6. McLuhan, Marshall. *Understanding Media: The Extensions of Man.* New York: McGraw-Hill, 1964.
7. Minsky, Marvin, and Seymour Papert. *Perceptrons: An Introduction to Computational Geometry.* Cambridge, Mass.: MIT Press, 1969.
8. Mumford, Lewis. *The Myth of the Machine.* 2 vols. New York: Harcourt, Brace and World, 1967–70.
9. _____. *Technics and Civilization.* New York: Harcourt, Brace and Company, 1934.
10. Piaget, Jean. *Judgment and Reasoning in the Child.* New York: Harcourt, Brace and Company, 1928.
11. _____. *The Language and Thought of the Child.* New York: Harcourt, Brace and Company, 1926.
12. _____. *The Origins of Intelligence in Children.* New York: International Universities Press, 1952.
13. Suzuki, Shinichi. *Nurtured by Love: A New Approach to Education.* New York: Exposition Press, 1969.

Ted Nelson, *Computer Lib/Dream Machines*

1. See T. H. Nelson, *The Snunking of the Heart: On the Psychology of Puns and Preterism in Carroll and Others,* 1980, unless a decent writing system comes along.

Alan Kay and Adele Goldberg, "Personal Dynamic Media"

1. Baeker, Ronald. "A Conversational Extensible System for the Animation of Shaded Images," *Proc. ACM SIGGRAPH Symposium,* Philadelphia, Pa., June 1976.
2. Goldberg, Adele, and Alan Kay, eds. *Smalltalk-72 Instruction Manual,* Xerox Palo Alto Research Center, Technical Report No. SSL 76-6, March 1976.
3. Goldeen, Marian. "Learning About Smalltalk," *Creative Computing,* September–October 1975.
4. Kay, Alan. *The Reactive Engine.* Doctoral dissertation, University of Utah, September 1969.
5. Learning Research Group. "Personal Dynamic Media," Xerox Palo Alto Research Center, Technical Report No. SSL 76-1, March 1976.
6. Saunders, S. "Improved FM Audio Synthesis Methods for Realtime Digital Music Generation," *Proc., ACM Computer Science Conference,* Washington, D.C., February 1975.
7. Smith, David C. *PYGMALION: A Creative Programming Environment,* Doctoral dissertation, Stanford University Computer Science Department, June 1975.
8. Snook, Tod. *Three-dimensional Geometric Modelling.* Master's thesis, University of California, Berkeley, September 1976.

Tim Berners-Lee, "Information Management: A Proposal"

1. T. H. Nelson, "Getting It Out of Our System," in Information Retrieval: A Critical Review, G. Schechter, ed., Washington, D.C.: Thomson Books, 1967, pp. 191–210. J. B. Smish and S. F. Weiss, "An Overview of Hypertext," in *Communications of the ACM,* vol. 31, no. 7, July 1988, and other articles in the same special "Hypertext" issue. B. Campbell and J. Goodman, "HAM: A General Purpose Hypertext Abstract Machine," in *Communications of the ACM,* vol. 31, no. 7, July 1988. R. M. Akscyn, D. McCracken, and E. A. Yoder, "KMS: A Distributed Hypermedia System for Managing Knowledge in Originations," in *Communications of the ACM,* vol. 31, no. 7, July 1988.
2. *Hypertext on Hypertext,* a hypertext version of the special Comms of the ACM edition, is available from the ACM for the Macintosh or PC.
3. J. Moline et al. eds. *Proceedings of the Hypertext Standardisation Workshop, January*

16–18, 1990, National Institute of Standards and Technology, pub. U.S. Dept. of Commerce.

4. Under unix, type man rn to find out about the rn command which is used for reading uucp news.
5. Under VMS, type HELP NOTES to find out about the VAX/NOTES system.
6. On CERNVM, type FIND DOCFIND for information about how to access the CERNDOC programs.

George Landow and Paul Delany, "Hypertext, Hypermedia and Literary Studies: The State of the Art"

1. Jay David Bolter, "Topographic Writing: Hypertext and the Electronic Writing Space," in *Hypermedia and Literary Studies,* George Landow and Paul Delany, eds., Cambridge, Mass.: MIT Press, 1991.
2. The term *Hypertext* was coined by Theodor H. Nelson in the 1960s.
3. Roland Barthes, *S/Z,* trans. Richard Miller, New York: Hill & Wang, 1974. We have generally used the term *blocks* here, though its connotations are unfortunate, given the actual fluidity and readiness to bond of electronic texts. See also discussion of *nodes* in John Slatin, "Reading Hypertext: Order and Coherence in a New Medium," in Landow and Delaney, op. cit.
4. Vannevar Bush, "As We May Think," *Atlantic Monthly,* no. 176, July 1945: pp. 101–8. James M. Nyce and Paul Kahn have shown that Bush had written a longer version of this essay by 1937: "Innovation, Pragmatism, and Technological Continuity: Vannevar Bush's Memex," *Journal of the American Association for Information Science,* no. 40, 1989, pp. 214–20.
5. In oral cultures, of course, the text had quite a different status in the mind, one that was closer in some respects to the hypertextual model. See Walter J. Ong, *Orality and Literacy: The Technologizing of the Word,* London: Methuen, 1982.
6. Lyrical poems are a special case: they are independent works, but have a close intertextual relationship with others in the same tradition, or by the same author.
7. We may here distinguish between hypertext and a related concept, Standard Generalized Mark-up Language (SGML). SGML concerns the marking out of units in an electronic text and then attaching to them a "tag" or descriptive category (such as "noun phrase" or "animal image"). A pro-

gram can then sort the tags to support grammatical or stylistic analysis. SGML thus sorts textual units by categories; hypertext links the units into larger structures.

8. Nicole Yankelovich, Norman Meyrowitz, and Andries van Dam, "Reading and Writing the Electronic Book," in Landow and Delany, op. cit.
9. For some speculations on these textual megastructures or "docuverses," see the discussion of the "New Alexandria" in the original, unedited version of this essay in Landow and Delaney, op. cit., and Yankelovich, "From Electronic Books to Electronic Libraries," in Landow and Delaney, op. cit.
10. A notable proto-hypertextual work is Humphrey Jennings, *Pandemonium: The Coming of the Machine as Seen by Contemporary Observers, 1660–1886,* New York: Free Press, 1985. This is a chronological series of short passages and images on the Industrial Revolution; Jennings's posthumous editor, Charles Madge, listed sixteen "Theme Sequences," each of them an alternate ordering of groups of passages. Jennings's remarkable work could very easily and usefully be adapted for hypertext presentation.
11. Examples of educational hypertext include Ben Schneiderman's Hyperties Holocaust materials, the many sets of HyperCard materials, including the Harvard Perseus Project (see Gregory Crane and Elli Mylonas, "Ancient Materials, Modern Media: Shaping the Study of Classics with Hypertext," in Landow and Delany, op. cit.), and the Intermedia materials developed at Brown. Among the applications of hypertext to poetry and fiction one can number Michael Joyce's, "Afternoon," Stuart Moulthrop's adaptation of Borges's "Forking Paths" to both HyperCard and StorySpace (see Stuart Moulthrop, "Reading from the Map: Metonymy and Metaphor in the Fiction of 'Forking Paths,' " in Landow and Delany, op. cit.), and William Dickey's mixed media poetry in HyperCard (see William Dickey, "Poem Descending a Staircase: Hypertext and the Simultaneity of Experience," in Landow and Delany, op. cit.).
12. Kenneth Morrell, "Teaching with *HyperCard.* An Evaluation of the Computer-Based Section in Literature and Arts C-14: The Concept of the Hero in Hellenic Civilization." Perseus Project Working Paper 3. Cambridge, Mass.: Department of Classics, Harvard University, 1988.
13. Marshall McLuhan, *The Gutenberg Galaxy: The Making of Typographic Man,* Toronto: University of Toronto Press, 1962; Elizabeth L. Eisenstein, *The Printing Press as an Agent of Change: Communications and Cultural Transformations in Early-Modern Europe,* 2 vols. Cambridge, Mass.: Cambridge University Press, 1979; and J. David Bolter, *Writing Space: The Computer, Hypertext and the History*

of Writing Hillsdale, N.J.: Lawrence Erlbaum, 1990. We would like to thank Professor Bolter for sharing a draft of his work with us before publication.

Morton Heilig, "The Cinema of the Future"

1. The nude eye actually has a vertical range of 180°, but this is reduced to approximately 150° by the brow and cheek of the head.
2. Vistavision, by photographing first on a negative frame twice the original size before printing on to normal size positive, has partially returned screen image to its usual sharpness.

Ivan Sutherland, "The Ultimate Display"

1. K. C. Knowlton. "A Computer Technique for Producing Animated Movies," *Proceedings of the Spring Joint Computer Conference* Washington, D.C.: Spartan, 1964.
2. I. E. Sutherland. "Sketchpad—A Man-Machine Graphical Communication System," *Proceedings of the Spring Joint Computer Conference.* Detroit, Mich. May 1963 Washington, D.C.: Spartan, 1964.

Scott Fisher, "Virtual Interface Environments"

Notes

1. The principal project team at NASA is C. Coler, S. Fisher, M. McGreevy, W. Robinett, and E. Wenzel. At Sterling Software: S. Bryson, J. Humphries, R. Jacoby, D. Kaiser, D. Kerr, and P. Stone.
2. The principal project team at VPL was T. Zimmerman, C. Blanchard, S. Bryson, and J. Grimaud.

References

1. Comeau, C., and J. Bryan. "Headsight Television System Provides Remote Surveillance." In *Electronics,* November 10, 1961, pp. 86–90.

2. Fisher, Scott S. "Telepresence Master Glove Controller for Dexterous Robotic End-Effectors." In *Advances in Intelligent Robotics Systems,* D. P. Casasent, ed. Proc. SPIE 726, 1986.
3. Fisher, S. S., M. McGreevy, J. Humphries, and W. Robinett. "Virtual Environment Display System." ACM 1986 Workshop on 3D Interactive Graphics, Chapel Hill, N.C. October 23–24, 1986.
4. Fisher, S. S., M. McGreevy, J. Humphries, and W. Robinett. "Virtual Interface Environment for Telepresence Applications." In *Proceedings of ANS International Topical Meeting on Remote Systems and Robotics in Hostile Environments.* J. D. Berger, ed., 1987.
5. Fisher, S. S., E. M. Wenzel, C. Coler, and M. W. McGreevy. "Virtual Interface Environment Workstations." In *Proceedings of the Human Factors Society 32nd Annual Meeting,* Anaheim, Calif., October 24–28, 1988.
6. Foley, James D. "Interfaces for Advanced Computing," *Scientific American,* vol. 257, no. 4 (1987), pp. 126–35.
7. Herot, C. "Spatial Management of Data." *ACM Transactions on Database Systems,* vol. 5, no. 4 (1980).
8. Knowlton, K. C. "Computer Displays Optically Superimposed on Input Devices." *The Bell System Technical Journal,* vol. 56, no. 3 (March, 1977), pp. 367–83.
9. Lippman, Andrew. "Movie-Maps: An Application of the Optical Videodisc to Computer Graphics." *Computer Graphics,* vol. 14, no. 3 (1980), pp. 32ff.
10. Negroponte, N. "Media Room." *Proceedings of the Society for Information Display,* vol. 22, no. 2 (1981), pp. 109–13.
11. Schmandt, C. "Spatial Input/Display Correspondence in a Stereoscopic Computer Graphic Work Station." *Computer Graphics, Proceedings of ACM SIGGRAPH '83,* vol. 17, no. 3 (1983), pp. 253ff.
12. Sutherland, I. E. "Head-Mounted Three-Dimensional Display." *Proceedings of the Fall Joint Computer Conference,* vol. 33 (1968), pp. 757–64.
13. Wenzel, E. M., F. L. Wightman, and S. H. Foster. "A Virtual Display System for Conveying Three-Dimensional Information." *Proc. Hum. Fac. Soc.* (1988).

Marcos Novak, "Liquid Architectures in Cyberspace"

Notes

1. Bruce Sterling has recently collected five: Gibsonian cyberspace, Barlovian cyberspace, Virtual Reality, Simulation, and Telepresence. The definition above is a concatenation of all five plus a sixth, set forth in "Making Reality a Cyberspace" by Wendy Kellog and others.
2. Thomas Kuhn discussed the notion of world lines with respect to scientific concepts in a talk on untranslatability between different intellectual frameworks (Kuhn, 1990).

References

1. Alberti, L. B. [1404–1472]. *The Ten Books of Architecture.* New York: Dover Publications, 1986.
2. Anderson, J. A., and E. Rosenthal, eds. *Neurocomputing: Foundations of Research.* Cambridge, Mass.: MIT Press, 1988.
3. Benedikt, M., ed. *Cyberspace: Collected Abstracts.* University of Texas at Austin, 1990.
4. Berman, M. *Coming to Our Senses.* New York: Simon and Schuster, 1989.
5. Bonner, J. T. *The Evolution of Complexity by Means of Natural Selection.* Princeton, N.J.: Princeton University Press, 1988.
6. Brisson, D. W., ed. *Hypergraphics: Visualizing Complex Relationships in Art, Science and Technology.* Boulder, Colo.: Westview, 1978.
7. Conrads, U. *Programs and Manifestoes on 20th-Century Architecture.* Cambridge, Mass.: MIT Press, 1970.
8. Coyne, R. D. "Tools for Exploring Associative Reasoning in Design." In *The Electronic Design Studio,* M. McCullough, W. J. Mitchell, and P. Purcell, eds. Cambridge, Mass.: MIT Press, 1990.
9. Davies. P., ed. *The New Physics.* Cambridge, U.K.: Cambridge University Press, 1989.
10. Dawkins, R. *The Blind Watchmaker.* New York: W. W. Norton & Company, 1986.
11. Feyerabend, P. *Against Method.* London: Verso, 1988.
12. Gregory, B. *Inventing Reality: Physics As Language.* New York: Wiley Science Editions, 1988.

13. Helsel, S. K., and J. P. Roth, eds. *Virtual Realities: Theory, Practice and Promise.* London: Meckler, 1991.
14. Hillier, B., and J. Hanson. *The Social Logic of Space.* Cambridge, U.K.: Cambridge University Press, 1984.
15. Hollier, D. *Against Architecture: The Writings of George Bataille.* Cambridge, Mass.: MIT Press, 1989.
16. Kostoff, S. *A History of Architecture.* Oxford: Oxford University Press, 1985.
17. Kuhn, T. "Untranslatabilty," lecture notes, UCLA, 1990.
18. Küppers, B. O. *Information and the Origin of Life.* Cambridge, Mass.: MIT Press, 1990.
19. Langton, C. G. *Artificial Life: Proceedings of an Interdisciplinary Workshop on the Synthesis and Simulation of Living Systems Held September 1987.* Reading, Mass.: Addison-Wesley, 1989.
20. Lorca, F. G. *Poet in New York.* New York: Noonday Press, 1989.
21. McClelland, J. L., and D. E. Rumelhart, eds. *Parallel Distributed Processing: Exploration in the Microstructure of Cognition,* vols. 1 and 2. Cambridge, Mass.: MIT Press, 1986.
22. Negroponte, N. *Soft Architecture Machines.* Cambridge, Mass.: MIT Press, 1975.
23. Pérez-Gómez, A. *Architecture and the Crisis of Modern Science.* Cambridge, Mass.: MIT Press, 1983.
24. Pickover, C. *Computers, Pattern, Chaos and Beauty: Graphics from an Unseen World.* New York: St. Martin's Press, 1990.
25. Sartre, J. P. *Being and Nothingness.* New York: Washington Square Press, 1956.
26. Scarry, E. *The Body in Pain.* New York: Oxford University Press, 1985.
27. Sterling, B. "Cyberspace (TM)," *Interzone,* November 1990.
28. Tzonis, A. *Classical Architecture: The Poetics of Order.* Cambridge, Mass.: MIT Press, 1986.
29. Vidler, A. "The Building in Pain." In *AA Files.* London: AA Press, 1990.
30. Woods, L. *OneFiveFour.* New York: Princeton Architectural Press, 1989.

Bill Viola, "Will There Be Condominiums in Data Space?"

1. A. R. Luria, *The Mind of the Mnemonist,* New York: Basic Books, 1968.
2. The Greeks perfected a system of memory that used the mental imprint-

ing of any objects or key points to be remembered onto specific locations along a pathway previously memorized from an actual temple. To recall the points in their proper order, one simply had to take the walk through the temple in one's mind, observing the contents left at each location along the way.

3. A. K. Coomaraswamy, *The Transformation of Nature in Art,* New Yok: Dover Publications, 1956; reprint of the original Harvard University Press edition, Boston, 1934.
4. A Landmark interactive laserdisc project by MIT Media Lab, in the late 1970s, that mapped the city of Aspen, street by street, with moving cameras so that the viewer could take a "ride" through the city, going anywhere at will—one of the first visual-mapping database moving-image projects related to data space ideas and today's virtual reality technology.

Roy Ascott, "Is There Love in the Telematic Embrace?"

1. The neologism "telematique" was coined by Simon Nora and Alain Minc in *L'Informatisation de la societe,* Paris: La Documentation Francaise, 1978.
2. The German word "*Gesamtkunstwerk*" was used by Richard Wagner to refer to his vision of a "total artwork" integrating music, image, and poetry.
3. Humberto R. Maturana and Francisco J. Varela, *The Tree of Knowledge: The Biological Roots of Human Understanding,* Boston: Shambhala, 1987.
4. Roland Barthes, "From Work to Text," in *Image-Music Text,* trans. Stephen Heath, New York: Hill & Wang, 1977.
5. C. E. Shannon and W. Weaver, *The Mathematical Theory of Communication,* Urbana: University of Illinois Press, 1949.
6. Heinz von Foerster, *Observing Systems,* New York: Intersystems, 1981.
7. James Gleick, *Chaos,* New York: Heinneman, 1987.
8. J. A. Wheeler and W. H. Zurek, *Quantum Theory and Measurement,* Princeton, N.J.: Princeton University Press, 1983.
9. See Michel Sanouillet, ed., *Salt Seller: The Writings of Marcel Duchamp* (Marchand du Sel) New York: Oxford University Press, 1973.
10. Charles Fourier (1772–1837). His "system of passionate attraction" (elaborated in *Theorie des quatres mouvements et des destinees generales* [Paris: Bureaux de la Phalange, 1841]) sought universal harmony.
11. Peter Russell, *The Awakening Earth,* London: Routledge, 1982.

12. James L. McClelland, David E. Rummelhart, and the PDP Research Group, *Parallel Distributed Processing,* vol. 1, Cambridge, Mass.: MIT Press, 1986.
13. The term was first proposed in Roy Ascott, "Art and Telematics: Towards a Network Consciousness," in *Art Telecommunications.* Heidi Grundmann, ed., Vancouver: Western Front, 1984.
14. Paul J. Curran, *Principles of Remote Sensing,* New York: Longman, 1985.
15. Stephen R. Graubard, ed., *The Artificial Intelligence Debate,* Cambridge, Mass.: MIT Press, 1988.
16. Koji Kobayashi, *Computers and Communications,* Cambridge, Mass.: MIT Press, 1986.
17. Roy Ascott, "Art, Technology and Computer Science," in XLII Esposizione Internazionale d'Arte (Venice: Biennale di Venezia, 1986). The international commissioners for the 1986 Biennale were Roy Ascott, Don Foresta, Tom Sherman, and Tomaso Trini.
18. Roy Ascott, "Gesamtdatenwerk: Konnektivitat, Transformation and Transzendenz," in *Kunstforum* 103 September/October 1989. Project by Roy Ascott in collaboration with Peter Appleton, Mathias Fuchs, Robert Pepperell, and Miles Visman.
19. Norbert Wiener, *Cybernetics or Control and Communication in the Animal and Machine,* Cambridge, Mass.: MIT Press, 1948.
20. VPL Research, Inc., of California demonstrated "shared virtual reality" and "walk-through cyberspace" at Texpo '89 in San Francisco, June 1989.

Pavel Curtis, "Mudding: Social Phenomena in Text-Based Virtual Realities"

1. Forster, E. M., "The Machine Stops," in *The Science Fiction Hall of Fame,* vol. IIB, Ben Bova, ed., Avon, 1973. Originally in E. M. Forster, *The Eternal Moment and Other Stories,* New York: Harcourt Brace Jovanovich, 1928.
2. Eric S. Raymond, ed., *The New Hacker's Dictionary,* Cambridge, Mass.: MIT Press, 1991.
3. In fact, these two commands are so frequently used that single-character abbreviations are provided for them. The two example commands would usually be typed as follows:

Can anyone hear me?

:smiles.

4. Forster, op cit.
5. The "MOO" in "LambdaMOO" stands for "MUD, Object-Oriented." The origin of the "Lambda" part is more obscure, based on my years of experience with the Lisp programming language.
6. Van Gelder, Lindsy, "The Strange Case of the Electronic Lover," in Dunlop and Kling, op. cit.
7. Sara Kiesler et al., "Social Psychological Aspects of Computer-Mediated Communication," in *Computerization and Controversy,* Charles Dunlop and Robert Kling, eds., San Diego, Calif.: Academic Press, 1991. Kiesler and her colleagues have investigated the effects of electronic anonymity on the decision-making and problem-solving processes in organizations; some of their observations parallel mine given here.
8. Van Gelder, op cit.
9. Morningstar, Chip, and F. Randall Farmer, "The Lessons of Lucasfilm's Habitat," in *Cyberspace,* Michael Benedikt, ed., Cambridge, Mass.: MIT Press, 1991.
10. Forster, op cit.
11. McLuhan, Marshall, *Understanding Media,* New York: McGraw-Hill, 1964.
12. Forster, op cit.

Pierre Lévy, "The Art and Architecture of Cyberspace"

1. William Gibson, *Neuromancer,* New York: Ace Science Fiction Books, 1984.

<<a note on the web site>>

artmuseum.net
http://www.artmuseum.net

Multimedia: From Wagner to Virtual Reality is a unique hybrid publishing project that joins W.W. Norton & Company with our media partner, Intel Corporation's ArtMuseum.net, to present an untold history of multimedia.

The book and the Web site are meant to work in tandem. On-line, *Multimedia: From Wagner to Virtual Reality* is a dynamic, growing resource featuring hyperlinked texts and a wealth of multimedia documentation—vintage photographs, rare video clips, and an interactive timeline chronicling the medium's pioneers. It also includes "in-depth" profiles on several of the artists and scientists featured in the book, extending its scope to include illustrated descriptions of the artworks and technological invention that underlie the evolution of the medium.

The site was developed in close collaboration with ArtMuseum.net and its Arts and Education program. It is our hope that the Web site will serve as a valuable resource for students, teachers, artists, and critics probing the interactive media arts, as well as those attempting to make sense of this burgeoning medium that is transforming our art and our culture.

<< acknowledgments >>

A project of this scope doesn't come to fruition without the assistance and insistence of a good many people. Randall Packer has explored the roots of multimedia for over a dozen years as research for his work as a composer, media artist, and educator. These investigations would have been far more challenging were it not for the support of the following people and institutions: Zakros Interarts and the talented performers, musicians, and artists who participated in reconstructions of seminal new music theater productions that inspired the research for this project; the San Francisco State Multimedia Studies Program; Josi Callan and the San Jose Museum of Art; the University of California, Berkeley, Department of Art Practice; and David Ross and the San Francisco Museum of Modern Art.

Thanks to Eric Mendlow for bringing Randall Packer and Ken Jordan together in the San Francisco spring of 1998. The seeds for this project were planted during that meeting in Randall's Mission loft, as questions about the history of multimedia prompted book after book to be taken down from the library shelves, and piled in increasingly ungainly stacks on every available desk space. After that evening, this anthology seemed inevitable.

We would like to acknowledge the inspiration of the critics, theorists, and historians who laid much of the groundwork for this investigation, as well as friends whose conversation helped shape ideas expressed in this book. They include: Robert Atkins, Steward Brand, Paul Ceruzzi, Tim

Druckrey, Richard Foreman, Peter Frank, Rose Lee Goldberg, Cynthia Goodman, Katie Hafner, Michael Heim, E. T. Kirby, Michael Kirby, Richard Kostelanetz, Brenda Laurel, Sylvere Lotringer, Margot Lovejoy, Matthew Lyon, Marshall McLuhan, Margaret Morse, Nicholas Negroponte, Frank Popper, Lillian Schwartz, Kristine Stiles, Calvin Tomkins, Sherry Turkle, Paul Virilio, Peter Weibel, Robert Wilson, and Gene Youngblood. A special note of appreciation for Howard Rheingold, whose 1985 history of personal computing, *Tools For Thought,* was particularly influential.

The early support of our editor, Alane Salierno Mason of W.W. Norton, helped tremendously to get this book underway, just as her fierce editorial eye kept the prose taut and accessible; thanks also to Stefanie Diaz at Norton for her assistance in multiple arenas.

Critical to the project was the support of Kevin Teixeira of Intel Corporation, who believed in this book from the get-go, and saw its potential as a dynamic, multimedia on-line destination. His enthusiastic backing, as well as the expert advice and technical support of Vince Thomas and the ArtMuseum.net staff, including Annie Rodkins and Irena Rogovsky, not only made the Web component to this project possible, they made it a pleasure. In addition, Phyllis Hecht, Web Manager and Art Director of the National Gallery of Art, offered stalwart support and guidance in the design of the Web site.

We would particularly like to thank a few of the authors included in this volume for their encouragement and instruction over the years: Douglas Engelbart, Alan Kay, Lynn Hershman, Ted Nelson, Billy Klüver, and Scott Fisher. Our intent has been to present one of the possible histories of multimedia. Our hope is for this encounter to mark the beginning of the reader's exploration of this field, not the end, and that it will lead to further encounters with these authors and their colleagues. Some articles in this book have been edited down for reasons of space (with edits marked by the use of an ellipse); throughout, references to illustrations or passages not reproduced in this volume have been deleted.

The preparation of this book has been a deeply rewarding, stimulating experience, in large part due to the breadth and insight of these remarkable thinkers. We'd like to extend our gratitude to the pioneers of multimedia, not only for graciously providing us with material when they were approached, but, most important, because their work and vision has been the inspiration fueling our efforts.

permissions

Grateful acknowledgment is made to the following for permission to reprint copyrighted materials:

Richard Wagner, "Outlines of the Artwork of the Future," *The Artwork of the Future*, originally published in 1849. Reprinted in English, *Prose Works*, vol. 1, translation by William Ashton Ellis, originally published in 1895 by Kegan Paul, Trenchy, Trübner & Co., Ltd. Reprinted by University of Nebraska Press, 1993.

F. T. Marinetti et al., "The Futurist Cinema," *Futurist Manifestos*, edited by Umbro Apollonio, originally published by L'Italia futurista (Milan), 1916. Translation by R. W. Flint, copyright © 1973 Thames and Hudson, Ltd. (London). Reprinted by permission of Thames and Hudson, Ltd.

László Moholy-Nagy, "Theater, Circus, Variety," *The Theater of the Bauhaus*, edited by Walter Gropius and Arthur S. Wensinger, translated by Arthur S. Wensinger (Middletown, CT: Wesleyan University Press, 1929, 1961). Reprinted by permission of Wesleyan University Press.

Dick Higgins, "Intermedia," *A Dialectic of Centuries*, originally published by Something Else Press (New York), 1966. Reprinted by Printed Editions, copyright © 1978 Richard C. Higgins. Courtesy of The Estate of Dick Higgins.

Billy Klüver, "The Great Northeastern Power Failure," paper presented at annual meeting of the College Art

Association, New York, January 28, 1996. Published as "Det Stora Strömavbrottet I New York," *Dagens Nyheter* (Stockholm), March 6, 1966; to be included in the forthcoming autobiographical anthology of the author's writings. Courtesy of the author.

Nam June Paik, "Cybernated Art," *Manifestos,* originally published in Great Bear Pamphlets (New York: Something Else Press, 1966). Reprinted in *Theories and Documents of Contemporary Art: A Sourcebook of Artists' Writings* (Berkeley, CA: University of California Press, 1996). Courtesy of the Artist and Holly Solomon Gallery.

Norbert Weiner, "Cybernetics in History," excerpt from *The Human Use of Human Beings* by Norbert Weiner. Copyright © 1950 by Norbert Weiner, renewed 1977 by Margaret Weiner. Reprinted by permission of Houghton Mifflin Company. All rights reserved.

J.C.R. Licklider, "Man-Computer Symbiosis," copyright © 1960 IRE (now IEEE). Reprinted with permission of IEEE from *Transcriptions on Human Factors in Intelligence,* vol. HFE-1, pp. 4–11, March 1960.

Douglas Engelbart, "Augmenting Human Intellect," originally published in *The Augmentation Papers,* Bootstrap Institute, 1962. Copyright © 1993 Bootstrap Institute. Courtesy of Douglas C. Engelbart.

John Cage, "Diary: Audience 1966," *A Year from Monday* (Middletown, CT: Wesleyan University Press, 1966). Copyright © 1967 John Cage. Reprinted with permission of Wesleyan University Press.

Roy Ascott, "Behaviourist Art and the Cybernetic Vision," *Cybernetica,* International Association for Cybernetics, Namur, vol. IX, 1966, pp. 247–64; vol. X, 1967, pp. 25–56. Copyright © 1966 Roy Ascott. Courtesy of the author.

Myron Krueger, "Responsive Environments," originally published as "Responsive Environments," American Federation of Information Processing Systems 46 (June 13–16, 1977), pp. 423–33. Reprinted in *Theories and Documents of Contemporary Art: A Sourcebook of Artists' Writings* (Berkeley, CA: University of California Press, 1996). Copyright © 1977 Myron W. Krueger. Courtesy of the author.

Alan Kay, "User Interface: A Personal View," *The Art of Human-Computer Interface Design,* edited by Brenda Laurel (Reading, MA: Addison-Wesley Publishing Company, 1989). Courtesy of the author.

Vannevar Bush, "As We May Think," originally published in *The Atlantic Monthly,* vol. 176, no. 1 (July 1945), pp. 101–8. Copyright © 1945 Vannevar Bush.

Ted Nelson, excerpts from *Computer Lib/Dream Machines.* Copyright © 1974, 1987 Theodor H. Nelson. Courtesy of the author.

Alan Kay and Adele Goldberg, "Personal Dynamic Media," IEEE (The Institute of Electrical and Electronics Engineers, Inc.) *Computer*, 10 (March 1977), pp. 31–41. Courtesy of Alan Kay.

Marc Canter, "The New Workstation: CD ROM Authoring Systems," *CD-ROM: The New Papyrus: The Current and Future State of the Art*, edited by Steve Lambert and Suzanne Ropiequet (Redmond, WA: Microsoft Press, 1986). Courtesy of the author.

Tim Berners-Lee, "Information Management: A Proposal," CERN (Geneva, Switzerland), 1989. Courtesy of the author.

George Landow and Paul Delany, "Hypertext, Hypermedia and Literary Studies: The State of the Art," originally appeared in *Hypermedia and Literary Studies*, edited by George Landow and Paul Delany. Copyright © 1991 by Massachusetts Institute of Technology. Reprinted by permission of Georges Borchardt, Inc., for the authors.

Morton Heilig, "Cinema of the Future," originally appeared in Spanish as "El cine del futuro," *Espacios* (Mexico), no. 23–24, (January–June 1955). Translated by Uri Feldman of the Massachusetts Institute of Technology Media Lab and reprinted in *Presence: Teleoperators and Virtual Environments*, Massachusetts Institute of Technology Press, vol. 1, no. 3, (1992).

Ivan Sutherland, "The Ultimate Display," Proceedings of the IFIP (International Federation for Information Processing) Congress, 1965, pp. 506–8. Copyright © 1965 Ivan Sutherland.

Scott Fisher, "Virtual Interface Environments," *The Art of Human-Computer Interface Design*, edited by Brenda Laurel (Reading, MA: Addison Wesley Publishing Company, 1989). Copyright © 1990 Scott Fisher. Courtesy of the author.

Daniel Sandin, Thomas DeFanti, and Carolina Cruz-Neira, "Room with a View." Copyright © IEEE (The Institute of Electrical and Electronics Engineers, Inc.). Reprinted, with permission from *Spectrum*, October 1993.

William Gibson, "Academy Leader," *Cyberspace, First Steps*, edited by Michael Benedikt (Cambridge, MA: Massachusetts Institute of Technology Press, 1991). Copyright © 1991 William Gibson. Courtesy of the author.

Marcos Novak, "Liquid Architectures in Cyberspace," *Cyberspace: First Steps*, edited by Michael Benedikt (Cambridge, MA: Massachusetts Institute of Technology Press, 1990). Reprinted by permission of Massachusetts Institute of Technology Press.

Excerpt from Federico García Lorca's poem "Sleepwalking Ballad," translated by Will Kirkland. Copyright © Will Kirkland and Herederos de Federico García Lorca. Excerpt from Federico García Lorca's lecture, *A Poet in New York*,

<<index>>

Page numbers in *italics* refer to illustrations.